Pathology of Thermal Injury: A Practical Approach

PATHOLOGY OF THERMAL INJURY: A PRACTICAL APPROACH

Thomas W. Panke, M.D.

Associate Clinical Professor, Pathology
University of Cincinnati
School of Medicine
and
Good Samaritan Hospital
Cincinnati, Ohio

Charles G. McLeod, Jr., D.V.M.

Lieutenant Colonel, U.S. Army Veterinary Corps
Chief, Comparative Pathology
U.S. Army Medical Research Institute of Chemical Defense
Aberdeen Proving Ground
Edgewood, Maryland

Grune & Stratton, Inc.

(Harcourt Brace Jovanovich, Publishers)

Orlando   San Diego   New York
London   Toronto   Montreal   Sydney   Tokyo

Library of Congress Cataloging in Publication Data

Panke, Thomas W.
  Pathology of thermal injury: A practical approach

  1. Burns and scalds—Complications and sequelae.
2. Physiology, Pathological. I. McLeod, Charles,
1941–   . II. Title. [DNLM: 1. Burns—complications.
2. Burns—pathology. WO 704 P193p]
RD96.4.P38 1985   617′.1101   85-12670
ISBN 0-8089-1754-4

Grune & Stratton, Inc.
Orlando, FL 32887

Distributed in the United Kingdom by
Grune & Stratton, Ltd.
24/28 Oval Road, London NW 1

Library of Congress Catalog Number 85-12670
International Standard Book Number 0-8089-1754-4

Printed in the United States of America
85 86 87 88   10 9 8 7 6 5 4 3 2 1

CONTENTS

FOREWORD

Burn injury not only destroys the cutaneous barrier protecting the patient from a hostile environment, but it also causes pathophysiological changes in every other organ system. The extent and duration of organ function alterations are directly related to the extent of thermal injury. These same changes occur to some degree in all injured patients, which makes the burn patient the universal trauma model. Since the severity of thermal injury is easily quantifiable and can be expressed as a percentage of total body surface, treatment results can be assessed by making patient-to-patient and institution-to-institution comparisons.

Multidisciplinary research involving surgeons, internists, pathologists, veterinarians, physiologists, and biochemists has defined the multisystem consequences of thermal injury. Recent major clinical advances have been accompanied or preceded by careful pathologic studies describing the pathogenesis of burn wound sepsis, the natural history of stress ulcers, the morphologic changes of inhalation injury, and the development of reproducible animal models, to name but a few of the benefits that have accrued from the integration of pathologists into the burn team. Pathologists have been members of the burn team at our Institute for Surgical Research for over 20 years, and their contributions have played a major role in the development of current burn care, which has resulted in improved patient survival.

Thomas Panke, who is certified in anatomic and clinical pathology and has special expertise in hematopathology, has made original observations of post-burn changes in the cardiovascular, gastrointestinal, endocrine, and hemotopoietic systems as detailed in this volume. Charles McLeod, a certified veterinary pathologist with special expertise in inhalation injury and in animal models of infectious diseases, has developed new models of post-burn gastrointestinal stress lesions and post-injury opportunistic infection. These models have permitted the development of treatment techniques to minimize post-burn complications.

In *Pathology of Thermal Injury: A Practical Approach* we have for the first time a compendium of burn pathology that emphasizes the multisystem consequences of thermal injury. The authors have made clinical pathologic correlations to aid in our understanding of this most severe form of trauma.

Basil A. Pruitt, Jr., M.D., F.A.C.S.
Colonel, U.S. Army Medical Corps
Commanding Officer and Director
Institute of Surgical Research
Brooke Army Medical Center
Fort Sam Houston, Texas

ACKNOWLEDGMENTS

Pathology of Thermal Injury: A Practical Approach was prepared with the assistance of many people. First, and most important, was the patient and generous guidance received from the Commander and Director of the U.S. Army Institute of Surgical Research, Colonel Basil A. Pruitt, Jr. He unselfishly reviewed and edited each chapter and provided invaluable guidance through his experience and personal knowledge of the burn literature. Without the assistance and encouragement of Colonel Pruitt, and the facilities at the U.S. Army Institute of Surgical Research, this text would not exist.

There were many other friends upon whom we called for medical and editorial assistance in their areas of expertise. We would like to thank them all. They include Norbert Goebel, Drs. A. D. Mason, Jr., William McManus, Albert McManus, Gerald Penn, M. S. N. Murthy, Nariosang Kandawalla, and Phyllis Leist.

We are especially grateful to our secretary, Margaret I. Cooper, who deserves as much credit as anyone in preparing this manuscript. Mrs. Cooper edited and typed the many copies of each chapter in addition to her many other duties. Jery Williams placed some of the finishing touches on several chapters, and we wish to thank her as well.

Paulette Langlinais prepared our electron photomicrographs, and Tony Dorsaneo, Frank Rodriques, Ray Newman, Jane Norton, and Eric Sourvere assisted in our preparation of figures and illustrations and deserve our gratitude.

Finally, with great love, we wish to thank our understanding wives, Liz and Joan, who encouraged our efforts and somehow managed to persevere in our absence through many late evenings and weekends during the last few years. To them and our families, we wish to dedicate this book.

PREFACE

"Why write a burn book?" We have heard this question repeatedly from our peers, our spouses, and occasionally from each other.

During our assignment in pathology at the Institute of Surgical Research, Fort Sam Houston, Texas, we found that information about burn pathology, unlike most other specialities of pathology had to be gathered from scattered sources. There were many valuable publications, but a practical guide for the *burn* pathologist was not available. Visiting pathologists who had a sudden need to interpret biopsies, perform autopsies, or determine cause of death in burn victims had also expressed this concern to us.

What began as an attempt to write a simple monograph got out of hand. As we wrote the early chapters, it became evident that we would have to address each of the major body systems. It would have been an oversimplification and misleading to present only the mechanics of burn pathology when the reality is that thermal trauma directly or indirectly affects every system and organ in the body. In no way have we attempted to encompass the entire burn literature, but it is our intent and hope that this work will help pathologists and other burn specialists to better understand this complex disease process.

Pathology of Thermal Injury:
A Practical Approach

Chapter 1

INTRODUCTION TO BURN PATHOLOGY

We wish to introduce the reader to the pathology and physiologic consequences of burn trauma via a brief historical review of the treatment and study of burn injury, followed by a short discourse on the current "state of the art." Finally, we wish to discuss the future development and needs in burn pathology.

The treatment of major thermal trauma remains one of the most challenging problems in the field of medicine today. Annually it affects more than 2 million patients who utilize more than 11,000 hospital beds/day at an expenditure of nearly 1.5 billion dollars. Over 12,000 deaths each year are attributed to thermal injury.[1]

HISTORICAL ANECDOTES

From prehistoric times, humans have utilized fire to prepare food, to warm their bodies and destroy their enemies, and have often suffered injuries because of this close association. In contrast to the many mysterious infectious diseases that plagued the world, burns were of obvious etiology. Every civilization has no doubt been concerned with the treatment of this painful, disfiguring, and often fatal injury.

The emphasis of early efforts was directed at topical therapy rather than descriptions of pathological findings. Artz[2] and Scarborough[3] have presented excellent and entertaining reviews of this topic. Dr. Artz mentions the first burn specialist, a Neanderthal man who used plant extracts in burn therapy. The early Egyptians treated burns with incantations and mixtures of gum, goat's hair, and milk from women who had given birth to sons. In the fifth and sixth centuries BC, oriental physicians advocated tinctures and extracts of tea leaves for treating burns. Hippocrates, around 430 BC, suggested, ". . . having old swine's seam (fat) and mixed it with resin and

1

bitumen, and having spread it on a piece of cloth and warmed it at the fire, apply a bandage." He also described several tanning methods as those used in the Orient, although solutions of bark were used rather than tea.[2] Early Romans utilized topical mixtures of honey and bran (behold, another benefit of fiber!), and Pliny the Elder felt that exposure of the wound to open air rather than bandaging was desirable. Galen prescribed vinegar or wine. An old Arabian physician, Rhazes, in the ninth century recommended ice-cold water for burn injuries. Some of these early remedies were embarrassingly similar to those "discovered" in modern times.[4,5]

On the advice of a knowledgeable country girl, Ambrose Pare tried and apparently had success with onions and salt in topical burn therapy. This treatment was used for years and was referenced as late as 300 years after discovery.[6] Although these agents are now used most often for improving the palatability of burned hamburgers, it may be time to "rediscover" their usefulness in burn treatment.

In summary, it appears that everything from pigeon dung to chopped leeks to fire itself has been utilized in topical burn therapy. With the introduction of the hypodermic syringe, opiates, anesthetics, and grafting, the trend changed towards development of technology directed at early closure of the burn wound.[7]

An important part of this process was based on the descriptive classification of burns. The first classification of burns was conceived by Fabry in 1607, who wrote of it in his *De Combustionibus*.[9] His system was based on the appearance of the wound surface in addition to a few other signs. This morphologic method of diagnosing the depth of burn injury was used for nearly 200 years[8] and included (1) redness with blistering of the skin; (2) withering of the skin without charring; and (3) eschar formation and charring. Other early authors refined this early description of the burn wound by a more exacting analysis of the burn wound depth and additional pathologic features.[9,10]

Our understanding of burn trauma has had its greatest advancement in the last 50 years.[11] Earlier, the burn wound had been considered primarily on the basis of local tissue damage. With the improvement of effective resuscitative techniques, the many early deaths due to shock and renal failure were markedly decreased. Postburn survival in patients with large burns was increased and lengthened, and new problems developed for the clinician, surgeon, and pathologist. It was apparent that burn trauma influenced virtually all body systems. The concept of the "burn syndrome" appeared.

THE CURRENT "STATE OF THE ART"

As is true with other fields of medicine, the use of the laboratory for monitoring the burn patient's clinical course has steadily increased. In addition to standard laboratory tests (BUN, glucose, electrolytes, liver function tests, coagulation studies, etc.), a variety of sophisticated tests and other techniques are carried out. The limulus lysate assay and other tests which may quantitate endotoxin are used to delineate early phases of infection. Pulmonary scintiphotography following inhalation and intravenous injection of Xenon 133 is a qualitative and semiquantitative means to assess pulmonary parenchymal injury in patients who have an inhalation injury. Microbiological techniques quantitate microbiologic flora in the burn wound. Phage typing determines whether a nosocomial contagious source is present.

Recently, histological evaluation of the burn wound biopsy has provided an accurate and rapid (within 4 hr) technique for diagnosing burn wound infection.[12] Since certain histological patterns are typical for specific bacteria (*Pseudomonas aeruginosa*), morphological interpretation may often provide a strong presumptive diagnosis on which specific treatment may be initiated. Histologic evaluation is often more useful and definitive in the diagnosis of invasive mycotic infection. Invasive infections by Phycomycetes, *Aspergillus* sp., and *Candida* sp. may often be diagnosed based on the morphology of the microorganism present in unburned tissue rather than the histologic pattern of necrosis. The electron microscope is utilized for evaluating atypical and nondiagnostic burn wound lesions which appear to be viral in origin.

Intravenous catheterization has become a vital adjunct to clinical treatment of severely burned patients. Unfortunately, thrombophlebitis (sterile or infectious) and infectious cardiac valvulitis may complicate such use. Suppurative thrombophlebitis involving a peripheral vein is clinically evident in less than half of the patients having such a complication, and cultures and biopsy procedures are often necessary to make that diagnosis. Central vein (subclavian or jugular vein or superior vena cava) infectious thrombophlebitis and infectious valvulitis are often insidious and are potentially life-threatening. Careful evaluation of the use of central venous catheters has led to several useful proposals which may reduce the incidence of such infectious thrombophlebitis and valvulitis.

We are currently in the era of successful management of what were previously termed "lethal burns."[13] Use of microbiologically controlled environments, topical antimicrobial agents, improved nutritional support, and early surgical removal of burned tissue in burn patient management, and extensive use of biochemical and physiological monitoring, have improved the survival of patients with burns of up to 60 percent of the total body surface.

Each new drug, surgical technique, or change in supportive management of burn patients may present another group of problems for the professional burn team. These problems will be effectively managed only by the "team approach," as emphasized by Artz, Moncrief, and Pruitt.[14]

THE FUTURE IN BURN PATHOLOGY

We visualize a variety of exciting future developments which will directly involve laboratory investigators in their evaluation and implementation. Physicians now recognize that the skin is a vital, protective, complex glandular organ. Destruction of skin by thermal injury removes an effective barrier to infection and provides multiple portals for bacterial, mycotic, and viral entry. Cutaneous burns (thermal injury) also cause pathological and physiological changes in all other organ systems. Indeed, extensive burns are considered by some to be the "universal trauma model."[11] Prevention of infection and closure of the burn wound are prime objectives in caring for the burn patient. Use of adjuvants to accelerate removal of eschar and employment of biological or synthetic materials to cover the burn wound are obvious directions toward which therapy will progress. Some current efforts to improve the care of severely burned patients have utilized early resection of the burn wound, immune suppression, application of related donor allograft, and artificial wound coverings to the burn wound.[15–19] The pathologist will play a vital role in analyzing possible untoward local or systemic effects of such agents employed in the care of the burn wound.

Microbial contamination of the burn wound progressively increases as time elapses in the postburn period. Topical antimicrobial agents only suppress the growth of burn wound microorganisms, which remain an ever-present threat for burn wound invasion. Isolation of severely burned patients has decreased or delayed "cross-contamination" between patients.[17] Significant wound infection is particularly uncommon when eschar is slow to separate, but occurs with greater frequency after eschar separation when healing is delayed. Early burn wound excision may produce as yet unrealized improved survival of severely burned patients.[17,20] Other centers have demonstrated that enzymatic treatment accelerates debridement of eschar and permits, in selected patients, earlier closure.[21,22] However, patients referred to our institute who have been treated with enzymes have frequently suffered conversion of partial thickness burns to full thickness skin loss, and others have reported an association between use of enzymatic debriding agents and sepsis.[23] Additional, well-controlled trials are needed to determine the benefit of enzymatic treatment. These clinical trials should include close clinical evaluation of the burn wound coupled with wound biopsies to identify extension of the enzymatic activity beyond the burn wound. Successful use of enzymes for eschar debridement may depend on judicious patient selection (e.g., perhaps the thin skin of the elderly patient is not appropriate for enzyme therapy) and careful control of the dose and exposure period of the enzyme.

The development of septicemia marks a pivotal point in the postburn course. Delay in specific therapy of the disseminating infectious process may be fatal. In the future, gas chromatography may provide a rapid technique for identifying the offending bacteria within a few hours. This will be aided by better and more reliable techniques for diagnosing early endotoxemia, which will permit institution of therapy even before significant bacteremia occurs.

Rapid identification of burn wound infection is currently carried out by use of the burn wound biopsy, and the pattern for invasive infection appears to be most specific for *Pseudomonas aeruginosa*. Other gram-negative bacterial organisms and some cases of *Pseudomonas* infection produce a very nonspecific pattern of invasive infection. Although broad-spectrum antibiotic treatment is initiated as soon as invasive sepsis is identified, the antibiotic regimen may have to be modified when cultural identification and antibiograms of the specific microorganisms have been determined. Obviously, an alternate diagnostic method is necessary for these atypical infections. Early studies in the use of immunofluorescent-tagged specific microbes are promising and suggest that specific identification may be possible even in the absence of a characteristic histologic pattern of invasive infection. Although *Pseudomonas* toxoids have failed to give protection in animal experiments,[24] production of specific antibacterial vaccines may reduce the incidence of infection in the burn patient,[25] although a variety of vaccines have also proved ineffective to date.

Depressed humoral and cellular immune function in the burn patient suggests that prophylactic replacement of these deficient factors in severely burned patients will be an important adjunct in future patient care. Currently, immunotherapy of infectious disease in the burn patient is in an experimental stage and focuses on specific and nonspecific gammaglobulin therapy.[26] Monitoring of the T-lymphocytic system is currently employed for those severely burned patients undergoing immunosuppression with antithymocyte globulin to prolong allograft survival.[17] Additional evaluation of reticuloendothelial and neutrophilic function will also be essential. Adjuvant therapy with hematopoietic transplants and granulocytic transfusions may all have utility in improving defense mechanisms in the burn patient.[27]

In summary, it is apparent that much remains to be accomplished in the areas of surgery, clinical therapy, and diagnostic pathology of thermal injury. We anticipate that the pathologist will continue to be a vital contributor to the fund of scientific knowledge concerned with the care of the thermally injured patient. New, successful modes of therapy can only be developed with the aid of the concerned and knowledgeable pathologist who is cognizant of current developments and capable of detecting and defining new complications of burn therapy.

References

1. Lloyd, J. R.: Thermal trauma: Therapeutic achievements and investigative horizons. Surg. Clin. N. Amer. 57:121–138, 1977.

2. Artz, C. P.: Historical aspects of burn management. Surg. Clin. N. Amer. 50:1193–1200, 1970.

3. Scarborough, J.: On medications for burns in classical antiquity. Clin. Plast. Surg. 10:603–610, 1983.

4. Blumenfeld, A.: Cold-water immersion for burns (Editorial). New Engl. J. Med. 290:58, 1974.

5. Demling, R. H., Mazess, R. B., and Wolberg, W.: The effect of immediate and delayed cold immersion on burn edema formation and resorption. J. Trauma 19:56–60, 1979.

6. Lewis, S. R.: The Controversy of 2000 years—Closed or open therapy. *In* Burns: A Symposium. L. Goldman and R. E. Gardner, eds. Springfield, Ill., Charles C. Thomas, 1965, pp 61–69.

7. Cockshott, W. P.: The history of the treatment of burns. Surg. Gynec. Obstet. 102:116–124, 1956.

8. Sonnenburg, E., and Tschmarke, P.: Die Verbrennungen und die Erfrieryngen. Stuttgart, 1915, pp 6, 18.

9. Jackson, D. MacG.: A historical review of the use of local physical signs in burns. Brit. J. Plast. Surg. 23:211–218, 1970.

10. Harkins, H. N.: Acute ulcer of the duodenum (Curling's ulcer) as a complication of burns; relation to sepsis. Report of a case with survey of 107 cases collected from the literature, 94 with necropsy, 13 with recovery; experimental studies. Surgery 3:608–641, 1938.

11. Pruitt, B. A., Jr.: Forces and factors influencing trauma care: 1983 A.A.S.T. presidential address. J. Trauma 24:463–470, 1984.

12. Teplitz, C.: The pathology of burns and the fundamentals of burn wound sepsis. *In* Burns. A Team Approach., 1st ed. C. P. Artz, J. A. Moncrief, and B. A. Pruitt, Jr., eds. Philadelphia, W. B. Saunders Co., 1979, pp 45–94.

13. Monafo, W. W., Robinson, H. N., Yoshioha, T., and Ayvazian, V. H.: "Lethal" burns progress report. Arch. Surg. 113:397–401, 1978.

14. Artz, C. P., Moncrief, J. A., and Pruitt, B. A.: Burns. A Team Approach. Philadelphia, W. B. Saunders Co., 1979.

15. Burke, J. F., May, J. W., Jr., Albright, N., Quinby, W. C., and Russell, P. S.: Temporary skin transplantation and immunosuppression for extensive burns. N. Engl. J. Med. 290:269–271, 1974.

16. Burke, J. F., Quinby, W. C., Jr., Bondoc, C. C., Cosimi, A. B., Russell, P. S., and Szyfelbein, S. K.: Immunosuppression and temporary skin transplantation in the treatment of massive third degree burns. Ann. Surg. 182:183–195, 1975.

17. Burke, J. F., Quinby, W. C., and Bondoc, C. C.: Early excision and prompt wound closure supplemented with immunosuppression. Surg. Clin. N. Amer. 58:1141–1150, 1978.

18. Burke, J. F.: Observations on the development of an artificial skin. J. Trauma 23:543–551, 1983.

19. Wise, D. L.: Burn Wound Coverings. CRC Press, Ames. Iowa: Iowa State, 1984.

20. Jackson, D. MacG.: Second thoughts on the burn wound. J. Trauma 9:839–862, 1969.

21. Dimick, A. R.: Experience with the use of proteolytic enzyme (Travase[R]) in burn patients. J. Trauma 17:948–955, 1977.

22. Zawacki, B. E.: The effect of Travase on heat-injured skin. Surgery 77:132–136, 1975.

23. Hummel, R. P., Kautz, P. D., MacMillan, B. G., and Altemeier, W. A.: The continuing problem of sepsis—Following enzymatic debridement of burns. J. Trauma 14:572–579, 1974.

24. Walker, H. L., McLeod, C. G., Jr., Leppla, S. H., and Mason, A. D., Jr.: Evaluation of *Pseudomonas aeruginosa* Toxin A in experimental rat burn wound sepsis. Infection and Immunity 25:828–830, 1979.

25. Wasserman, D., Schlotterer, M., Paul, P., and Rieu, M.: Systematic utilization of an anti*Pseudomonas aeruginosa* vaccine in a severe burn unit. Scand. J. Plast. Reconstr. Surg. 13:81–84, 1979.

26. Jones, R. J., Roe, E. A., and Gupta, J. L.: Low mortality in burned patients in a *Pseudomonas* vaccine trial. Lancet 2:401–403, 1978.

27. Workman, R. D., Faville, R. J., Strate, R. G., Quie, P. G., Jager, R. M., and McCullough, J.: Granulocyte transfusions for patients with severe thermal burns. Transfusion 18:142–148, 1978.

Chapter 2

CLASSIFICATION OF BURN WOUND INJURY AND MECHANISMS OF REPAIR

At postmortem, the pathologist lacks certain clinical parameters which are vital for assessing the depth of burn injury, e.g., the degree of cutaneous sensitivity and integrity of the circulation. Careful review of premortem burn wound evaluation, combined with postmortem histologic examination, yields the most accurate information as to the depth and type of thermal injury. At autopsy, the pathologist should also make his own estimation of the percentage of body surface burned (see Chapter 4) and compare this with clinical estimates.

TYPES OF BURNS

Flame burns are the most common type of thermal injury and usually result from the ignition of the victim's clothing. Borders of flame burns tend to be irregular and the depth variable (Fig. 2–1).

Flash burns result from sudden explosion of gases, finely particulate material suspended in air, or from an electric arc. Exposed body surfaces are burned uniformly (Fig. 2–2) (usually a partial thickness burn). Flash burns may have mixed zones of flame injury if clothing is ignited.

At the skin surface, *electric* contact and exit burns are usually well-circumscribed, although there may be massive destruction of deeper structures (Fig. 2–3). *Arc* burns often occur with electric shock injury and may be indistinguishable from flame burns. As with flash burns, there may be secondary flame burns if clothing is ignited.

Scald burns usually involve infants and young children. "Spill" scalds

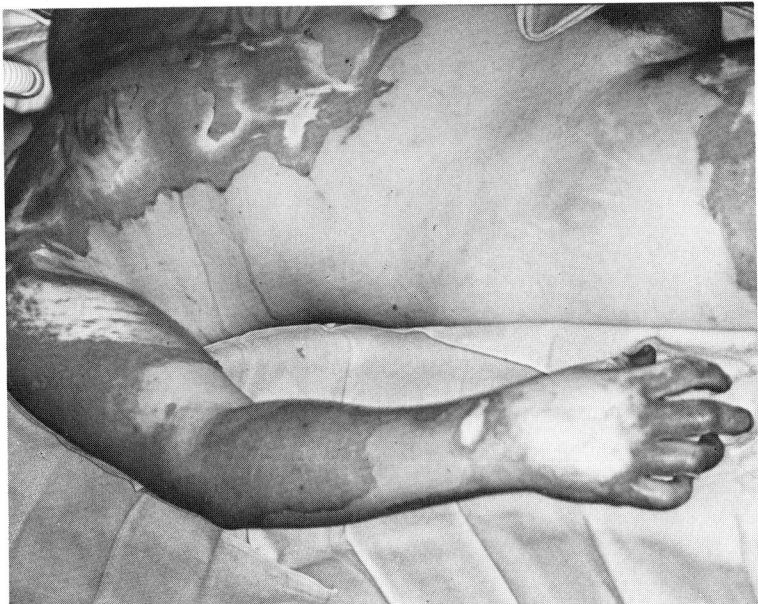

Fig. 2–1. A third degree flame burn usually has irregular borders and a variable depth of injury.

have characteristically irregular borders and may involve the side of the head, ipsilateral shoulder and upper trunk.[1] The wounds are usually superficial and have vesicles and bullae with a red moist surface. "Immersion" scalds are characterized by straight, uniform margins demarcating the depth of immersion (Fig. 2–4). The burn is at least partial thickness and is often full thickness. The red appearance of a deep scald burn wound is caused by the presence of hemoglobin fixed in the tissues. The hard, dry, inelastic, and transparent form of scald injury is usually full thickness.

It is often possible in the acute stages to differentiate flame from scald burns. Flame injuries will often have the epidermis "seared on," whereas scald burns cause the epidermis to be uplifted by edema.[2] In addition, the zone between coagulated tissue and subjacent viable tissue is more regular and linear in scald burns.[3]

DEPTH OF SURFACE BURNS

This area of burn pathology has undergone extensive evaluation.[4] Our intent is to review briefly the morphologic changes seen in burns of various depths. The mildest change following thermal injury occurs in *superficial*

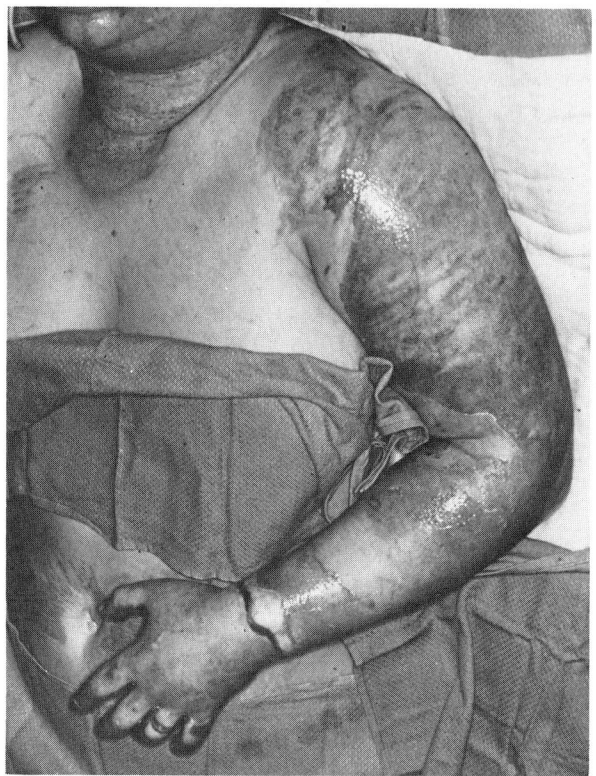

Fig. 2–2. Flash burns tend to have a uniform depth of injury and are often of partial thickness.

partial thickness (first degree) burns and consists of a mild erythematous alteration of the skin sometimes with vesicle formation. Prominent edema causes the wound surface to be elevated above the surrounding unburned tissue. Histologically, there may be hydropic swelling of the epithelial nuclei and superficial adnexa, and chromatin is displaced toward one side of the nucleus. More severe lesions may lead to actual nuclear pyknosis and cytoplasmic vacuolization. The erythema is due to hyperemia of superficial dermal capillaries with occasional extravasation of erythrocytes (Fig. 2–5). Collagen fibers may be separated by edema.

The *deep partial thickness* (second degree) burn is characterized by a soft, dry, waxy, white appearance after the devitalized material is removed. The tissue is not initially edematous and is not sensitive to pinprick, but perception of deep pressure is still intact. Histologically, necrosis occurs in both the dermis and epidermis. The epithelial nuclei show marked pyknosis and cytoplasmic alterations. Development of intercellular edema causes detachment of the basal cells from the epidermal basement membrane and leads to formation of a subepidermal bulla (Fig. 2–6). The basal cells soon become detached from surrounding cells and some may be observed within the vesicular or bullous fluid. The dermal injury consists of progressive eosinophilia, swelling, and fusion of collagen fibers. Blood vessels, especially venules within the dermis, are occluded by thrombi which are predominantly erythrocytic in composition. Deep partial thickness burns may destroy some of the adnexal structures, but the capacity for regeneration is still present.

Full thickness (third degree) burns yield a white-to-black, hard, "leathery," inelastic eschar that may have a glistening, apparently translucent surface. The wound is insensitive to all but deep pressure. Coagulative necrosis affects the entire thickness of the epidermis and dermis and usually extends into subcutaneous fat. These changes result in prominent shrinkage of both the nucleus and cytoplasm of each epithelial, fat, and connective tissue cell within the affected area. Nuclei of epithelial cells may retain their distorted pyknotic appearance for days, and this heat coagulation of the epithelium has been compared to a crude form of "histologic fixation" (Fig. 2–7).[2] Similar changes of coagulative necrosis occur in the eccrine sweat

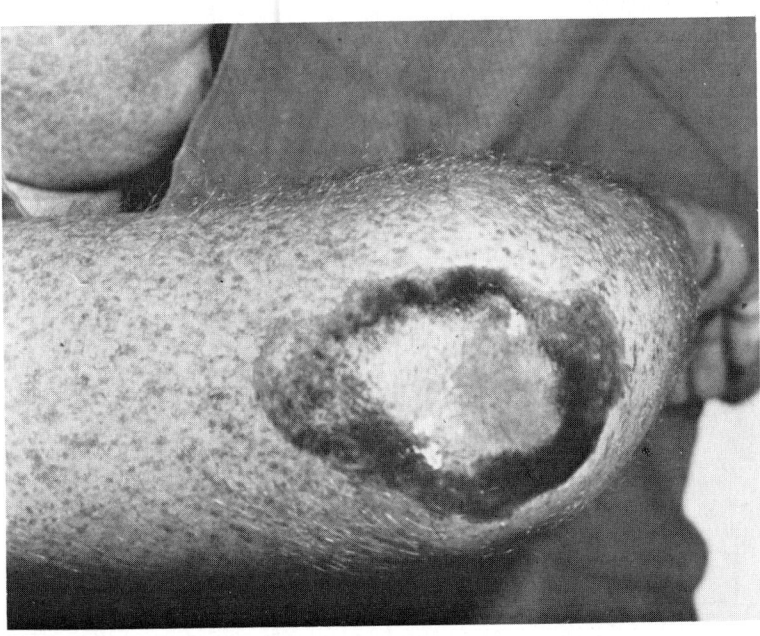

Fig. 2–3. An electric burn usually has well circumscribed contact and exit sites of burn injury with massive destruction of deep tissues.

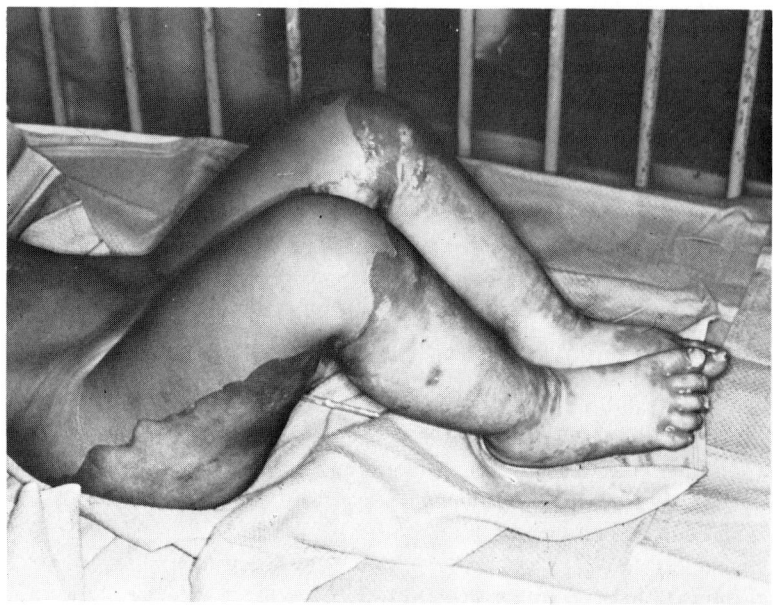

Fig. 2–4. Immersion scald burns usually have straight uniform margins which are indicative of the depth of immersion.

glands and sebaceous glands. Deep dermal and subcutaneous blood vessels often contain thrombi.

Because of the varying sensitivity of different tissues to injury, collagen which appears viable may be intermixed with necrotic adnexal structures in the early postburn period.[5,6] This arrangement complicates differentiation of deep partial-thickness burns from full thickness burns. Necrotic changes in dermal collagen may be subtle. Edema is usually conspicuous at the junction between coagulated and viable tissue;[3] and bundles of homogenized, coagulated collagen fibers have a striking affinity for acid dyes such as orange G and analine blue.[2] Masson's trichrome stain may be useful in discriminating dead collagen (pink to red) from viable collagen (blue).[6]

FUNCTIONAL ZONES OF THE BURN WOUND

Close observation of the burn wound demonstrates three separate "functional" zones in the immediate postburn period: the zone of immediate coagulation, surrounded by a rim of capillary stasis, which in turn is surrounded by a region of hyperemia (Fig. 2–8).[5,7] The outer *zone of hyperemia* is characterized by an almost complete loss of epidermis but no apparent changes (with the exception of hyperemia) in the dermis. Digital pressure on this area causes blanching of the tissue; release of pressure results in return of the former erythematous appearance. A pressure cuff used to constrict the venous supply of an extremity burn wound causes this zone of hyperemia to become cyanotic.

The intermediate *zone of vascular stasis* is similarly erythematous in the early postburn period and blanches with digital pressure; color reappears when the pressure is released. Cuff pressure occluding venous outflow causes the immediate area to turn white instead of cyanotic. This zone of vascular stasis undergoes progressive vascular occlusion during the first 24 hr postburn and terminates in necrosis of tissue in this intermediate area.

The central *zone of coagulated tissue* appears pale white and is unaffected by digital pressure or venous occlusion. The progressive development of necrosis in the intermediate zone within the first 24 hr may explain the apparent increment of third degree burn that is often noted on the second postburn day. The changes that occur in this intermediate "functional" zone may be related to a hypercoagulable state (vascular stasis and injured blood vessels) in the local microvasculature which leads to thrombosis. It is well known that there is a gradual and progressive disappearance of small blood

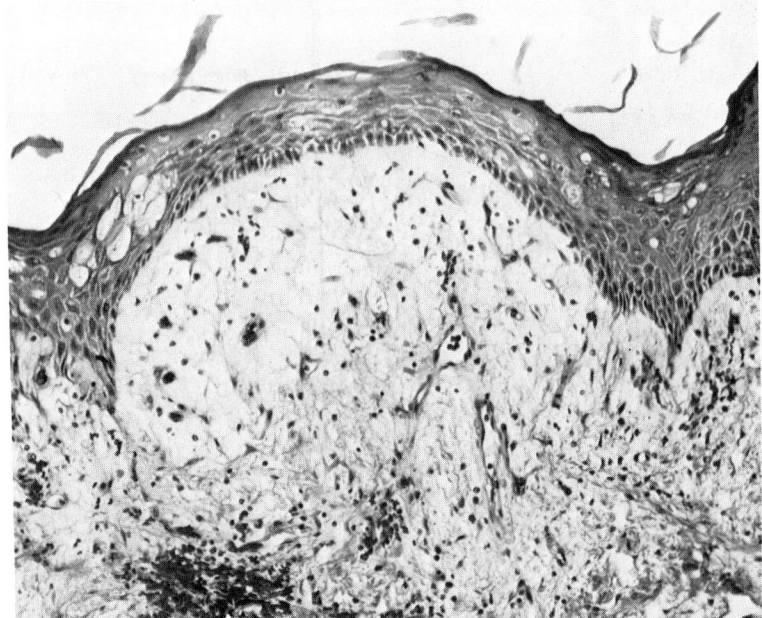

Fig. 2–5.   Partial thickness burns have mild vesiculations, erythema, and very minimal nuclear pyknosis. [hematoxylin and eosin (H & E), × 126]

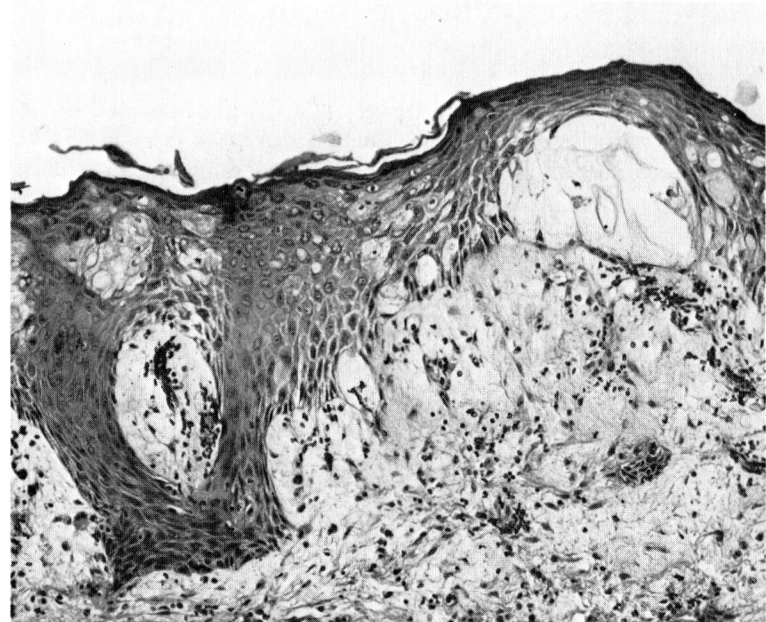

Fig. 2–6. Subepidermal and intraepidermal vesicles and bulla are seen in a deep partial thickness (second degree) burn. Superficial adnexal structures may be focally damaged, but deep hair follicles remain viable. (H & E, × 126)

vessels in the periphery of the burn wound. With wound infection or other stresses, deep partial thickness burns may progress to full thickness through a reduction of patent blood vessels. The reported improvement in burn wound healing and prognosis in patients treated early with anticoagulants may be an effect of the reduction of both microvascular thrombosis and occlusion in the burn wound.[8,9] Similarly, the progression of deep partial thickness to full thickness burns which occurs following wound infection may be a result of increased vascular thrombosis. Experimental data suggest that corticosteroids retard this progression of partial to full thickness wounds by maintaining vascular microcirculation.[10]

A report by Massiha and Monafo[11] has demonstrated that the development of dermal ischemia which occurs in the zone of vascular stasis is related to occlusive events in the venous system and not the arterioles or capillaries. Order et al.[12] stressed the importance of arteriolar occlusion in their experimental study in burned rats; however, as Massiha and Monafo[11] suggest, these arteriolar abnormalities may be secondary to the occlusive venular thrombosis which restricts[6] capillary outflow. The thin-walled venules would logically be more susceptible to thermal injury than arterioles. Jackson[5] has shown microscopically patent arterioles adjacent to thrombosed venules in skin 24 hr postburn.

In a more recent experiment, preservation of the microcirculation in the zone of vascular stasis was extended from 24 hr to 72 hr through the use of anticoagulants.[9]

The polymorphonuclear leukocytic response seen at the base of full thickness burns is variable and may not appear for 8–10 days. Some investigators report a heightened acute inflammatory response to infection;[6] however, the acute inflammatory response may also be nonspecifically elevated in response to necrotic tissue. The most severe infections of the burn wound by *P. aeruginosa* often have a sparse to absolute absence of inflammatory response.[2]

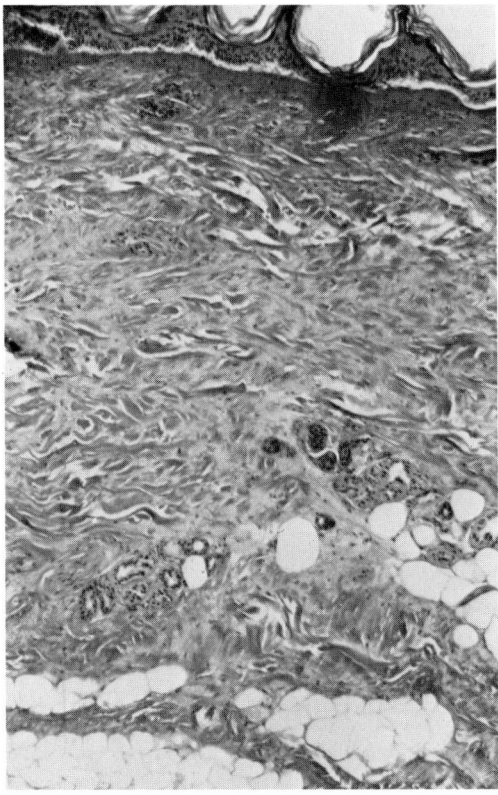

Fig. 2–7. Full thickness (third degree) thermal injury results in total necrosis of the epidermis and all adnexal structures. Dermis and epidermis have a "fixed" appearance with no evidence of vesiculation or edema. Nuclei of rounded fat cells are absent or pyknotic. (H & E, × 90)

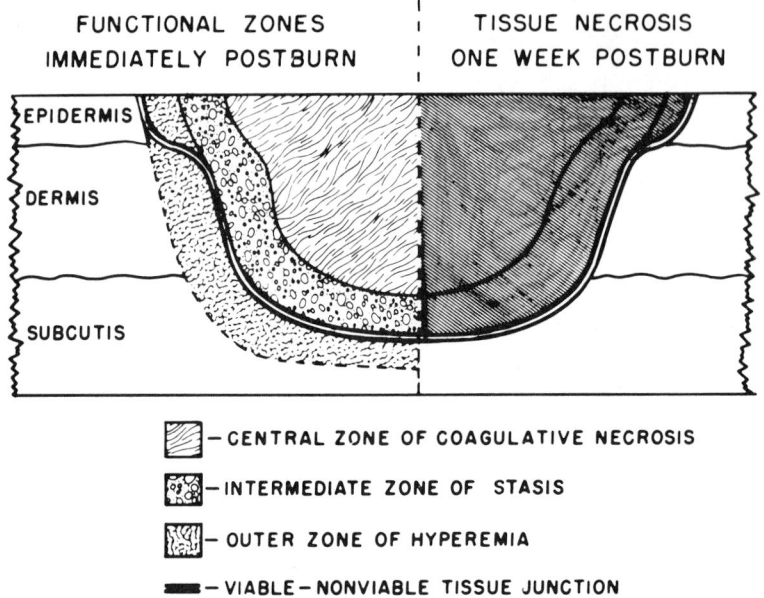

FUNCTIONAL ZONES
IMMEDIATELY POSTBURN

TISSUE NECROSIS
ONE WEEK POSTBURN

EPIDERMIS

DERMIS

SUBCUTIS

▨ — CENTRAL ZONE OF COAGULATIVE NECROSIS

▨ — INTERMEDIATE ZONE OF STASIS

▨ — OUTER ZONE OF HYPEREMIA

▬ — VIABLE – NONVIABLE TISSUE JUNCTION

Fig. 2–8.   Functional zones of a burn wound.

HEALING BURN WOUND

Classically, granulation tissue usually appears at the base of the burn wound within the first postburn week (Fig. 2–9A). In severe burns, the appearance of granulation tissue may be delayed for two or three weeks. Delayed appearance of granulation tissue may also be seen in patients with septicemia and its associated severe metabolic abnormalities.[3]

The granulation tissue is composed of a complex network of fibroblasts, capillaries, inflammatory cells, and reticular and collagen fibers. The initial event in the development of healthy granulation tissue appears to be the proliferation and ingrowth of capillaries at the margin of viable tissue. Fibrin and fibronectin are consistent components of healing wounds and appear to serve as a framework for the development and ingrowth of both neovascular and fibroblastic tissues (Fig. 2–9B).[13] Capillary proliferation is closely followed by the appearance of fibroblasts, macrophages, chronic inflammatory cells, and mast cells. Factors and conditions which stimulate and mediate the repair mechanism are not well understood, but include tissue hypoxia, hypotension, and tissue metabolic changes.

As repair continues in the dermis, it is characterized by a continued

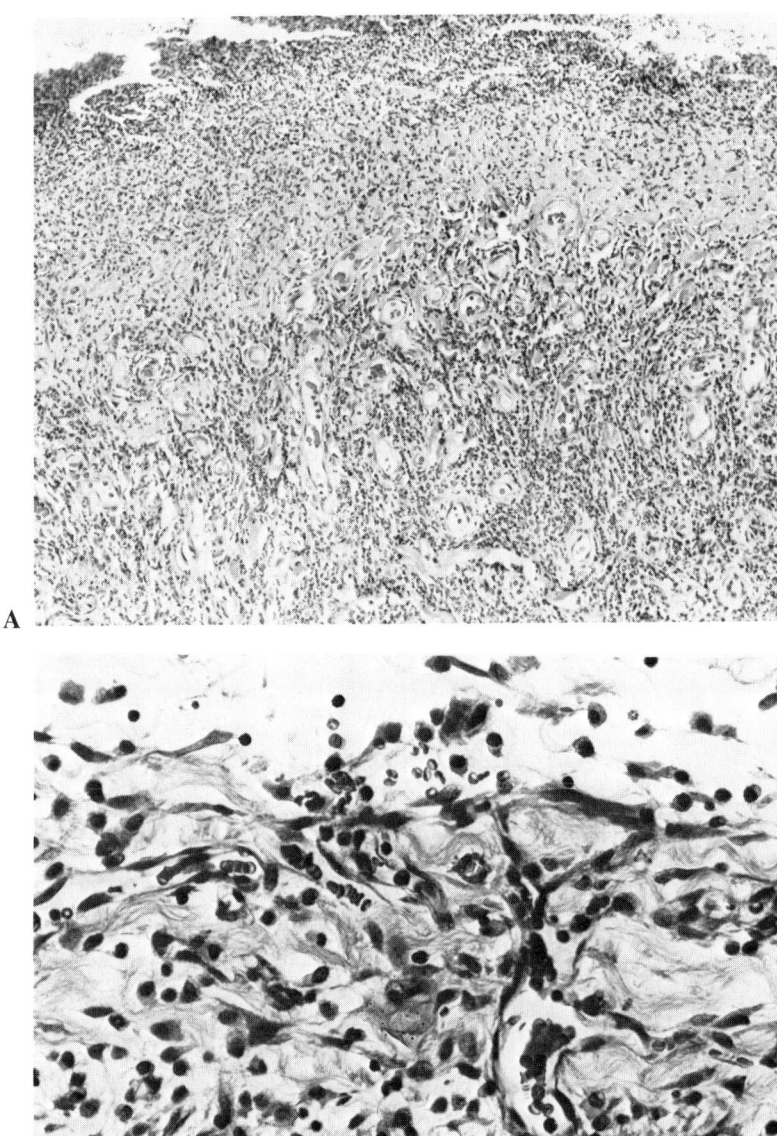

Fig. 2–9. (A): Granulation tissue develops within 5–7 days following injury. Proliferating new capillaries are seen at the base of this full thickness burn wound. Wound surface consists of necrotic leukocytes and fibrin. (H & E, × 126) (**B**): Higher magnification of the zone between viable tissue and the burn wound surface. Acellular fibrillar material is fibrin which appears to serve as a framework for budding capillaries and fibroblasts. Note macrophages and polymorphs which may be numerous, even in uninfected wounds. (H & E, × 270)

17

proliferation and maturation of fibroblasts, which eventually form a prominent immature collagen network (Fig. 2–10) containing abundant glycoproteins and mucopolysaccharides.[14] Capillaries become progressively compressed and reduced in size and number during wound repair and gradually diminish as the wound is covered by new epithelium (Fig. 2–11). Myofibroblasts appear to play an important role in wound healing, specifically contraction,[15] and their variances in distribution and numbers in different types of wounds (burns versus cold injury) have recently been studied.[16] These specialized fibroblasts appear in deep partial and full thickness burns at 3–5 days after trauma.[17] These functionally unique cells are attached to other cells and adjacent myofibroblasts by a "myofibroblast-anchoring substance" which is composed of collagen and fibronectin.[18] Although the number of myofibroblasts is initially small, it progressively increases and may comprise 50–100 percent of the fibroblast population several weeks after the burn.[17] The unidirectional alignment and contraction of these myofibroblasts is important in wound repair (Fig. 2–12). Pressure over the burn wound has been associated with more filamentous and longer myofibroblasts.[19] Exuberant granulation tissue is most often seen in patients

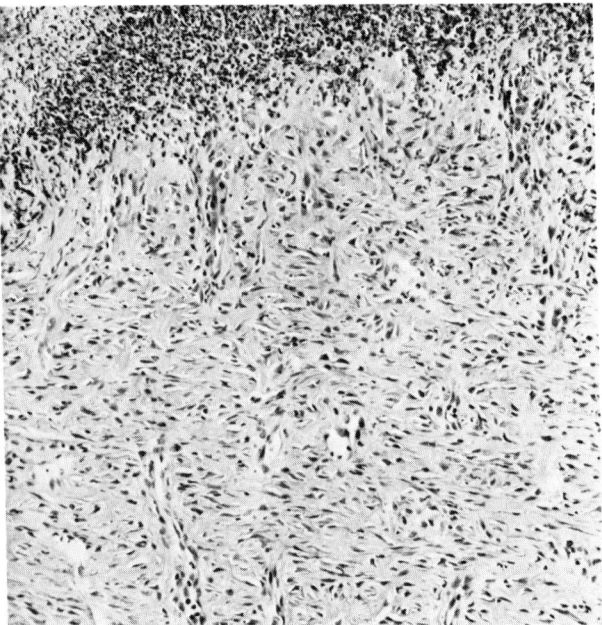

Fig. 2–10. Continued fibroblastic activity and maturation result in early collagen deposition. Zone of intense hypercellularity represents lower border of nonviable tissue. (H & E, × 126)

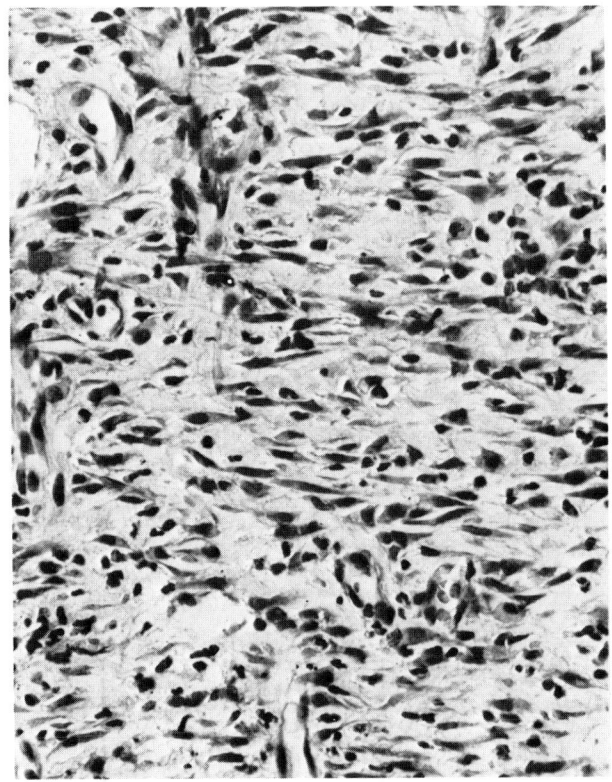

Fig. 2–11. Normal wound maturation results in reduction in size and number of capillaries. This change is most evident directly beneath newly formed epidermis. (H & E, × 216

with deep partial thickness burns, and it is wounds of this type that may lead to hypertrophic scars in children and young adults. (Fig. 2–13) As collagen matures, the number of fibroblasts decreases and the plump fibroblast is replaced by slender fibrocytes (Fig. 2–14) which are presumably less active. Collagen fibers tend to be oriented parallel to the skin surface, and the amount of ground substance within the wound decreases with maturity.[14]

Regeneration of the epidermis in a partial thickness burn starts within 42 hr of thermal injury.[17] Increased mitotic activity of the basal cells of the stratum spinosum is identified in squamous epithelium at the margin of the burn wound and within viable remnants of skin appendages in the burn wound. A diurnal rhythm has been demonstrated in regenerating epithelium which appears most active during rest or sleep.[20] During migration, squamous cells penetrate blood and fibrin by production of proteolytic en-

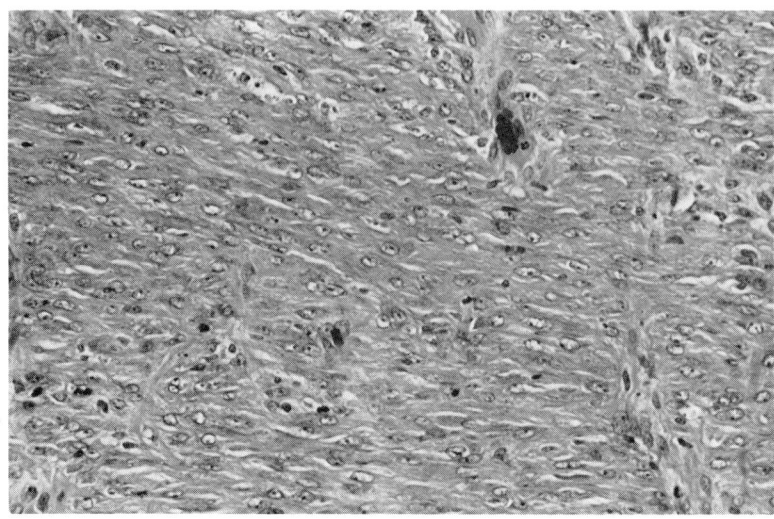

Fig. 2–12. Specialized fibroblasts or "myofibroblasts" are present in large numbers in healing full thickness wounds. These cells are arranged in bundles parallel to the wound surface, and their contractile activity may be important in closure of small wounds. (H & E, × 216)

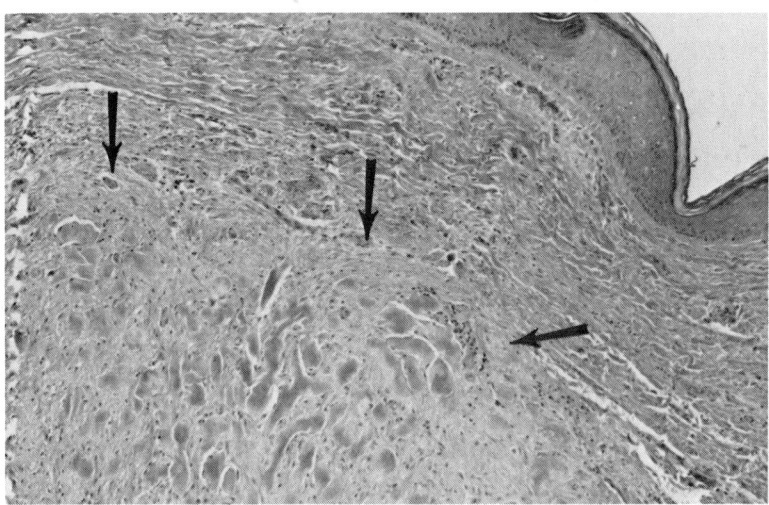

Fig. 2–13. Abnormal organization and differentiation of fibroblasts in some patients result in whorling masses (arrows outline margin) of abnormal thickened collagen bundles and subsequent hypertrophic scars. Overlying epidermis has hyperplastic changes. (H & E, × 126)

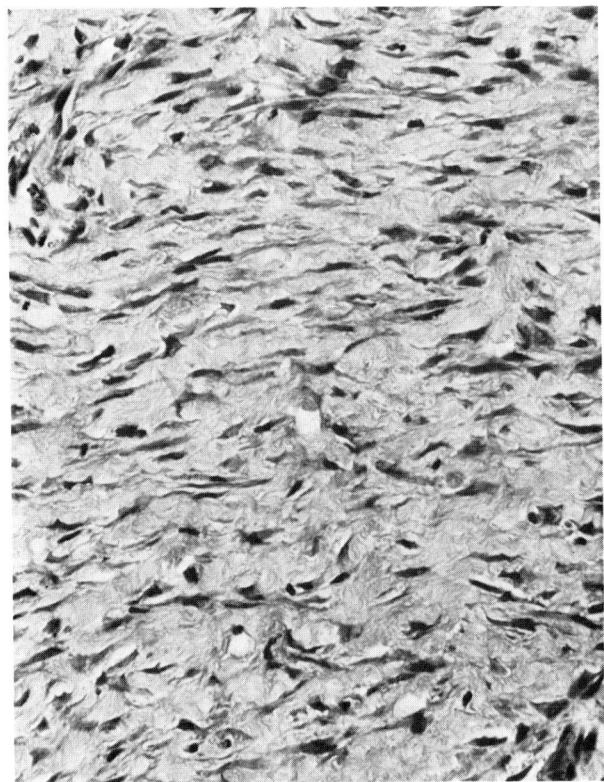

Fig. 2–14. Fibroblasts become slender and less prominent as mature collagen develops within the wound. (H & E, × 252)

zymes.[6,21] The rate of epithelialization under the bulla of a second degree burn appears to be twice that of similar desiccated areas. Partial epithelialization appears to precede basement membrane formation (Fig. 2–15). In third degree burns, epithelialization usually progresses no further than 1 cm from viable tissue adjacent to the burn wound.[17] This complex process of burn wound healing may be disrupted by a number of pathologic states that commonly occur in burn patients (Table 2–1).

ELASTIC TISSUE IN SCARS

Scar tissue lacks elastic tissue in the early phases of healing, and with healing, fine reticular elastic fibers appear. Hypertrophic scars (keloids) often lack elastic fibers and have a moderate chronic inflammatory (lympho-

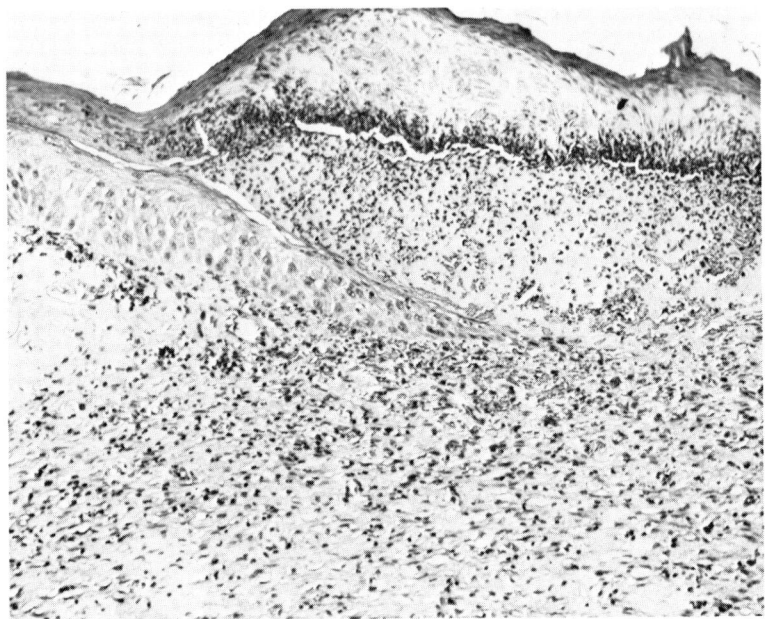

Fig. 2–15. Advancing new epidermis undermines necrotic debris on the wound surface. (H & E, × 126)

cytes and plasma cells) infiltrate and scattered giant cells. When elastic fibers appear in keloids, they are clumped and irregularly distributed in contrast to nonhypertrophic scars.[22] The explanation for the initial lack of elastic fibers and the subsequent clumped and frayed appearance of elastin in keloids is unknown. Antibodies against elastin have been indentified in serum and may play a role in the retarded growth of elastic tissue in scars.[22,23] Also, multinucleated giant cells appear to have a prominent role in elastolysis.[24] Keloids are said to have depressed levels of copper which may adversely affect elastin biosynthesis.[25]

TECHNICAL DIFFICULTIES IN PROCESSING
BURN WOUND BIOPSIES

Sizeable technical difficulties must be surmounted by the pathologist and his technician before the histologic sections of some burn wound biopsies are ready for interpretation. Marked dehydration of thermally injured skin occurs with the passage of time. The resultant superficial tissue is hardened eschar which has been compared to shoe leather by many

Table 2–1. *Conditions Which Impair Wound Healing*

Incomplete excised necrotic tissue from initial injury or subsequent rough handling
Delayed or incomplete eschar separation
Foreign material—residual nidus of inflammation/infection
Shock—poor perfusion
Septicemia
Abnormal metabolic/electrolytic milieu
Acidosis with or without hypoxia
Hepatic failure—leading to abnormal protein synthesis, etc.
Renal failure—uremia
Prolonged steroid administration—associated with nitrogen and potassium
    depletion
Immune suppression—renders the immune system ineffective against bacteria,
    viruses, fungi
Excessive mobilization—damages wound bed, tears off grafts
Starvation—lack of adequate caloric, protein, mineral, and vitamin intake

---

frustrated histotechnologists. The contrast between hard desiccated eschar and the soft malleable subcutaneous tissue often results in "tearing" and "crush" artifacts during the sectioning of the tissue block. These tissue alterations may severely limit the pathologist's ability to interpret selected specimens.

EVALUATION OF THE BURN WOUND BIOPSY

The pathologist must be aware of all the useful information available in his burn wound biopsies. Of course, the surgeon and clinician are immediately interested in the presence or absence of "invasive" infection,[26] but the following histologic determinations should also be made when possible:

1. Burn wound depth—presence or absence of viable adnexal structures.
2. Nature and degree of the inflammatory response at the junction of viable and nonviable tissue.
3. State of repair—epithelial regeneration, maturity of granulation tissue.
4. Presence of intravascular thrombi—disseminated intravascular coagulopathy.
5. Presence of colonizing bacterial or fungal organisms—these may be of significance in forthcoming biopsies.

References ————————————————————————————————

1. Moncrief, J. A.: Burns I. Assessment. J. A. M. A. 242:72–74, 1979.
2. Teplitz, C: Pathology of burns. *In* Artz, C. P., and Moncrief, J. A.: The Treatment of Burns, Chapter 2, 2nd. ed. Philadelphia, W. B. Saunders Co., 1969, pp 22–88.
3. Teplitz, C: The pathology of burns and the fundamentals of burn wound sepsis. *In* Artz, C. P., Moncrief, J. A., and Pruitt, B. A., Jr.: Burns. A Team Approach. Philadelphia, W. B. Saunders Co., 1979, pp 45–94.
4. Galbraith, S., and Foad, M.: Burn appearance and spontaneous healing: A prospective study. Burns 8:317–320, 1982.
5. Jackson, D. M.: Second thoughts on the burn wound. J. Trauma 9:839–862, 1969.
6. Winter, G. D.: Histological aspects of burn wound healing. Burns 1:191–196, 1975.
7. Jackson, D: The diagnosis of the depth of burning. Brit J. Surg. 40:588–596, 1953.
8. McCleery, R. S., Schaffarzick, W. R., and Light, R. A.: An experimental study of the effect of heparin on the local pathology of burns. Surgery 26:548–564, 1949.
9. Noble, H. G. S., Robson, M. L., and Krizek, T. J.: Dermal ischemia in the burn wound. J. Surg. Res. 23:117–125, 1977.
10. Robson, M. C., Kucan, J. O., Paik, K. I., and Eriksson, E.: Prevention of dermal ischemia after thermal injury. Arch. Surg. 113:621–625, 1978.
11. Massiha, H., and Monafo, W. W.: Dermal ischemia in thermal injury: The importance of venous occlusion. J. Trauma 14:705–711, 1974.
12. Order, S. E., Mason, A. D., Jr., Walker, H. L., Lindberg, R. F., Switzer, W. E., and Moncrief, J. A.: The pathogenesis of second and third degree burns and conversion to full thickness injury. Surg. Gynec. Obstet. 120:983–991, 1965.
13. Grinnell, F., Billingham, R. E., and Burgess, L.: Distribution of fibronectin during wound healing in vivo. J. Invest. Derm. 76:181–196, 1981.
14. Parks, D. H., Evans, E. B., and Larson, D. L.: Prevention and correction of deformity after severe burns. Surg. Clin. N. Am. 58:1279–1289, 1978.
15. Gabbiani, G., Ryan, G. B., and Majno, G.: Presence of modified fibroblasts in granulation tissue and their possible role in wound contraction. Experimentia 27:549–550, 1971.
16. Ehrlich, H. P., and Hembay, R. M.: A comparative study of fibroblasts in healing freeze and burn injuries in rats. Am. J. Path. 117:218–224, 1984.
17. Baur, P. S.: The healing of burn wounds. Tex. Soc. Electron Micros. Newsletter 8:15–23, 1977.
18. Baur, P. S., Jr.: Myofibroblasts: Their attachments and consequences in human burn wounds, granulation tissues and scars. J. B. C. R. 3:214–219, 1982.
19. Bauer, P. S., Barrott, G. F., Linares, H. A., Dobrkovsky, M., de la Houssaye, A. J., and Larson, D. L.: Wound contractions, scar contractures and myofibroblasts: A classical case study. J. Trauma 18:8–22, 1978.

20. Van Winkle, W.: The epithelium in wound healing. Surg. Gynecol. Obstet. 127:1089–1115, 1968.
21. McMinn, R. M. H.: Wound healing. The Cell in Medical Science, Vol. 4. New York, Academic Press, 1976, pp 321–356.
22. Linares, H. A., and Larson, D. L.: Elastic tissue and hypertrophic scars. Burns 3:4–7, 1976.
23. Stein, F., Pezess, M. P., Robert, L., and Poulain, N.: Anti-elastin antibodies in normal and pathological human sera. Nature 207:312–313, 1965.
24. Shionoya, S., Tsunekawa, S., and Kamiya, K.: Elastolysis and giant cell reaction against disintegrated elastic fibers. Nature 207:311–312, 1965.
25. Carnes, W. H.: Brief communication: Failure of elastogenesis in copper deficiency. Clin. Sci. 38:9, 1970.
26. Pruitt, B. A., Lindberg, R. B., McManus, W. F., and Mason, A. D., Jr.: Current approach to prevention and treatment of *Pseudomonas aeruginosa* infections in burned patients. Reviews of Infectious Diseases 5:889–897, 1983.

Chapter 3

THE BURN WOUND BIOPSY

Therapy for burn wound infection is based on the results of histological examination of wound biopsies, as well as quantitative bacterial cultures.[1,2,3] Specimens are selected from areas of the burn wound suspected of having invasive infection using the criteria listed in Table 3–1. A portion of the biopsy is submitted to the microbiology laboratory for quantitative cultures and the remaining tissue is retained for histologic examination. The presence of desiccated eschar as well as the need to evaluate subcutaneous tissue for microbiologic invasion precludes the use of frozen sections. Selected suspicious skin biopsies are rapidly processed on the Ultra Autotechnicon (Appendix 3–1) and sections are subsequently stained with hematoxylin and eosin, periodic acid-Schiff, and tissue Gram's stains (Appendix 3–2). If indicated, the mycotic and bacterial stains are combined by staining a single section with PAS and Giemsa (Appendix 3–3). The entire histologic procedure, including special stains, can be performed in 4 hr or less.

As with the other areas of pathology, full evaluation of the specimen requires knowledge of certain clinical aspects which include

1. Site of biopsy and whether the biopsy is from the burn wound or another site
2. Postburn day of biopsy
3. Extent of burn wound at site of biopsy
4. General appearance of the burn wound—color, nature of exudate, hemorrhage
5. Presence (or absence) of topical antibiotics
6. Use of autograft or homograft at biopsy site
7. History of previous tangential or scalpel excision of the burn wound
8. Use of enzymatic debriding agents
9. Status of coagulation system

Table 3–1 *Selection of Sites for Evaluation of Invasive Infection of the Burn Wound*

Direct gross observation of the burn wound by the pathologist

Hemorrhagic areas—*Pseudomonas aeruginosa*
Zygomycetes
Advancing violaceous wound margins
Metastatic abcesses (lesions) in unburned tissue
Excavation (small) in burn wound, especially in second degree burn areas
Vesicular lesion in healing or healed burn wound
Necrosis and discoloration of subcutaneous fat
Skin grafting sites, especially where there is poor autograft acceptance, or accelerated homograft rejection
Clinical history
Conversion of second degree burn to third degree burn
Unexpectedly rapid eschar separation
Lysis and degeneration of granulation tissue with "neoeschar" formation

To insure that all of these clinical data are available, we have devised a form which is completed by our surgical staff prior to submission of the biopsy (Appendix 3–4).

EVALUATION OF THE BIOPSY

Since invasion of the burn wound is defined as extension of organisms with necrosis beyond (deeper than) the thermally injured tissue, an adequate biopsy must contain viable dermal (second degree burn) or subcutaneous (third degree burn) tissue. If the burn is severe, it may be necessary to include deep subcutaneous tissue. Bacteria and/or fungi even deep in the eschar of a superficial biopsy specimen, albeit suspicious, represent "colonization," and only the inclusion of viable tissue in the specimen will permit one to assess whether there has been "invasion."

Prior to a search for microorganisms, one must establish the depth of coagulative necrosis due to the thermal injury. Although the depth of necrosis related to the burn may vary somewhat, any significant deep extension into viable tissue should alert one to the possibility of invasive infection (Fig. 3–1A). The pathologist may actually choose to mark the depth of burn on the hematoxylin-and-eosin stained sections. Special stains often yield poor differential staining between viable and nonviable tissue; there-

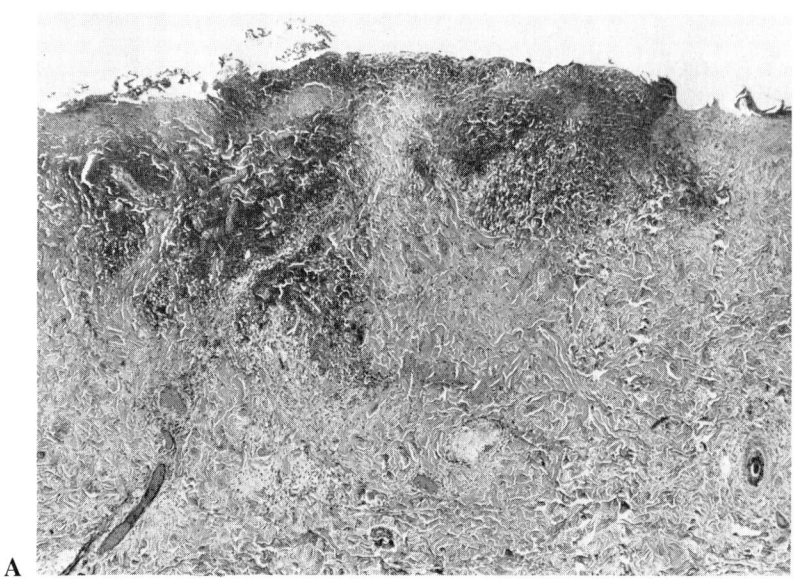

A

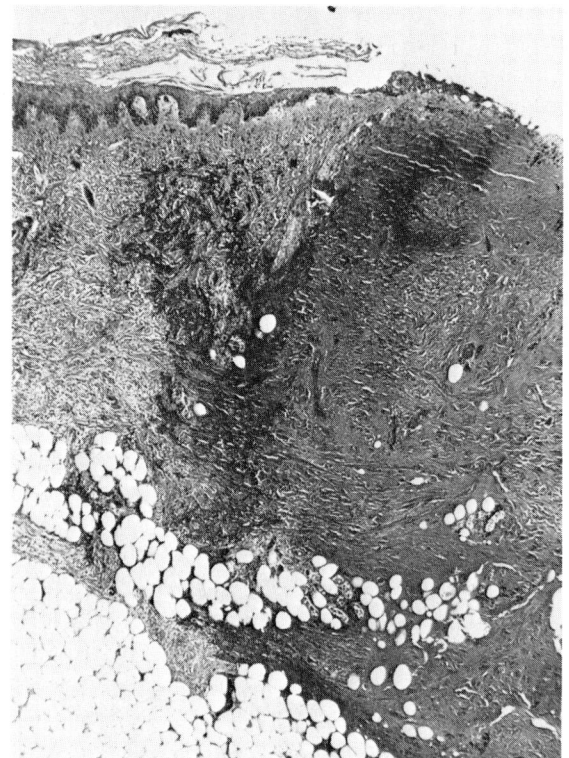

B

Fig. 3–1 A, B.

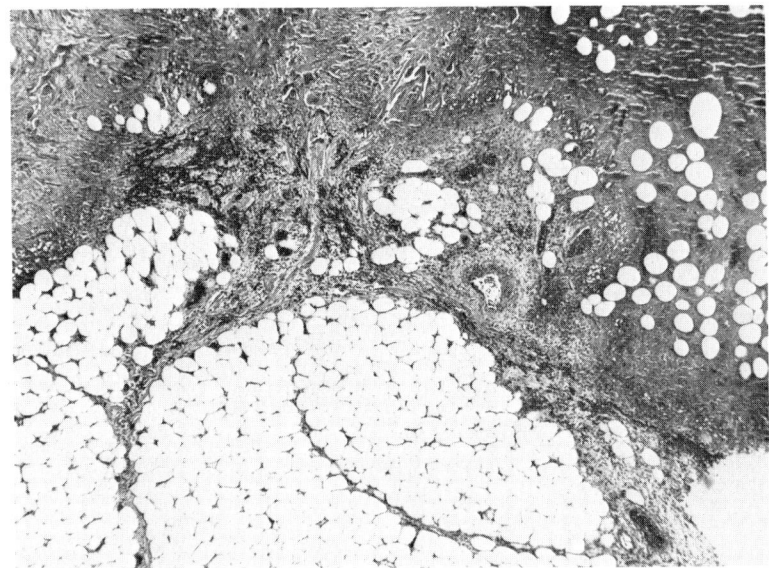

Fig. 3–1. (A): Full thickness thermal injury with recent extension of necrosis into viable collagen. (H & E, × 31.5) (**B**): "Invasive" infection may be seen as extension of necrosis at the base or lateral border of the burn wound margin, as in this case of invasive *Pseudomonas aeruginosa* infection. (H & E, × 31.5) (**C**): In this section, an irregular junction between the burn wound and viable fat are highly suggestive of infection. Hemorrhage and suppurative inflammation are also present. (H & E, × 31.5)

fore, it is wise to refer to the hematoxylin-and-eosin sections to establish morphologic landmarks such as blood vessels, nerves, etc., that may serve to identify the deep border of viable tissue in the sections with special stains.

Microbiologic invasion of the burn wound starts focally (or multifocally) and imparts an exaggerated, irregular margin to the base or lateral border of the burn wound (Fig. 3–1A–C, 3–2A). The presence of microorganisms in unburned tissue is the "sine qua non" for invasive infection. Bacteria (*Pseudomonas aeruginosa*) or fungi (usually Zygomycetes or *Aspergillus* sp.) within blood vessels also indicate invasive infection and the possibility of hematogenous dissemination.

Occasionally, one will encounter wound biopsies with early infection. In these biopsies, the pattern of necrosis may suggest an invasive infection, but the microorganisms are obscured by the suppurative inflammatory process and the basophilic background material. In such cases, the number of bacteria may be very low in the necrotic foci (< 10^6/g tissue) and even fewer microorganisms may be present in adjacent viable tissue. It is virtually

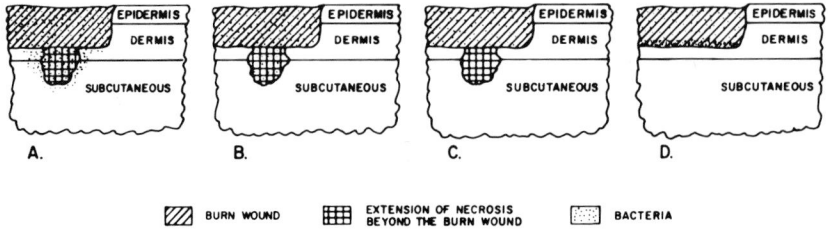

Fig. 3–2. Superimposed on these diagramatic cross-sections of skin in a partial thickness (second degree) burn with different stages of invasive infection. **(A):** Classical invasive bacterial infection of the burn wound. Necrosis with bacteria extends beyond the burn would and bacteria also invade viable tissue. **(B):** Diagnostic, but not classical, invasive bacterial infection. Necrosis with bacteria extends into viable tissue, but no bacteria have invaded beyond the advancing necrotic margin. **(C):** "Highly suspicious for invasive infection." The pattern of necrosis suggests invasive infection, but the microorganisms are confined to eschar. **(D):** "Highly suspicious for invasive infection." No evidence of new necrosis is seen, but microorganisms are numerous in the eschar and specifically the subeschar space immediately adjacent to viable tissue.

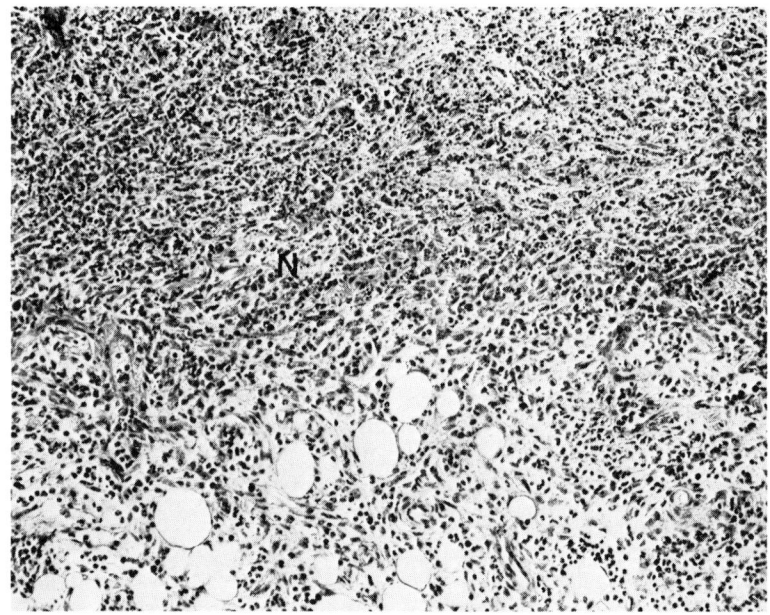

Fig. 3–3. Zones of recent necrosis (N) within granulation tissue of a burn wound should be examined closely for invasive microorganisms. (H & E, × 126)

impossible to detect bacteria histologically when fewer than 10^5 microorganisms/g tissue are present. Identification of bacteria in areas of necrosis which have obviously extended beyond the depth of the burn (i.e., necrosis of viable tissue) is sufficient for a diagnosis of invasive infection despite the lack of bacteria in viable tissue (Fig. 3–2B). However, such cases are very uncommon, and in general, bacteria should be present in viable tissue for one to render a diagnosis of invasive infection. If no microorganisms are seen in lesions which present a pattern of necrosis strongly suggestive of an invasive infection, the diagnosis should be designated as "highly suspicious for invasive infection" and additional biopsies should be submitted for histological evaluation (Fig. 3–2C).

In contrast to the above cases in which there is a paucity of microorganisms, some biopsies of the burn would will present with numerous microorganisms in eschar with extension to the subeschar area. If no suspicious pattern of necrosis (irregular foci of necrosis extending beyond the eschar) are seen and no microorganisms are seen in viable tissue, the biopsy is highly suspicious for, but *not* diagnostic of, invasive infection of the burn wound (Fig. 3–2D).

In most cases, burn wounds more than 7–10 days old have a distinct line of granulation tissue at the border of viable and nonviable tissue. This layer of granulation tissue is usually so well delineated that areas of bland necrosis resulting from an invasive infection are accentuated by the highly cellular and vascular granulation tissue (Fig. 3–3). Special stains for bacteria and fungi will usually demonstrate the causative agent.

Surgical reports should also describe other findings that may be of value to the clinician and surgeon, such as maturity of the granulation tissue on the wound surface, the presence and type of superficial microbial flora, hemorrhage, and presence of intravascular fibrin–platelet thrombi suggesting disseminated intravascular coagulopathy. Conversion of a second degree burn to third degree is also important to emphasize in the report. This may represent additional evidence for an invasive infection.

ACUTE INFLAMMATORY CELLS

Some investigators have mentioned that a heightened acute inflammatory process may occur in areas of invasive infection.[4] Rough handling of the burn wound and foreign material, such as necrotic hair fragments and other extraneous debris, nonspecifically induce a neutrophilic response. In fact, polymorphonuclear leukocytes may be found at variable levels in the eschar and viable tissue until the wound is healed, and are not specific indicators of invasive infection (Fig. 3–4).

Prior to the common usage of effective topical antibiotics against *Pseudomonas aeruginosa,* a pattern of diffuse burn wound sepsis was

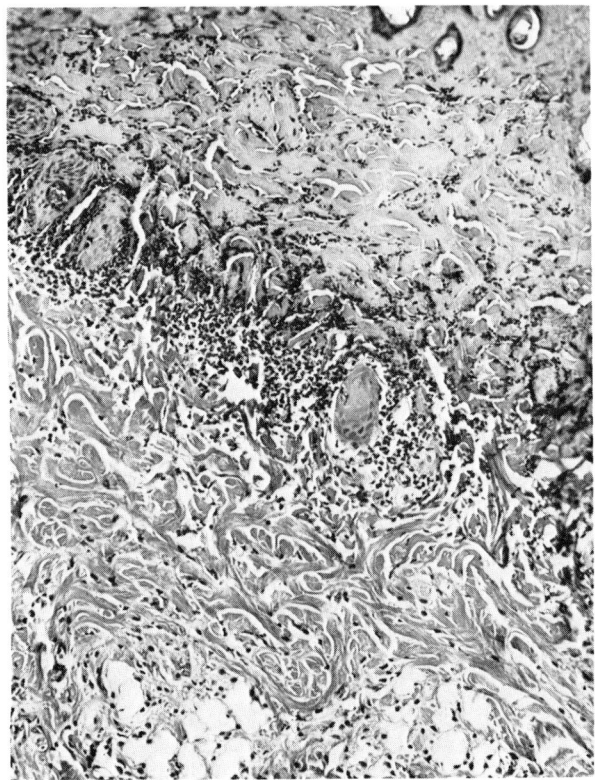

Fig. 3–4. Suppurative infiltrate at the junction of burned and viable tissue in a healthy 9-day-old burn wound. Suppurative inflammation is not always indicative of invasive infection. (H & E, × 126)

common. In these cases, there was a consistent gross and microscopic absence of suppurative inflammation.[5] Currently, some burn wound biopsies with an extensive invasive infection by *Pseudomonas aeruginosa* are likewise associated with a sparse acute inflammatory infiltrate. This has also been noted in current studies utilizing an animal model. In summary, the polymorphonuclear response is not an invariable accompaniment of invasive infection (Fig. 3–5).

HEMORRHAGE

One of the most frequent markers for invasive infection is the presence of acute hemorrhage (especially common in infections due to *Pseudomonas aeruginosa* and Zygomycetes) in the burn wound. However, small hemorrhages as well as thrombi are frequent in the early postburn phase and are

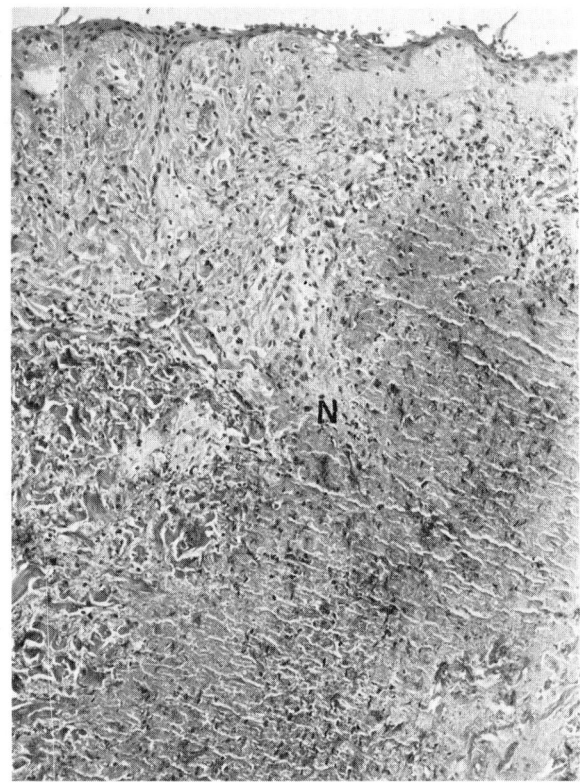

Fig. 3–5. Invasive *Pseudomonas aeruginosa* Infection in a 10-day-old second degree burn wound. Note the absence of acute inflammation at the junction of viable and necrotic (N) tissue. (H & E, × 126)

associated with the initial trauma (fire, explosions, etc.), and later may be related to minor trauma associated with handling of the burn patient. In a small number of patients, hemorrhages are related to a mild ongoing disseminated intravascular coagulopathy. These are readily recognized histologically by the presence of fibrin–platelet thrombi in the small blood vessels, usually microvasculature, of viable tissue adjacent to the burn wound.

MODE OF EXTENSION OF BACTERIA
INTO BURNED SKIN

Superficial bacterial colonization of eschar (supraeschar colonization) is usually followed by periadnexal extension of these microorganisms along the path of hair follicles (perifollicular colonization) into deep dermis.[5] Gram-negative bacilli as well as gram-positive cocci may utilize this route (Fig.

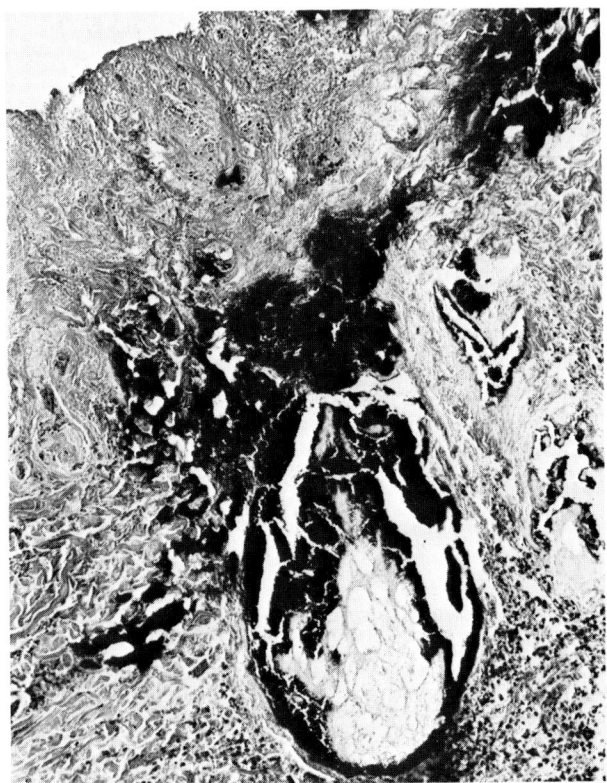

Fig. 3–6. Colonizing bacteria may be present in large numbers within necrotic hair follicles and adjacent adnexae. Extension into viable dermis and subcutis may follow. (*Staphylococcus*, Gram's stain, × 126)

3–6). In second degree burns, extension of these bacteria along necrotic hair follicles may simulate invasive infection; but without the finding of organisms in viable portions of hair follicles or adjacent debris, the diagnosis of invasive infection should not be made. However, susbsequent development of invasive infection in periadnexal tissue may occur. Close proximity of microorganisms within the dead hair follicle to the viable tissue and the direct communication of periadnexal structures via a ductal system explain the common spread of microorganisms to these structures in partial thickness (second degree) burns.

PSEUDOINVASIVE INFECTION

The following conditions may obscure or confuse the histologic picture and lead to an erroneous diagnosis of invasive infection.

EXTENSION OF MICROORGANISMS ALONG THE HAIR SHAFT. This has been described above. The critical distinction between this finding and that of invasive infection is the restriction of microorganisms to that necrotic portion of hair follicle within the eschar in a noninvasive lesion.

INADEQUATE (SHALLOW) BIOPSY. A biopsy may consist of eschar and lack underlying viable tissue. This problem arises frequently with deep third degree burns in which microbiologic organisms colonize superficial eschar and extend into underlying necrotic tissue. A biopsy that is no deeper than the eschar may be suspicious but yields no information about the depth of, or presence or absence of invasive infection in, the burn wound.

ENZYMATIC DEBRIDEMENT. The recent reappearance of certain topical proteolytic agents for debridement has provided a new problem for the pathologist. These agents, in some patients, may irregularly extend the burn wound margin into viable tissue. This extension closely simulates the histologic appearance (hemorrhagic necrosis) of invasive infection. In such cases, one must rely completely upon the observation (with special stains) of microorganisms in viable tissue for a diagnosis of invasive infection of the burn wound. Because enzymatic therapy holds the promise of earlier escharectomy with decreased morbidity and mortality, it is likely to become a therapeutic modality. Its effect on eschar and viable tissue must be further elucidated histologically so it may be clearly separated from "invasive infection" of the burn wound.[6]

ASPERGILLUS NIGER. Superficial colonization of the burn wound by *Asperillus niger* may create lesions which clinically simulate an invasive infection.[7] The dark brown colonies of *Aspergillus niger* are usually smaller (1–4 mm) than lesions commonly seen with an invasive infection by *Pseudomonas aeruginosa* or Zygomycetes (greater than 4 mm). The colonies of *Aspergillus niger* are usually confined to the superficial eschar and are often noninvasive (Fig. 3–7 A & B).

SEVERAL CONDITIONS SIMULATING THE BASOPHILIC (LOW POWER) APPEARANCE OF BACTERIA AS WELL AS OBSCURING THE PRESENCE OF BACTERIA IN THE BURN WOUND. Of the entities which may simulate bacteria, some are readily eliminated. Mineralization (blue calcium deposits) and karyorrhexic nuclear fragments usually can be differentiated on the basis of size and morphology from bacteria which are smaller and very uniform in size. However, injured tissue of the burn wound often has an amorphous, patchy, basophilic alteration which is identical to the appearance of colonies of "invading" bacteria when viewed at low power. At high power ($\times$ 500; $\times$ 1000), the absence of discrete microorganisms is conspicuous and the pattern of distribution of this patchy amorphous basophilic material simulates a fluid. It is likely that protein and/or carbohydrate molecules are

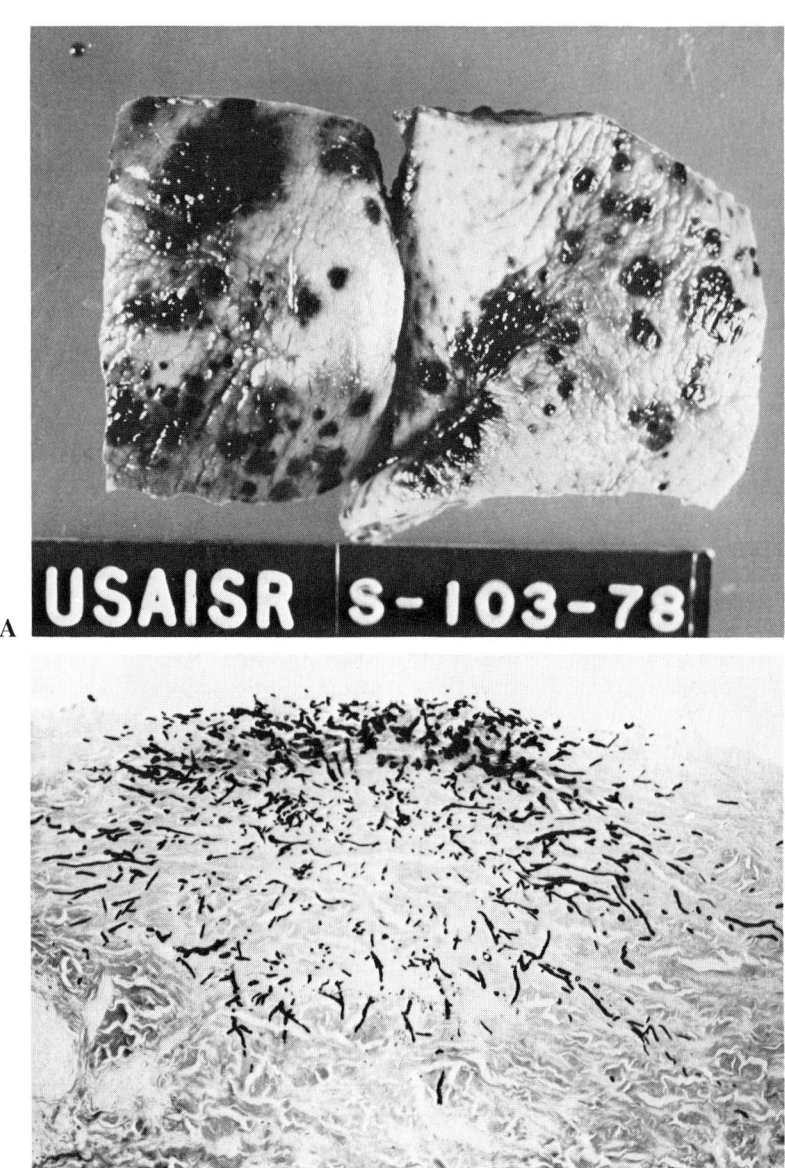

Fig. 3–7. (A): Pigmented colonies of *Aspergillus niger* may be clinically suggestive of invasive or metastatic *Pseudomonas* infection. In this burn wound surgical specimen, the small dark lesions were found to represent *Aspergillus niger* eschar colonization. The larger dark lesions were found to be invasive *Aspergillus niger* infection with associated hemorrhage and necrosis. **(B)**: Profuse superficial growth of *Aspergillus niger* in a burn wound eschar. Colonizing fungi and bacteria may be numerous on the burn wound eschar. [Gomori's methenamine silver (GMS), × 64.8]

dissolved in the fluid medium and extend into the eschar to provide this amorphous basophilic alteration.

PROFUSE FUNGAL AND BACTERIAL GROWTH CONFINED TO THE ESCHAR. A pathologist who is unfamiliar with the burn wound biopsy and does not carefully delineate the border of viable and nonviable tissue on sections with special stains will make a falsely positive diagnosis when he encounters profuse intraeschar fungal or bacterial growth (Fig. 3–7B).

CONTAMINATION OF SPECIMEN. There are probably a variety of ways a specimen may be contaminated, but we have experienced only two. Skin biopsies of the burn wound mistakenly sent in culture media often contain mumerous colonies of microorganisms in eschar. Specimens negative for invasive infection at the time of biopsy may, in fact, become positive due to microbial growth in the unburned portion of the biopsy while the specimen is in the culture media. Such a biopsy may lack acute inflammatory cells; however, rampant invasive infection may also have very few, if any, polymorphonuclear leukocytes (Fig. 3–5).

The second problem we experienced was mixed bacterial growth within our water bath. The latter condition was easily recognized by the variety of microorganisms seen (diplococci, cocci, and bacilli) and the observation of microorganisms surrounding, as well as lying on the surface of, tissue sections. More frequent changing of the water bath has eliminated this problem.

QUANTITATIVE CULTURES

Attempts have been made to use quantitative bacterial cultures alone to detect invasive infection of the burn wound or suitability of the wound bed for grafting.[8–16] These methods vary from culture "swabs" from the burn wound surface to utilizing the tissue biopsy itself for culture. One rapid and simple technique calls for immediate slide evaluation of the diluted tissue homogenate under the microscope. The homogenate is then cultured. When the slide test is positive, antibiotic sensitivities are set up simultaneously. However, this method as well as those above are associated with false positive and negative results. In addition, standard cultures require 24–48 hr before identification is complete. Superficial colonization of the eschar by a prominent number of organisms may totally mislead the clinician and falsely suggest an "invasive infection." Surface swab cultures often contain quantitatively more microorganisms than deep wound cultures and may wrongly indicate an invasive infection. In some cases, surface organisms are completely different from those invading the depth of the wound. Positive blood cultures may not be seen in patients with extensive, early, burn wound infection. By the time the blood culture is positive, the patient's burn wound may, and often does, contain extraordinary quantities of bacteria (10^8 or greater/g tissue). Quantitative wound cultures are needed to establish the diagnosis of significant wound infection $> 10^5$/g tissue) and may also simultaneously suggest a poor prognosis when bacteria exceed 10^8/g tissue. However, Woolfrey et al.[17] have demonstrated the importance of selecting more than one biopsy for culture. Twenty-five percent of the cultures failed to identify the significant microorganism, and bacterial densities in paired biopsy specimens were in agreement (had the same log 10 bacterial density) in only 38 percent of the samples.

The obvious ideal method for diagnosis of burn wound infection is rapid histologic examination coupled with quantitative bacterial cultures of the burn wound. A method for bacterial quantitation is found in Appendix 3–5. Attempts to use frozen section to accelerate histologic examination of infected tissue have unfortunately met with significant technical problems that preclude interpretation of such specimens.[2,18] But rapid, "same-day" tissue processing on an Ultra Autotechnicon does yield acceptable sections within 2–4 hr. Unexplainable discrepancies will be seen between histologic and microbiologic findings. But these follow-up comparisons, especially when multiple specimens are examined, serve to give confidence to both the pathologist and the microbiologist.

Reference

1. Pruitt, B. A., Jr., and Foley, F. D.: The use of biopsies in burn patient care. Surgery 73:887–897, 1973.
2. Pruitt, B. A., Jr., and McManus, A. T.: Opportunistic infections in severely burned patients. Am. J. Med. 76S:146–154, 1984.
3. Teplitz, C.: Pathology of burns. *In* Artz, C. P., and Moncrief, J. A.: The Treatment of Burns, Chapter 2, 2nd ed. Philadelphia, W. B. Saunders Co., 1969, pp 22–88.
4. Pruitt, B. A., Jr.: Infections caused by *Pseudomonas* species in patients with burns and in other surgical patients. J. Infect. Dis. 130(Suppl):8, 13, 1974.
5. Teplitz, C.: The pathology of burns and the fundamentals of burn wound sepsis. *In* Burns. A Team Approach, 1st ed. C. P. Artz, J. A. Moncrief, and B. A. Pruitt, Jr., eds. Philadelphia, W. B. Saunders Co., 1979, pp 45–94.
6. Makepeace, A. R.: Enzymatic debridement of burns:A review. Burns 9:153–157, 1983.
7. Panke, T. W., McManus, A. T., and Spebar, M. J.: *Aspergillus niger* infection of the burn wound: Gross appearance simulating ecthyma gangrenosa. Am. J. Clin. Pathol. 72:230–232, 1979.
8. Robson, M. C., and Krizek, T. J.: Predicting skin graft survival. J. Trauma 13:213–217, 1973.

9. Shuck, J. M., and Moncrief, J. A.: The management of burns. 1. General considerations and the sulfamylon methods. Curr. Probl. Surg. 3:1–52, 1969.

10. Heggers, J. P., Robson, M. C., and Ristroph, J. D.: A rapid method of performing quantitative wound cultures. Mil. Med. 134:666–667, 1969.

11. Robson, M. C., and Heggers, J. P.: Bacterial quantification of open wounds. Mil. Med. 134:19–24, 1969.

12. Edlich, R. F., Rodeheaver, G. T., Spengler, M., Hilbert, J., and Edgerton, M. T.: Practical bacteriologic monitoring of the burn victim. Clin. Plast. Surg. 4:561–569, 1977.

13. Loebl, E. C., Marvin, J. A., Heck, E. L., Curreri, P. W., and Baxter, C. R.: The use of quantitative biopsy cultures in bacteriologic monitoring of burn patients. J. Surg. Res. 16:1–5, 1974.

14. Bharadwaj, R., Phadke, S. A., Joshi, B. N.: Bacteriology of burn wound using the quantitative full thickness biopsy technique. Ind. J. Med. Res. 78:337–342, 1983.

15. Tahlen, R. N., Keswani, R. K., Saini, S., and Miglani, O. P.: Correlation of quantitative burn wound biopsy culture and surface swab culture to burn wound sepsis. Burns 10:217–224, 1984.

16. Bharadwaj, R., Joshi, B. N., and Phadke, S. A.: Assessment of burn wound sepsis by swab, full thickness biopsy culture and blood culture—A comparative study. Burns 10:124–126, 1984.

17. Woolfrey, B. F., Fox, J. M., and Quall, C. O.: An evaluation of burn wound quantitative microbiology. I. Quantitative eschar cultures. Am. J. Clin. Pathol. 75:532–537, 1981.

18. Pruitt, B. A., Jr.: Biopsy diagnosis of surgical infections. N. Engl. J. Med. 310:1737–1738, 1984.

Appendix 3–1 *Processing of the Rapid Section*

The biopsy secured from the wound of the burn patient is immediately placed in formalin and transported to the pathology laboratory. The specimen is then promptly evaluated by the pathologist. Appropriate measurements and photographs are taken and the specimen is subsequently cut into (no thicker than) 1 mm sections which are heat-fixed in warm 10% buffered formalin (60°–70° C) for 15–20 min. The specimen is processed on the Ultra Autotechnicon where the 2-hr disk is utilized and the time periods are listed below. The important feature of this autoprocessor is that a partial vacuum is utilized to accelerate penetration of the tissue by reagents.

Ethanol	95%	3 min
Ethanol	100%	2½ min
Ethanol	100%	2½ min
Ethanol	100%	5 min
Ethanol	100%	5 min
Ethanol	100%	7½ min
Xylene		2½ min
Xylene		5 min
Xylene		5 min
Paraffin		5 min
Paraffin		5 min

The specimens are paraffin-embedded and cut at 5–8 μ. Because of the hardness of eschar, gelatin is used in the water bath and the slides are coated with albumin. Three slides are made; this allows for staining with Harris' Hematoxylin and Eosin, McManus PAS and modified Humberstone Gram's stain (Appendix 3–2). Alternative stains for microbiologic flora include PAS-Giemsa (Appendix 3–3) and Gomori's methenamine silver.

Appendix 3–2   *A Modified Humberstone Gram's Stain*[a]

Solutions

1. Aqueous crystal violet, 1% solution

2. Gram's iodine

   Iodine- - - - - - - - - - - - - - - - - - -   1.0 g
   Potassium iodide - - - - - - - - - -   2.0 g
   Distilled water - - - - - - - - - - - -   300.0 ml

3. Cellosolve (ethylene glycol monoethyl ether)

4. Aqueous basic fuchsin, 0.5% solution

5. Gallego's differentiator

   Distilled water - - - - - - - - - - - - - -   50 ml
   40% formaldehyde - - - - - - - - - - -   1 ml
   Glacial acetic acid - - - - - - - - - - -   0.5 ml

6. Aqueous tartrazine, 1.5% solution

Method

1. Deparaffinize and bring to water through graded alcohols.

2. Stain in 1% aqueous crystal violet for 2 min.

3. Rinse in distilled water.

4. Mordant in Gram's iodine for 5 min.

5. Rinse in distilled water.

[b]6. Differentiate in Cellosolve until blue color no longer streams away from the section (approximately 5–10 sec).

[b]7. Quickly wash with distilled water.

8. Stain in basic fuchsin for 5 min.

9. Wash in distilled water.

---

[a] From Brown, R. C., and Hopps, H. C.: Staining method for gram positive and gram negative bacteria. *In* Manual of Histologic Staining Methods of the Armed Forces Institute of Pathology, 3rd. ed. New York, McGraw-Hill Book Co., 1968, pp 224–225.

[b] Critical steps in regard to timing.

10. Differentiate and "fix" basic fuchsin in Gallego's differentiator for 5 min.

11. Wash *thoroughly* in distilled water.

12. Blot lightly, only to remove excess water (not to dryness).

13. Apply 1.5% aqueous tartrazine and immediately blot away excess, but not to dryness.

[b]14. Quickly pass the slide through 3 changes of Cellosolve (6 quick dips each).

15. Pass through 3 changes of xylene, approximately 10 quick dips each; the slide can remain in the last xylene until convenient to mount.

16. Mount in Permount.

Note:

Preferably, all stains and solutions are applied to the slide which is in a horizontal position, except for steps 6, 14, and 15, in which case the slide is dipped into the solutions contained in Coplin jars.

These times have been thoroughly checked and are best followed, using an interval timer.

The stain works satisfactorily on tissues fixed in 10% formalin (sodium phosphate or sodium acetate buffered), gluteraldehyde (2.5% with phosphate buffer), and Formol-Zenker's fluid. It does *not* work well on tissues fixed in Bouin's solution.

Appendix 3–3 *PAS-Giemsa Stain*[a]

Reagents

0.5% Periodic acid
0.5 g periodic acid
100 ml distilled H_2O

Schiff Reagent

Harleco Schiff Reagent #2818 or make Coleman-Feulgen Reagent

Stock 2.6 pH buffer

24.2 ml h 0.2 N HCl
50 ml h 0.2 M glycine
Qs to 200 ml with distilled H_2O. Check and adjust pH to 2.6. Store in refrigerator.

Stock Giemsa Stain

1 g h Giemsa Stain (Allied Chemical)
66 ml h glycerine
Heat overnight in 60°C oven. Cool to room temperature, then add 66 ml h methanol. Filter and store in refrigerator.

Working Giemsa (Use only 1 day)

3 ml h stock buffer
3.5 ml h Stock Giemsa
45 ml distilled H_2O
Heat to 60°C in oven prior to use. pH 3.3 to 3.4.

Procedure

1. Deparaffinize and hydrate through alcohols to distilled H_2O.

2. Place in 0.5% periodic acid for 5 min.

3. Rinse in distilled H_2O

4. Place in Schiff's Reagent for 15 min.

---

[a] Teplitz, C., and Davis, D.: Modified buffered Giemsa method—A one-step general cytological and bacterial tissue section stain, RCS-MEDDG-288. Fort Sam Houston, Tex, United States Army Surgical Research Unit, pp 54-1–54-6, 1963.

5. Rinse in running tap water for 5 min.

[b]6. Stain in Harris' Hematoxylin for 1 min.

[b]7. Rinse in tap water.

[b]8. Differentiate quickly in acid alcohol and blue in tap water.

9. Stain in heated working Giemsa in 60°C oven for 40 min.

10. Rinse in distilled H_2O.

11. Differentiate in absolute acetone.

12. Clear in equal parts acetone–xylene, then two changes xylene.

13. Mount in Permount.

---

[b] Steps 6, 7, and 8 may be omitted for routine use, if desired. Staining with Hematoxylin gives more intensive nuclear staining, useful for photographing. Results: Bacteria stain a bright blue. Fungi stain magenta.

Appendix 3–4 *Tissue Examination—Burn Wound[a]*

CLINICAL RECORD

Submitted by_____ Date & time specimen obtained_____

Date burned_____ %TBSB_____ %3°_____

Site of biopsy_____ (burned_____ unburned_____ skin)

Rx Hx: Topical antibiotic, type_____ Enzymatic Rx_____

Subeschar antibiotic infusion_____

Excision (type)_____ Grafting (type)_____ DIC_____

Appearance of wound at Bx site_____

Clinical Dx:_____

_____
(Signature & title)

PATHOLOGICAL REPORT

Laboratory_____ Accession No._____

Signature of Pathologist   Date_____

Patient Name: Age____ Sex____ Race____ SSAN_____
                                       Register #_____ Ward_____

---

[a] Key: %TBSB—% of total body surface with second and third degree burns; % 3°—% of body surface with third degree burns; Rx Hx—therapeutic history; Enzymatic Rx—enzymatic treatment; DIC—disseminated intravascular coagulopathy, Bx—biopsy; Dx—diagnosis.

Appendix 3–5 *Processing of Burn Wound Specimens for Microbiologic and Histologic Evaluation*[a]

To insure uniform handling, quantitative cultures of tissue must be performed in a standardized manner; the following method is suggested.

1. The topical antibacterial cream should be meticulously removed from the burn wound site. During life, biopsy sites are most expeditiously selected immediately following the removal of topical antibiotic at the start of the day, and prior to reapplication of such cream. At postmortem examination, all antibiotic cream must be removed from the body so that all suspicious sites can be evaluated and selectively biopsied at the start of the autopsy and immediately submitted for culture.

2. The minimum size of tissue blocks from a surgical specimen should be 1.0×0.5 (rectangular biopsy) $\times 1.0$ cm depth. Tissue blocks from autopsy specimens should be at least 1.5 cm. Paired specimens should be obtained. One biopsy is used for histological preparation and microscopic exam, and the second specimen is processed as indicated below.

3. With flamed (sterile) forceps and scalpel previously dipped in 70% isopropyl alcohol, cut a block to slightly over 1 cm^3 (1 g). This specimen should be raised with a flamed forceps and sprayed with 70% isopropyl alcohol. The block is ignited and the flame is extinguished after 5 sec; this flaming procedure is repeated.

4. Using flamed scissors, the block is sectioned into small fragments.

5. The sample is transferred to a Ten Broeck grinder and sterile trypticase soy broth is added. The mixture is chilled in ice and the specimen is homogenized. (Eschar cannot be homogenized; but in a sense, it is "wrung out" to recover bacteria.) Trypticase soy broth is added to make a final volume of 2.0 ml.

6. Serial dilutions (10^{-8}, 10^{-6}, 10^{-4}, 10^{-2}) of the broth mixture are made. (Remember to calculate for the initial dilution with broth during homogenization.)

7. One ml of each dilution is added to appropriately labeled Petri dishes. The diluting agar (50°C) is promptly added and the contents of the Petri dish are gently swirled to distribute the sample. The culture plates are incubated for 48 hr and colony counts are performed.

[a] Adapted from Lindberg, R. B., Ph.D., unpublished procedure and from Teplitz, C.: The pathology of burns and the fundamentals of burn wound sepsis, 1st ed. *In* Burns. A Team Approach. C. P. Artz, J. A. Moncrief, and B. A. Pruitt, Jr., eds. Philadelphia, W. B. Saunders Co., 1979, pp 45–94.

Chapter 4

SPECIAL CONSIDERATIONS FOR THE BURN AUTOPSY

In contrast to the standard hospital autopsy, the burn autopsy findings are scrutinized not only by the primary physicians but also by future research physicians and others involved in medicolegal settlements of some cases. Data from the postmortem examination of burn patients serve several vital functions: (1) they provide immediate answers from problems that arose during the patient's hospital course; (2) in association with other cases, they may yield answers not readily perceived from analysis of any single case; (3) they are the vehicle to measure current and future therapy for the burn patient; and (4) their medicolegal importance is obvious to all in our litiginous society. Properly performed autopsies punctuated by meticulous observations at the autopsy table and the microscope maximize the benefits of the postmortem exam in achieving the benefits listed above.[1]

The burn autopsy in many ways is very similar to the postmortem examination performed on the nonburned patient; but variations occur that warrant consideration at the outset of this text. These variations will be only briefly mentioned here and more fully described in the appropriate sections of other chapters.

Evaluation of Extent of Burn Wound

In most clinical settings, burn wound size and depth are accurately measured early in the course of treatment because resuscitation is based on this information. The pathologist, however, should be familiar with the "Rule of Nines" (Table 4–1) and determine by his own estimation the percentage of body surface involved. In these determinations, it must also be considered that the body surface area of children is greater in respect to body weight than in adult patients. The extent of burns should be periodi-

Table 4–1 *Estimation of Percentage of Body Surface Burned—the "Rule of Nines"*

Burned Area	Percent of Body Surface
Head	9
Right Arm	9
Left arm	9
Right leg	18
Left leg	18
Anterior trunk	18
Posterior trunk	18
Neck	1

cally updated during the hospital course. At initial presentation, what appears to be a first or second degree burn may progress to second or third degree. In contrast, thermal injury may actually be less severe than initially determined. Armed with these evaluations, the pathologist becomes the final arbiter of extent of thermal injury.

Examination of Burn Wound for Infection

Depending on the number and size of areas suspected of having "invasive infection," three to six skins specimens are required to evaluate the burn wound for invasive infection by histologic means. It is often worthwhile to document photographically the appearance of the burn wound and several of the suspected invasive lesions. Photographs are invaluable both as medicolegal evidence and for educational future comparison with histologic sections (Figs. 4–1, 4–2, 4–3). It should also be remembered that the permit for postmortem examination may have to be modified to permit adequate burn wound examination.

Collecting Postmortem Cultures

Methods for obtaining cultures at autopsy are as variable as are the imaginations of the pathologists involved. In selecting cultures, the following may serve as a guide:

Obviously, the longer the interval between death and postmortem examination, the less reliable are the results of cultures. Intelligent use of postmortem cultures in conjunction with histologic examination and premortem cultures provides the most useful information. There will cer-

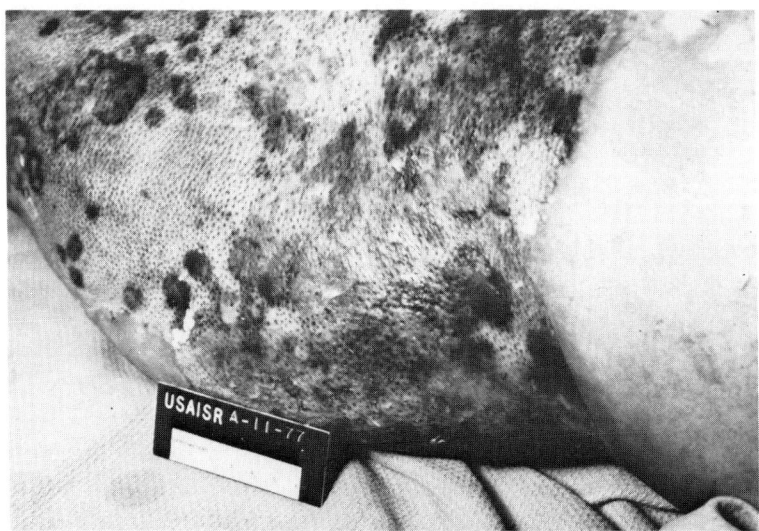

Fig. 4–1. The burn wound should be photographed and areas suspected of having invasive infection selected for histology and culturing.

tainly be instances in which the postmortem cultures may be meaningless, but these should always be performed if the autopsy is conducted within 24 hr following death. Careful selection of tissue for culture, whether it be skin or lung, is paramount.

Before the skin surface is evaluated, the topical antibiotic is completely removed. Tissue samples are collected in a sterile fashion (Chapter 3, Appendix 3–5). Samples should be 1.0 × 0.5 × 1.0 cm in depth and should be submitted for bacterial and mycotic cultures. An adjacent similar portion of tissue should be submitted for histological examination. Ideally, the latter specimen should include adjacent, peripheral, and deep unburned skin so that "invasive" infection can better be assessed. Careful review of the differential diagnosis for pseudoinvasive infection of the burn wound and the interpretation of special stains (Chapter 3) is vital for accurate histologic diagnosis.

Tissue is subsequently selected randomly from the spleen and liver, as well as from areas of the upper and lower lobes of the lung bilaterally. Liver and spleen specimens are often of more value when used in concert with blood cultures. The organ surfaces are seared with a hot spatula, and a sterile blade and forceps are used to obtain deep (greater than 1.5 cm below the surface of the organ) specimens. Blood is collected prior to opening the body. The skin over the subclavian vein is sterilized as one would to obtain blood from a living patient. A subclavian "stick" is probably the most valid

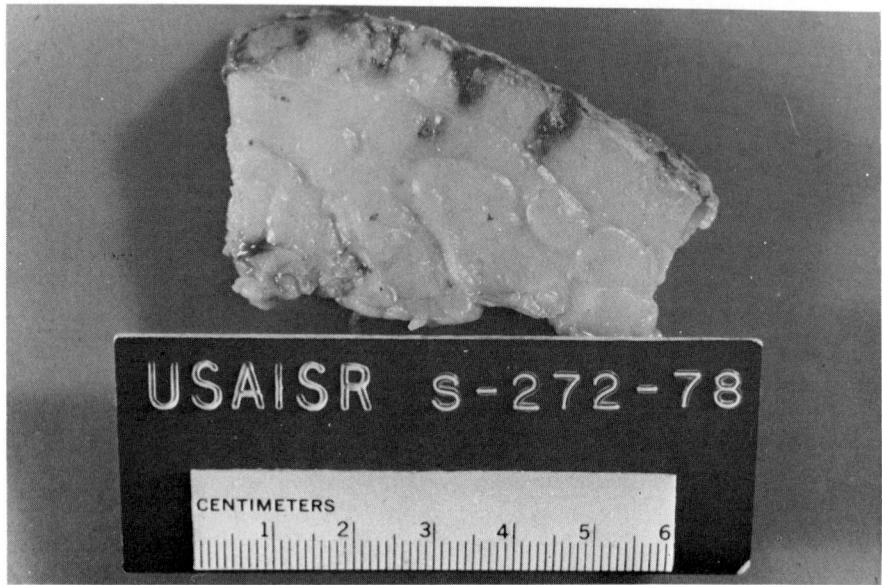

Fig. 4–2. A burn wound with two areas of hemorrhage extending into viable fat. These represent focal invasive infection.

technique through which to obtain meaningful postmortem blood cultures. During the performance of the postmortem examination, any additional areas suspected of having infection are prepared in a similar way for histology and culture. Special attention should be given to hemorrhagic lesions in or adjacent to the burn wound margins (Fig. 4–1).

Circumferential Burns of the Torso and Extremities

While it is well recognized that circumferential burns of the extremities are associated with vascular compression, the significance of similar circumferential burns of the chest and abdomen is less well appreciated. During both the acute postburn period and again during the healing phase of circumferential thoracic burns, a significant restriction of thoracic excursion has been shown to occur.[2] The probable mechanism for this process is that burn skin contracts,[3] becomes rigid, and with the development of tissue edema causes vascular, thoracic, or abdominal compression.

In the extremities, the arterial pulse is an excellent indicator of the degree of compression caused by circumferential burns. Escharotomies and occasionally fasciotomies are performed if these pulses are absent. On the torso, such indicators are not available to the clinician, and some patients

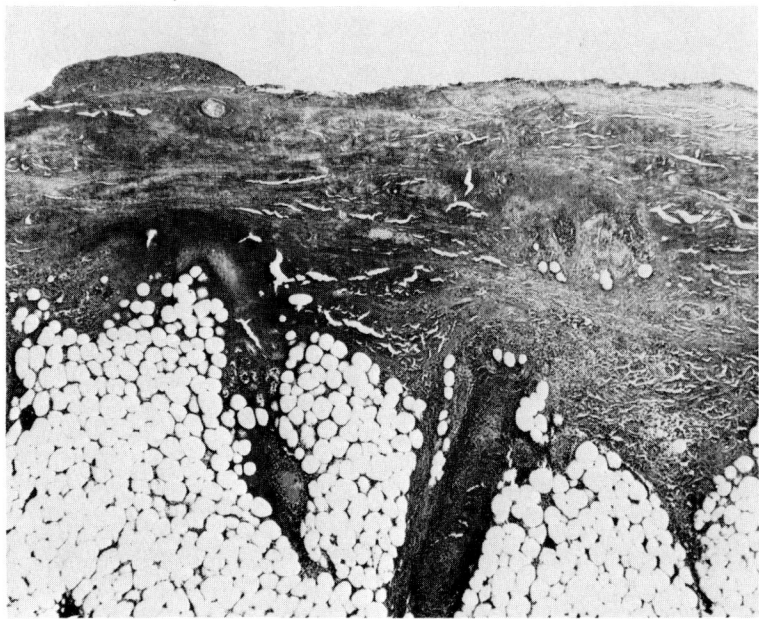

Fig. 4–3. Hemorrhage, necrosis, and an irregular zone of suppurative inflammation at the junction of burned and viable tissue suggest that invasive infection has occurred. Special stains of sections with these characteristics will usually reveal bacteria or fungi in the viable tissue. (H & E, × 9)

may exhaust their cardiopulmonary reserve before appropriate escharotomies are performed. It is possible that such circumferential burns of the torso may be the immediate cause of death in severely burned patients. At autopsy, a diagnosis of circumferential compression can be supported by observing the margins of long incisions made in the thoracic and abdominal wall. The edges of the incisions may retract 4–9 cm and the underlying tissue will be massively edematous in cases that have significant circumferential compression. Recently, Stone has described circumferential cervical burn would edema causing compression of the trachea.[4]

Venous Catheters

Central venous catheters are vital for monitoring the burn patient, providing large volumes of fluid in the immediate postburn course and hyperalimentation for the severely burned patient. Bland and infected thrombi are relatively common sequellae of cannula use. These may occur either in cannulated veins or in the heart (Fig. 4–4). Often these lesions are clinically ''silent,'' and only the presence of fever or positive *Staphylococ-*

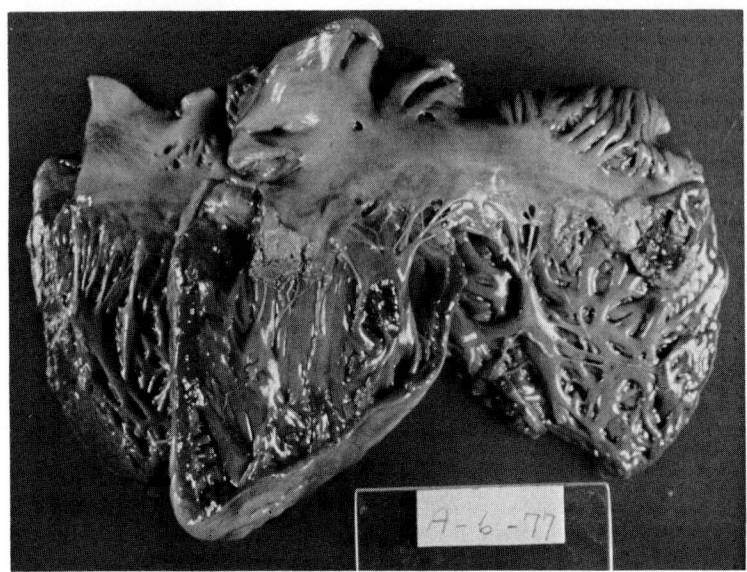

Fig. 4–4. Soft vegetations on the tricuspid valves of the heart contain *Staphylococcus*.

cus aureus blood cultures indicates the possibility of infected venous thrombi.

Selected peripheral veins should be routinely examined for these thrombi. In patients with unexplained septicemia, especially those with positive staphylococcal premortem blood cultures, the clinician should indicate on a diagram all veins catheterized and thus direct the pathologist toward a more productive search (Fig. 4–5).

Heart

Severely burned patients have a hyperdynamic cardiovascular physiologic response. In elderly patients, the coronary vessels are often severely compromised by atherosclerotic disease. The ischemic heart disease which results may easily be overlooked clinically because the typical clinical features are obscured by the overwhelming consequences of severe burn trauma.

Myocardial infarcts may also be seen in young burn patients with no apparent coronary vascular disease (see Chapter 9) and should be considered in any patient with multiple complications following severe burns.

Infectious and marantic thrombi involving cardiac valves are common in catheterized patients. They commonly occur in the right side of the heart

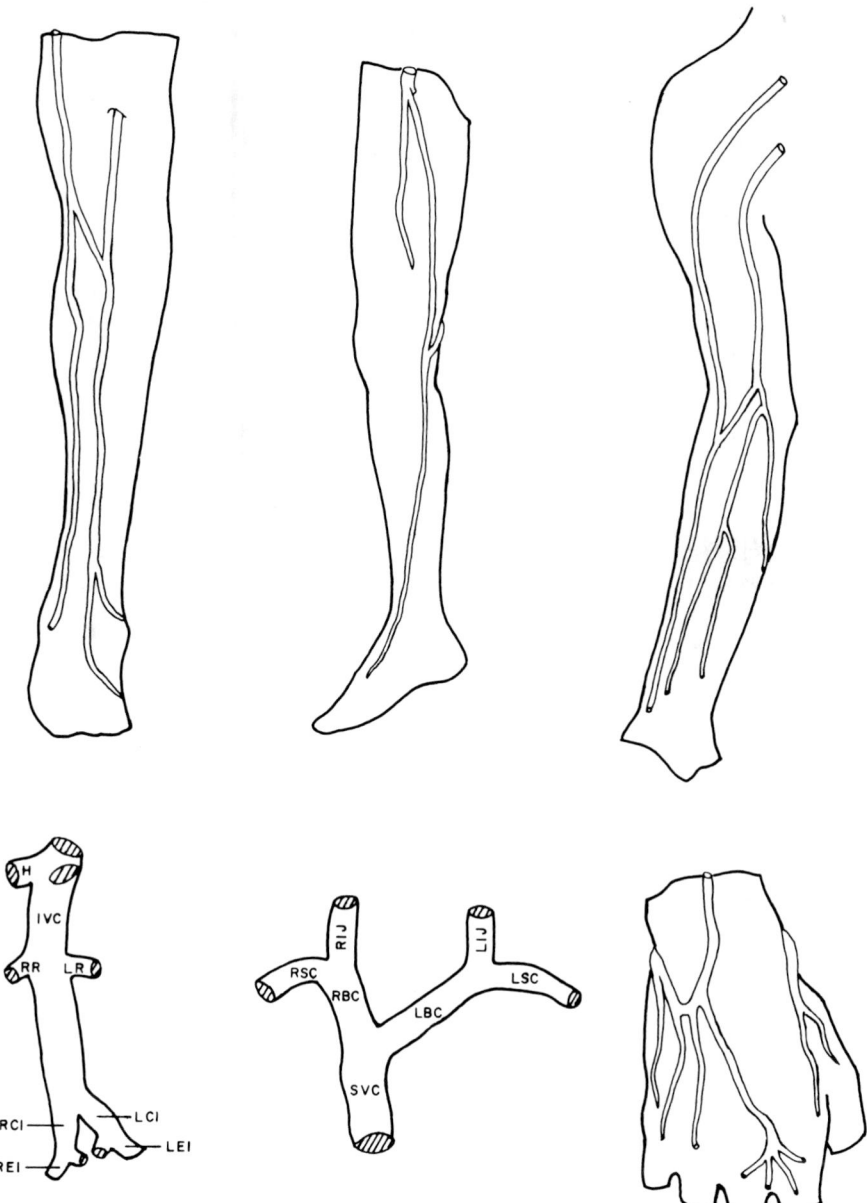

Fig. 4–5. Diagram used by clinicians to indicate to the pathologist sites of venous catheterization. H = Hepatic vein; IVC = Inferior vena cava; RR = Right renal vein; LR = Left renal vein; RCI = Right common iliac vein; LCI = Left common iliac vein; REI = Right external iliac vein; LEI = Left external iliac vein; RIJ = Right internal jugular vein; LIJ = Left internal jugular vein; RSC = Right subclavian vein; LSC = Left subclavian vein; RBC = Right brachiocephalic vein; LBC = Left brachiocephalic vein; SVC = Superior vena cava.

in those patients who have had a Swan-Ganz* catheter inserted to measure pulmonary wedge pressure.

Respiratory System

Careful inspection of the tracheobronchial tree for mucus plugs, aspirated gastric contents, or pulmonary tracheobronchial hemorrhage may demonstrate an immediate cause of death in the severely burned patient. Clinical evidence of "inhalation injury," especially in cases where there is a history of unconsciousness at the time of injury, should be documented pathologically by (1) prominent facial burns and "singed" nasal vibrissae; (2) burn injury to the oral and nasal cavity; (3) life-threatening laryngeal edema in the early postburn course; and (4) sloughed tracheobronchial mucosa replaced by a layer of mucus and "soot" which is attached to an erythematous tracheobronchial wall. If a patient with severe inhalation injury dies in the immediate postburn period, blood carbon monoxide levels may be very useful.

Inhalation injury in burn victims is more thoroughly evaluated by formalin inflation and fixation of the lungs. Both lungs are initially palpated externally for areas of consolidation, and samples of tissue are selected for culture. One or both lungs are then inflated with formalin and may be cut and examined immediately after the autopsy or on the following day.

Great care should be directed toward separating hematogenous pneumonia from a bronchopneumonia originating as a complication of an inhalation injury. Clinicians often question whether burn wound infection led to pneumonia (perivascular distribution) or burn wound therapy was effective and the infection arose in the lungs (bronchocentric following inhalation injury).

Careful dissection of the pulmonary vascular tree for the presence of thromboemboli is a vital part of any autopsy and is particularly important in the severely burned patient.

Gastrointestinal Tract

Curling's ulcers of the stomach and duodenum are a widely appreciated problem in the burn patient and must be carefully sought. Currently, however, ischemic enterocolitis has recently been shown to be a more common significant affliction of the gastrointestinal tract of the burn patient and may be an important portal for bacteremia. Pseudo-obstruction of the colon produces rather dramatic distention of the large bowel for inapparent reason.

---

* American Edwards Laboratories, Anasco, Puerto Rico

Liver

Conspicuous enlargement of the liver is common in the severely burned patient and often reflects "compensatory" hepatocellular hyperplasia. Significant hepatocellular necrosis occurs uncommonly and usually reflects terminal, dramatic, hypotensive episodes. The jaundice that occurs may reflect this but far more commonly is indicative of drug-related intrahepatic cholestasis, hemolytic anemia, or severe sepsis.

Acute acalculous cholecystitis is uncommon. It is often related to septicemia, but several other factors may be responsible.

Kidney

In patients with severe electrical injury, myoglobinuric nephrosis should be suspected. Carbon monoxide may also cause myonecrosis. As with any severely debilitated patients, acute renal tubular necrosis or renal cortical necrosis may occur.

Prostate

Necrotizing prostatitis is a relatively common finding in the burn patient and occurs in about 4–7 percent of our cases each year. It therefore represents another possible portal from which septicemia may occur.

Testes

Any patient with thermal injury to the inguinal area should have a histologic examination of the testes, which often demonstrates severe burn injury.

Endocrine System

Although relatively dramatic hormonal changes occur in most endocrine glands, conspicuous morphologic changes are most commonly identified in the adrenal glands. Severe stress is often associated with complete lipid depletion of the adrenal cortex, adrenal cortical "atrophy" and focal necrosis, and adrenal hemorrhage which occasionally is severe. Extensive pituitary infarcts have also been observed occasionally in severely burned patients. Significant morphologic lesions in the thyroid consist only of extensions of an infectious process from tracheostomy sites.

Child Abuse

The finding of cutaneous burns in children raises the possibility of child abuse. The physician and pathologist involved in such cases should be unusually sensitive to this problem. A discussion and guide for evaluation of suspected child abuse cases may be found in Chapter 17.

Miscellaneous

A number of lesions could easily escape the attention of the casual observer. The complete autopsy of the burn patient requires tissue sampling of infected osseous and cartilaginous lesions. Trauma and specifically bone fractures raise the possibility of life-threatening fat embolism. Ocular and auricular lesions, though not contributory to cause of death, should be graphically documented.

Assigning of Cause of Death

The plethora of events attending the thermal injury and complicating the hospital course may be perplexing especially to the neophyte physician (surgeon, internist, or pathologist) caring for the burn patient. However, changes in the burn wound as well as the numerous complications of thermal injury do occur at predictable points in the postburn course and clearly relate to the age of the patient, severity of thermal injury, and the presence of inhalation injury. A time-related framework of events and complications of burn injury is discussed at some length in the last chapter of this book. Various methods for predicting survival are presented. Evolution of disease processes are discussed. Certainly the review of that chapter would be an immeasureable aid for the conduction of any postmortem examination on the burn patient.

References ───────────────────────────────────

1. Sevitt, S.: A review of the complications of burns, their origin and importance for illness and death. J. Trauma 19:358–369, 1979.
2. Quinby, W. C., Jr.: Restrictive effects of thoracic burns in children. J. Trauma 12:646–655, 1972.
3. Wehr R. F., Smith, J. G., Jr., and Pirkle, D. E.: Early changes in rat skin following thermal injury. J. Trauma. 13:132–135, 1973.
4. Stone, H. H.: Pulmonary burns in children. J. Pediatr. Surg. 14:48–52, 1979.

Chapter 5

ELECTRIC INJURY

Electric injuries are common and are estimated to cause more than 1000 fatalities annually in the United States.[1] Many more people sustain serious but nonfatal electric injuries. Hence an understanding of this cause of thermal injury is important for the physician.

MECHANISM OF ELECTRIC INJURY

The electric injury (Fig. 5–1) is far more complicated for the surgeon than for the pathologist. Despite earlier publications that presented a complex and variable morphologic description of this injury, it is now generally accepted that an electric burn is a simple thermal injury.[2,3] When skin and adjacent soft tissues are examined histologically shortly after electric injury, they show changes similar to those of severe flame or scald burns. Deeper wounds caused by extreme high voltage resemble crush injury.[4] The physical laws that apply to electric burn injury have been elaborately discussed in other publications.[5,6] In general, the extent of heat produced and resulting tissue injury are proportional to the amount of current measured as amperes. Other factors that affect the degree of tissue injury are (1) voltage; (2) duration of contact; (3) type of current (alternating or direct); (4) the frequency (if alternating current); (5) resistance at points of contact; (6) path of the current through the body; (7) individual susceptibility; (8) intervening material between source of electricity and victim; and (9) direction in which current passed through the body.[6–8]

In animal experiments measuring tissue damage, the relationship between amperage, temperature, and duration of electric shock were studied.[2] Temperature rise within tissue paralleled the amperage increase. The highest temperature measurements were found adjacent to the skin contact points when the electrical shock was continued for a time period sufficient to allow the current to arc. Additionally, it was found that electrical burns appear to be self-limiting. If the duration of contact was sufficient to allow the current

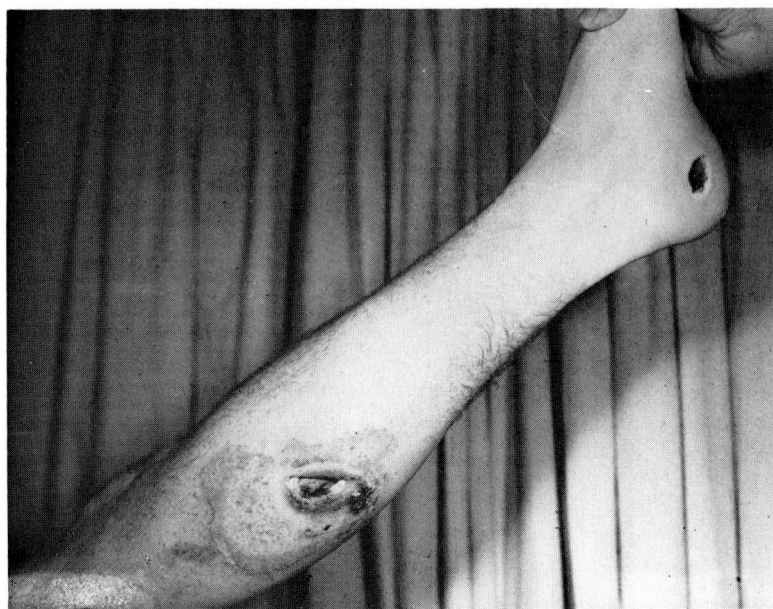

Fig. 5–1.   Acute electric burn contact site of the calf with a central charred zone and peripheral gray-white depressed zone of necrosis. Current exit wound is on the heel.

to arc, skin and muscle damage ceased because amperage then fell to zero. The mass of soft tissue penetrated by the electrical current was more closely related to the degree of tissue injury (heat generated) than to the internal resistance of individual tissues. Muscle injury appeared to be limited to that affected by the primary thermal injury.[2] These findings suggest that "progressive" muscle necrosis as seen clinically in electric burn patients may be due to the fact that some cells injured by heat take several days to degenerate or that nonviable tissue is initially inapparent and undiagnosed.

TISSUE RESISTANCE

The human body may be compared to a complex conductor in which tissues represent multiple circuits arranged in series and in parallel. Hence, tissues arranged in parallel with variable resistances will result in increased current flow through the low resistance pathway (e.g., in nerve instead of muscle; in muscle instead of fat). The following tissues are listed in a sequence of increasing resistance to electric energy: nerve, blood, muscle, skin, tendon, fat and bone.[9] Examples of variations in electric resistance can be found in the skin. A well-calloused hand has a resistance of 1 million ohms, while that of normal dry skin is approximately 5000 ohms, and of moist skin 1000 ohms.

TISSUE SENSITIVITY VARIATIONS

Of all tissues, bone has the greatest resistance (poorest conductor of electric current) and therefore generates the most heat. However, bone is also the most insensitive tissue to heat damage. High resistance at the point of contact and high voltage usually are associated with deeper and more extensive damage. Nerves and blood vessels most readily conduct electric current and generate only a minimal amount of heat; however, these structures are more sensitive to the heat generated than the other tissues of the body.[10]

ALTERNATING VS DIRECT CURRENT

As mentioned above, alternating current is more injurious to tissue. At the same levels of amperes, alternating current is fourfold more injurious to tissue than direct current[7] due to the former's tetanizing effect,[11] which prevents the victim from releasing the power source.[12]

PATHWAY OF ELECTRIC CURRENT AND
APPEARANCE OF INJURED TISSUE

At the site of contact, one sees a central charred area surrounded by a gray-white "leathery" depressed zone of necrosis (Fig. 5–1). These burns are dry, cold, and insensitive and are often without suppuration for days or weeks.[11] Skin contact points may develop progressively deeper ulcerations as the coagulated tissues slough or are surgically removed (Fig. 5–2) Soft tissues adjacent to the skin contact point show variable time-related changes, but in most cases of electric burn, the tissue necrosis extends into deep structures including tendons, muscles, nerves, cartilage and bone.[13] Soft tissue damage appears to be more severe when contact and exit points are on the same extremity.

Flow of current usually occurs along the line of least resistance. Sometimes, the whole body acts as a conductor and one sees distant isolated areas of vascular or neural necrosis. When the current "grounds," it usually

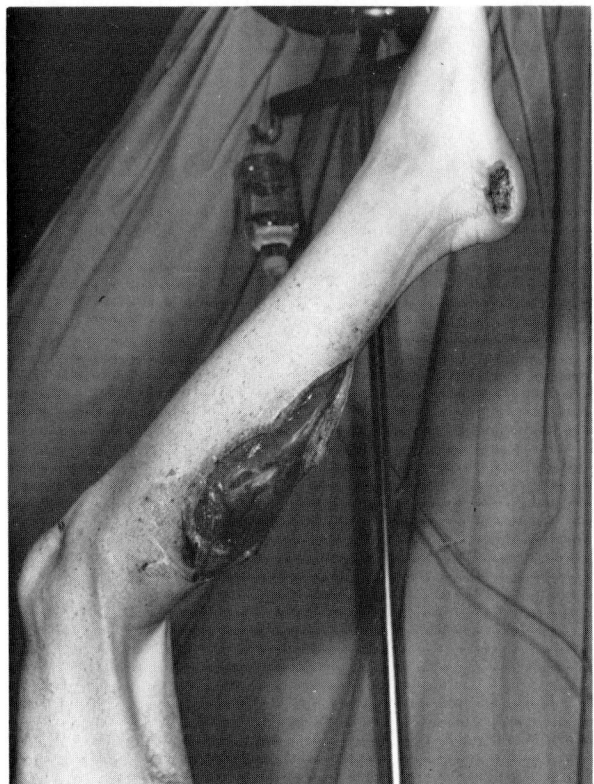

Fig. 5–2. Patient from Figure 5-1 after surgical excision of necrotic tissue.

concentrates and produces an extensive core of necrosis beneath the site where it again ruptures the skin (Fig. 5–3).[9,10] If no associated flame or arc burn is seen, the extent of internal damage (when good contact with a high voltage source has occurred) may not be recognized clinically. The clinical features of externally inapparent injury include tense brawny swelling and progressive neurological change (initially pain and later insensitivity) in an extremity.

An *arc burn* is produced by current coursing external to the body from the contact point to the "ground." It is produced across flexor surfaces (wrist, elbow) and has been called the "kissing" burn (Fig. 5–4). The electric arc is a visual manifestation of the extreme heat generated by the passage of electric current through the high resistance of air. The temperature of this arc may reach 2500–3000°C and may produce only a "flash" type burn in these patients.[1]

Electrical injuries caused by lightning are similar to other electrical burns, except that affected skin may have a characteristic "arborizing" or

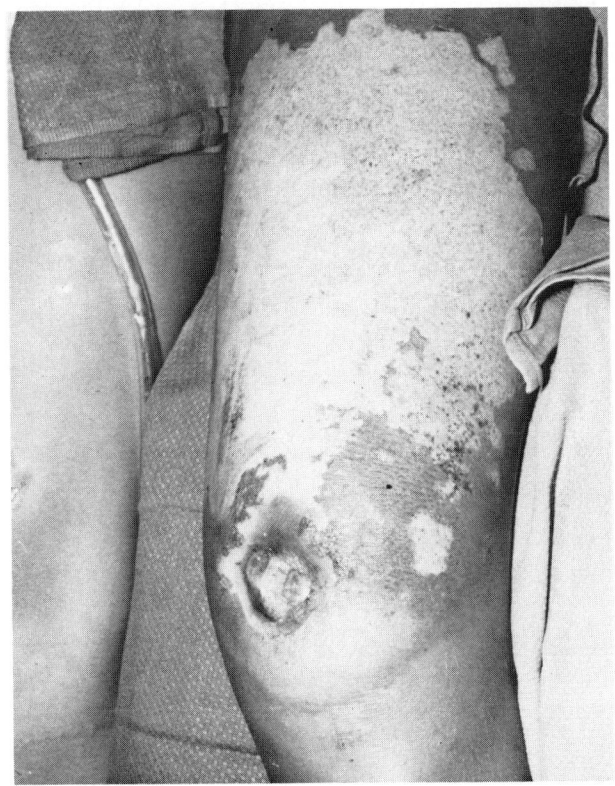

Fig. 5–3. An exit wound of an electric burn near the knee. The thigh in this patient received a secondary flame injury when the trousers were ignited at the current exit site.

"serpiginous" pattern. These peculiar burns are caused by the so-called "splash" effect of the arcing lightning bolt.[14]

Finally, a *thermal burn* may result when clothes are ignited by heat produced by the electric current (Fig. 5–3). This type of burn is important to distinguish from an arc burn which may result in deeper (unrecognized) injury. However, flame burns are often associated with arc burns,[9] and the two may be indistinguishable. Also, flame burns may mask small contact points where electric current has entered the body.

ELECTRIC INJURY, PATIENT EVALUATION

In patients with electrical injury, the points of electrical contact should be identified and exposed to demonstrate the depth and extent of tissue damage (Fig. 5–2). Whenever possible, selected biopsies of affected muscles

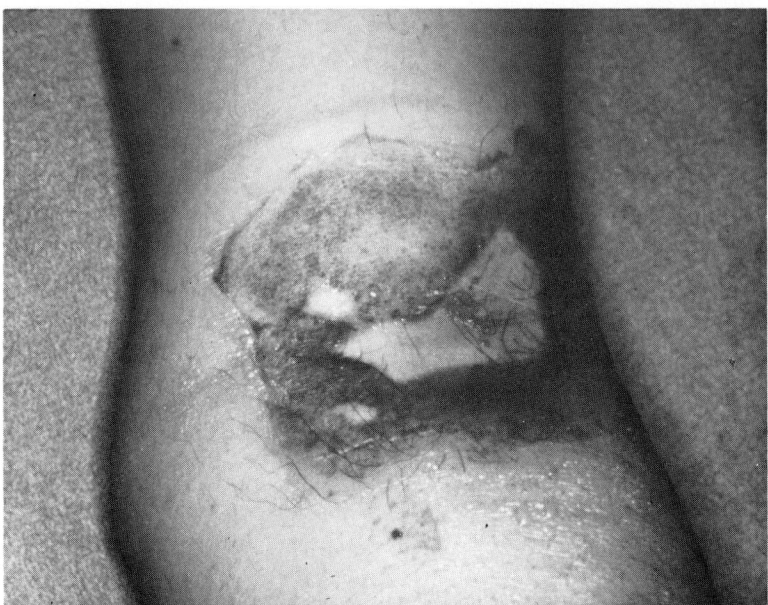

Fig. 5—4. Arc burns result from heat generated by current coursing along the skin surface, as occurred in the antecubital area of this patient.

along suspected pathways of injury should be examined histologically for necrosis and thrombi.[15] Electric burn wounds of the skull and left chest are of particular importance because of their anatomical relationship to vital structures.

Electric burns that appear relatively minor may actually have extensive areas of necrosis beyond the site of injury. This may be related to the electric conductivity and sensitivity of blood vessels which suffer injury to the tunica media and internal elastic lamina. Thrombosis and possibly acute vascular rupture may occur in sites far removed from the contact point.[15,16] Clinical evidence for such injuries may be delayed because of the time required for these vascular lesions to evolve.

MUSCLE BIOPSIES

Biopsies of muscle and other soft tissues have been utilized to determine viability of tissue in electrical burns.[17] Microscopically, electrical thermal injury produces a spiral or corkscrew-like appearance in the muscle fibers which have undergone coagulative necrosis (Fig. 5–5).[6,18] Pyknotic

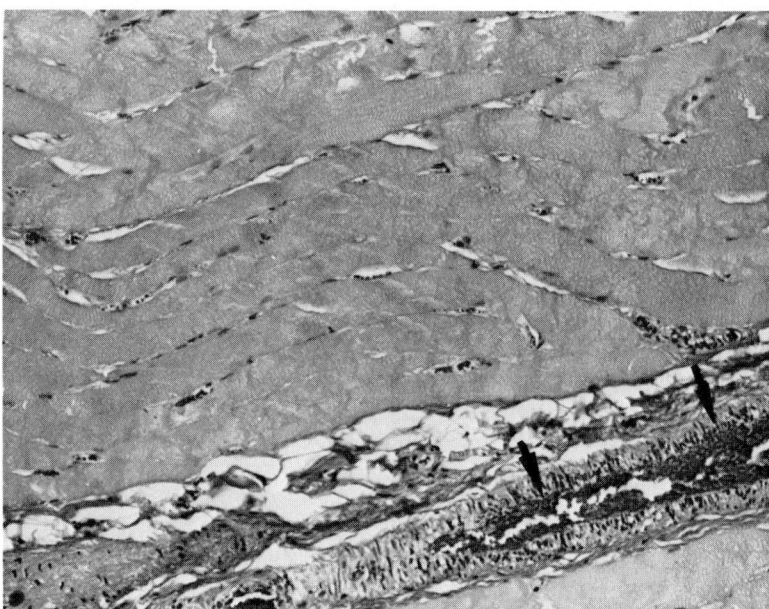

Fig. 5–5. The pathologist must be aware of the heat-fixed appearance of muscle following an electric burn. Muscle fibers have a hyalinized spiraled appearance and swelling. Striations may be accentuated or absent. Note the acute intramural hemorrhage in the small artery (arrows). (H & E, 126 ×)

sarcolemmal nuclei and eosinophilic changes in the muscle sarcoplasm also accompany necrosis (Fig. 5–6).[19] Unfortunately, 24 or more hours may be needed for such changes to be readily apparent. A physiologic approach (electromyographic stimulation) at the time of surgery may be more timely, more accurate, and easier to conduct. Recently ^{133}Xe washout determinations of muscle blood flow have been used for early detection of muscle ischemia.[19]

CARDIAC FINDINGS

Cardiac complications from electric injury usually consist of transient tachycardia, bradycardia, alterations in the S–T segment, and flattening or inversion of T waves. Acute myocardial infarction has also been reported.[11,14] Anginal attacks have been mentioned and, together with electrocardiographic changes, resolve within weeks to months following electrical injury. Findings at postmortem examination included dilated atria and

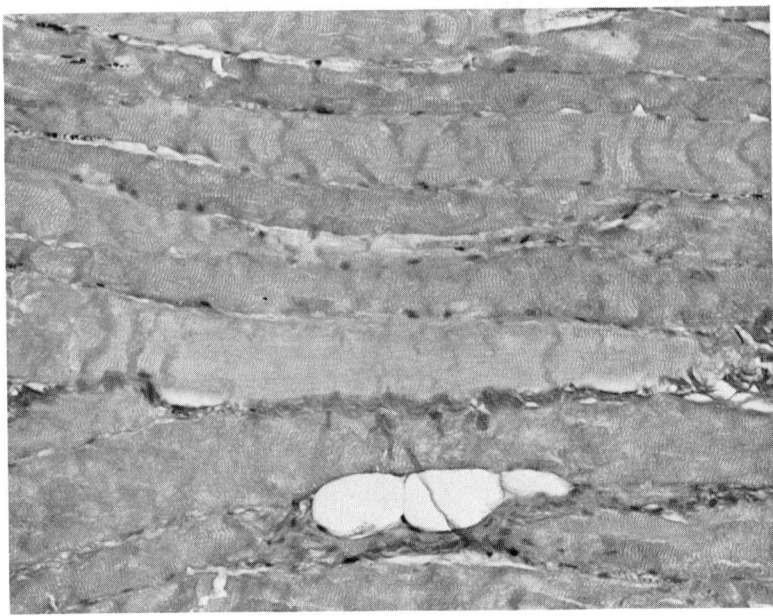

Fig. 5—6. Inflammation is usually absent in soft tissues following electric injury. In this section of skeletal muscle, muscle fibers show coagulative necrosis. Two necrotic fat cells are also present. (H & E, 126 ×)

hemorrhages of variable size in the pericardium, myocardium, and endocardium. Fragmented and twisted myocardial fibers, interstitial edema, and coagulative necrosis may be seen. By electron microscopy, detachment of mitochondrial membranes and rupture of myofibers have been described in rabbits subjected to low voltage current.[6]

NERVOUS SYSTEM

Some of the most common complications following electric injury involve the central nervous system, and their manifestations vary from transient unconsciousness to prolonged coma. Only about 20 percent of patients with central nervous system signs have electrical injury to the head.[20] Spinal cord lesions may present as a spastic paresis or as a deficit in motor function.[10] Symptoms referable to the electric injury may appear up to 3 years later. The most common signs and symptoms are those of incomplete spinal cord transection: muscle imbalance, abnormal gait, impotence, bladder dysfunction, and sympathetic nervous system overactivity which may be expressed by burning pain and hyperesthesia. Because of its high sensitivity and low resistance to electric current, nerve damage may extend to distant sites where no other tissue injury is present. Motor nerves appear to be more readily affected. Clinically, nerve damage is usually permanent.[9,21]

The resistance to the passage of electrical current through a nerve or its supporting vascular structures causes heat which destroys tissue. Disruption of vascular endothelium may cause thrombosis of smaller vessels.[22,23] Increased permeability of the microvasculature results in profound fluid losses which, when confined by fascial sheaths, causes dramatically increased hydrostatic pressure on nerves and vessels resulting in hypoxia. Because there may be no overlying injury, this problem may be easily overlooked. Deep escharotomy through fascial planes relieves the problem. Although the primary injury to the nerve by electric current cannot be repaired, the secondary hydrostatic compression injury to the nerve and its vascular structures are totally preventable.

VISCERAL LESIONS ASSOCIATED WITH
ACUTE ELECTRIC INJURY

Electric injury of the intestine is discussed in Chapter 12. Generally, the intestinal lesions associated with electric injury appear to be adjacent to sites of massive dissipation of electric (converted to thermal) energy.[20,24–26] Similar electric injury may involve the urinary bladder.[27] The incidence of gastrointestinal hemorrhage related to hemorrhagic gastritis or stress ulcers of the stomach or duodenum has been greatly reduced in most burn treatment centers by aggressive antacid therapy in all patients with significant thermal injury.[28]

Linear necrotic lesions or larger areas of necrosis may occur in the liver (Fig. 5–7) and probably represent passage of the electric current through the solid organ.[29] A case of acute gangrenous cholecystitis following electric injury has been reported. It was proposed that the gallbladder injury resulted from the high current created within the low resistance electrolyte medium of the gallbladder.[30] Another author has recognized a high incidence of cholelithiasis in patients with electric injury.[10]

Other visceral complications related directly to electric injuries include focal necrosis of the gallbladder[31,32] and pancreas.[33] Pancreatic injury appears to occur as a result of its proximity to areas of extensive muscle necrosis.

RENAL ABNORMALITIES

Myoglobinuric nephrosis and acute tubular necrosis are described in Chapter 14. Although myoglobinuria and/or hemoglobinuria are almost universal findings in high-voltage electrical injuries, associated renal failure

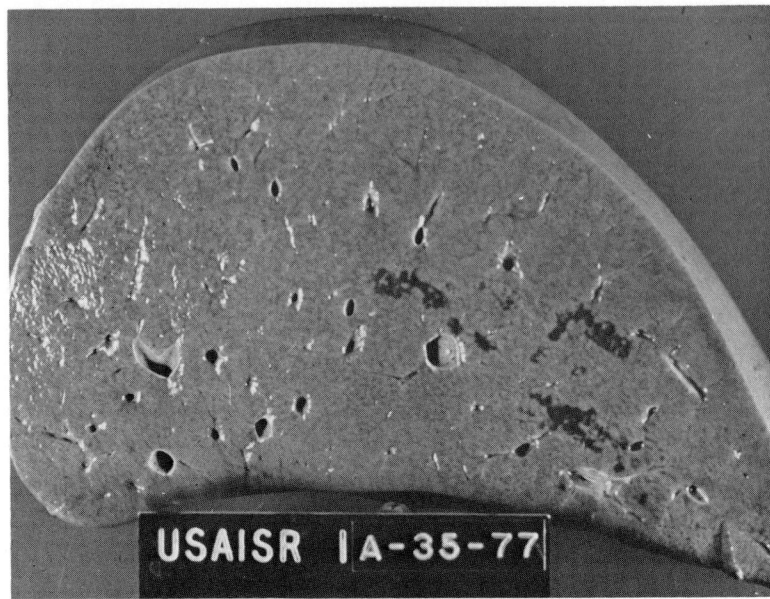

Fig. 5—7. Rarely, electric current may damage thoracic or abdominal viscera. The dark areas of necrosis within this liver represent sites of suspected electric current passage.

is not a major problem when timely and adequate resuscitation is implemented.[3]

SKELETAL SYSTEM

Skeletal fractures sustained during the electric shock occurred in 11 percent of one group of patients and principally involved long bones, but were found in the clavicle, skull, and spine (compression fractures). Usually the fractures are sustained as a result of a fall, but may be due to violent muscle contractions.[9,20]

INFECTIOUS COMPLICATIONS

Unrecognized foci of deep necrosis in muscle or other tissues are an everpresent threat as sites for aerobic and anaerobic infection.[20,34] Aggressive debridement of nonviable tissue appears to eliminate this problem.[17]

EYE INJURY

Cataracts may develop as immediate or late sequelae to electric burns. The incidence of eye injury is higher in patients that have an entry or electrical contact wound of the head, neck, or axilla.[10,35] Cataracts that result from electric burns are usually unilateral, whereas those resulting from lightning injury are usually bilateral.[6]

AUTOPSY FINDINGS, EARLY POSTBURN PERIOD

Electric burn patients who die immediately or shortly after injury may have few lesions other than those at the site of skin contact. Autopsy reveals little more than the expected nonspecific findings of pulmonary edema and mild acute petechial hemorrhages in the heart and pleura. Death in these patients is probably due to respiratory or cardiac arrest or both.

―――――――――――――――――――――――――――――― **References**

1. Wilkinson, C.: Electrical burns. Ariz. Med. 34:702–704, 1977.
2. Hunt, J. L., Masterson, T. S., and Pruitt, B. A., Jr.: The pathophysiology of acute electric injuries. J. Trauma 16:335–340, 1976.
3. Luce, E. A., and Gottlieb, S. E.: "True" high tension electrical injuries. Ann. Plast. Surg. 12:321–326, 1984.
4. Diamond, T. H., Twomey, A., and Myburgh, D. P.: High-voltage electrical injury. A case report and review of the literature. S. Afr. Med. J. 61:318–321, 1982.
5. Jaffe, R. H.: Electropathology. A review of the pathologic changes produced by electric currents. Arch. Path. 5:837–870, 1928.
6. Smogyi, E., and Tedeschi, C. G.: Injury by Electrical Force. *In* Forensic Medicine, Vol. 1—Mechanical Trauma. Philadelphia, W. B. Saunders, Co. 1977, pp 645–676.
7. Pearl, F. L.: Electric Shock. Arch. Surg. 27:227–249, 1933.
8. Kouwenhoven, W. B.: Effects of electricity on the human body. Elec. Engineering 68:199, 1949.
9. Moncrief, J. A., and Pruitt, B. A., Jr.: Electric Injury. Post Grad. Med. 48:189–194, 1970.
10. Baxter, C. R.: Present concepts in the management of major electric injury. Surg. Clin. N. Amer. 50:1401–1418, 1970.
11. Butler, E. D., and Gant, T. D.: Electrical injuries with special reference to the upper extremities. A review of 182 cases. Am. J. Surg. 134:95–101, 1977.

12. Lewis, G. K.: Electrical burns of the upper extremities. J. Bone Joint Surg. 40:27–40, 1958.

13. Sevitt, S.: Burns. Pathology and Therapeutics Applications, 1st. ed. London, Butterworth & Co, 1957, p 325.

14. Brian, A. W., McManus, W. F., Goodwin, C. W., Pruitt, B. A., Jr.: Lightning injury with survival in five patients. J.A.M.A. 253:243–245, 1985.

15. Robinson, D. W., Masters, F. W., and Forrest, W. J.: Electrical burns. A review and analysis of 33 cases. Surgery 57:385–390, 1965.

16. Sturmer, F. C., Jr.: Electrical burns: A case report. Ann. Surg. 154:120–124, 1961.

17. Burke, J. F., Quinby, W. C., Jr., Bondoc, C., McLaughlin, E., and Trelstad, R. L.: Patterns of high tension electrical injury in children and adolescents and their management. Amer. J. Surg. 133:492–497, 1977.

18. Quinby, W. C., Jr., Burke, J. F., Trelstad, R. L., and Caulfield, J.: The use of microscopy as a guide to primary excision of high tension electrical burns. J. Trauma 18:423–429, 1978.

19. Clayton, J. M., Hayes, A. C., Hammel, J., Boyd, W. C., Hartford, C. E., and Barnes, R. W.: Xenon-133 determinations of muscle blood flow in electrical injury. J. Trauma 17:293–298, 1977.

20. DiVincenti, F. C., Moncrief, J. A., and Pruitt, B. A., Jr.: Electrical injuries: A review of 65 cases. J. Trauma 9:497–507, 1969.

21. Levine, N. S., Atkins, A., McKeel, D. W., Jr., Peck, S. D., and Pruitt, B. A., Jr.: Spinal cord injury following electrical accidents: Case reports. J. Trauma 15:459–463, 1975.

22. Salisbury, R. E., and Dingeldein, G. P.: Peripheral nerve complications following burn injury. Clin. Orthop. Rel. Res. 163:92–97, 1982.

23. Rouse, R. G., and Dimic, A. R.: The treatment of electrical injury compared to burn injury: A review of pathophysiology and comparison of patient management protocols. J. Trauma 18:43–46, 1978.

24. Sturim, H. S.: The treatment of electrical burns. Surg. Gynec. Obstet. 128:129–133, 1978.

25. Sinha, J. K., and Roy, S. K.: Perforation of the caecum caused by an electrical burn. Brit. J. Plast. Surg. 29:179–181, 1976.

26. Marszalek, T., and Konieczng, L.: Electric spark injury of the intestine. Polski Przeglad Chirurgiczny (Warszawa) 45:1277–1278, 1973.

27. Chari, P. S., Bapna, B. C., and Balakrishnan, C.: Electrical burns causing a urinary bladder fistula. Case Report. Plast. Reconstr. Surg. 61:446–448, 1978.

28. McAlhaney, J. C., Jr., Czaja, A. J., and Pruitt, B. A., Jr.: Antacid control of complications from acute gastrointestinal disease after burns. J. Trauma 16:645–649, 1976.

29. Newsome, T. W., Curreri, P. W., and Eurenius, K.: Visceral injuries. An unusual complication of an electrical burn. Arch. Surg. 105:494–497, 1972.

30. Burke, J. F., Quinby, W. C., Jr., Bondoc, C., McLaughlin, E., and Trelstad, R. L.: Patterns of high tension electrical injury in children and adolescents and their management (from Discussion, p 497, Aliapoulious, M. A.). Amer. J. Surg. 133:492–497, 1977.

31. Smith, J., and Rank, B. K.: A case of severe electric burns with an unusual sequence of complications. Brit. J. Surg. 33:365–368, 1946.

32. Taylor, P. H., Pugsley, L. Q., and Vogel, E. H., Jr.: The intriguing electrical burn. J. Trauma 2:309–326, 1962.
33. Glazer, A. M.: Pancreatic necrosis in electrical shock. Arch. Pathol. 39:9–10, 1945.
34. Poate, W. J., and Macafee, A. L.: Gas gangrene following electrical burns: A report of two cases. Brit. J. Plast. Surg. 15:17–19, 1962.
35. Long, J. C.: A clinical and experimental study of electrical cataract. Am. J. Ophthalmol. 56:108–188, 1963.

Chapter 6

INFECTIOUS COMPLICATIONS— BACTERIAL

Several investigators have clearly demonstrated that bacterial and mycotic colonization of eschar may, and all too often do, lead to invasive infection of viable tissue beneath the burn wound.[1-3] This progressive process had been represented schematically in Fig. 6–1 and has been carefully reviewed in Chapter 3, in which this topic is further considered as it relates to the burn wound biopsy.

BACTERIAL FLORA OF THE BURN WOUND

A broad spectrum of bacteria colonize the burn wound, the most common including *Staphylococcus, Pseudomonas aeruginosa, Enterobacter cloacae, Proteus* sp., and *Escherichia coli.* Colonizing (noninvasive) gram-positive organisms frequently predominate in the wound cultures that are taken early in the burn course.[4] The incidence at which each microorganism acts as a pathogen varies from year to year, but *Pseudomonas aeruginosa* is currently the most common bacterial "invader" of the burn wound in many burn treatment centers.

Bacterial burn wound invasion is an uncommon complication before the fifth postburn day. This is clearly related to the progressive colonization of the burn wound eschar, which becomes evident by cultural examination after 48 hr and increases substantially over the next 72 hr. The burn wound is an excellent bacteriologic medium, and its capacity for supporting growth approaches that of trypticase soy broth.[5] Quantitative bacterial cultures should be routinely conducted on all surgical biopsies of the burn wound, as well as on selected skin specimens at postmortem examination. In reference to *Pseudomonas aeruginosa,* human autopsy and experimental animal studies indicate that bacterial counts of the eschar which exceed 10^5 microorganisms/g tissue are highly indicative of invasive bacterial infection

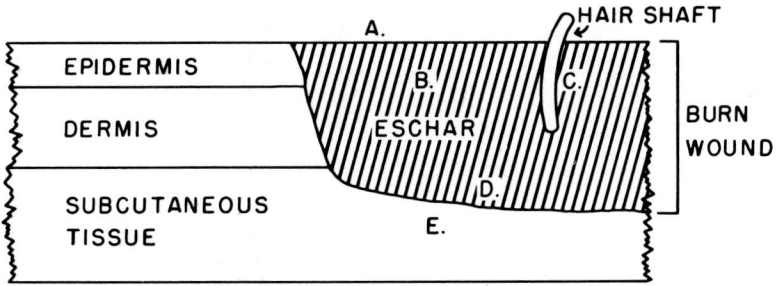

A. SUPERFICIAL COLONIZATION

B. INTRAESCHAR COLONIZATION

C. PERIFOLLICULAR/INTRAFOLLICULAR COLONIZATION

D. DEEP INTRAESCHAR COLONIZATION

E. INVASION OF VIABLE TISSUE, BURN WOUND INFECTION

Fig. 6—1. Schematic representation of sites of bacterial growth in the burn wound.

of the burn wound.[6] Such quantitative bacterial counts must be evaluated with caution, since reliability of single cultures have recently been questioned.[7] Although gram-negative bacteria other than *P. aeruginosa* probably have a similar growth phase within the eschar prior to invasion of viable tissue, quantitative cultures documenting the number of bacteria necessary for invasive infection to occur have been determined only for *P. aeruginosa*.

Staphylococcus sp. appear to colonize only the superficial eschar and extend along necrotic hair shafts into the eschar (Fig. 6–2). These bacteria are often seen in small rounded spaces or within microabscesses. Rarely, if ever, does *Staphylococcus* sp. invade viable tissue and cause a burn wound infection. It is thought that the primary mode of entry into the body for *Staphylococcus* sp. is via extension around a central venous cannula. (See Chapter 9).

The origin of the microorganisms which populate the burn wound is debated. Some authors, mainly on the basis of phage typing, suggest that the microbes originate from the patient's own skin or colonic flora. This concept of self-infection is further supported by the finding of innumerable phage types in different patients within the same burn ward. Sometimes,[8] however, the same strain (phage typing) of *P. aeruginosa* is cultured from the burn

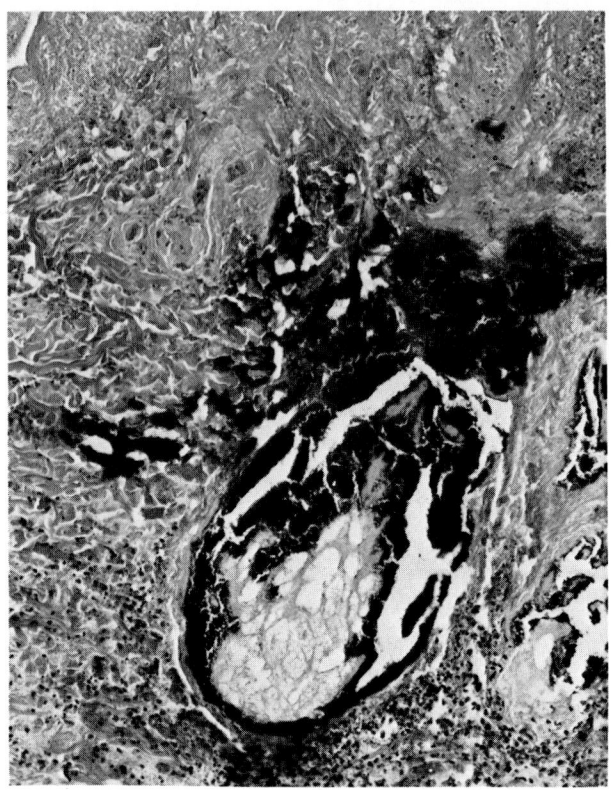

Fig. 6–2. Darkened areas around a necrotic hair shaft represent massive staphylococcal burn wound colonization. These bacteria are frequently found on or within the eschar but are seldom seen as true burn wound invaders. (H & E, × 135)

wound of several patients and suggests nosocomial, patient-to-patient spread.

Therapeutic handling of the burn patient provides yet another mechanism by which microorganisms gain entrance to the vascular system. A transient bacteremia, especially *Klebsiella* sp. and *Staphylococcus aureus,* may occur when the patient is manipulated during care of the burn wound.[9]

CHARACTERISTICS OF INVASIVE
Pseudomonas aeruginosa BURN WOUND
INFECTION

Prior to the use of effective topical antibiotics, *P. aeruginosa* often caused massive burn wound sepsis; the burn surface had characteristic broad black areas of necrosis which extended deep into subcutaneous fat

(Fig. 6–3). With the use of effective topical antibiotics, invasive lesions are more commonly focal, and currently may be undetected until clinical evidence of sepsis or a positive blood culture leads the clinician to examine the wound closely for lesions indicative of invasive infection (Fig. 6–4). It is these small lesions that are most often received by the pathologist for rapid processing and histologic evaluation (See Chapter Three).

The morphologic changes of invasive bacterial infection are most characteristic in *P. aeruginosa* wound sepsis. In addition to the nonspecific downward extension of necrosis beyond the base of the burn, there may be classically a prominent perivascular infiltrate of gram-negative bacilli (Fig. 6–5) with invasive infection by *P. aeruginosa*. A peculiar feature of this bacillary infiltrate is that it extends from perivascular tissue into the wall of the blood vessel but classically spares, or does not infiltrate, the endothelium until late in the disease course. Vascular involvement is usually confined to

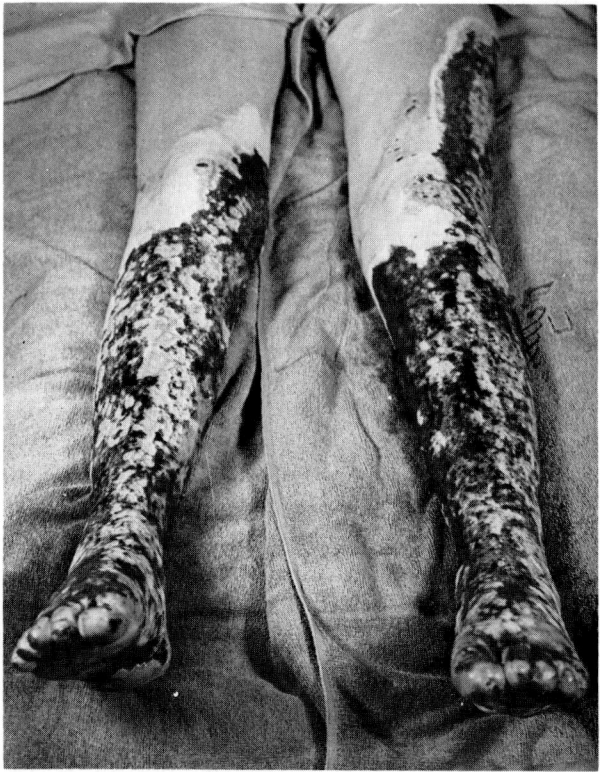

Fig. 6–3. Massive burn wound infection caused by *Pseudomonas aeruginosa*. Severe wound infections of this type have become less frequent with the use of topical antibacterials.

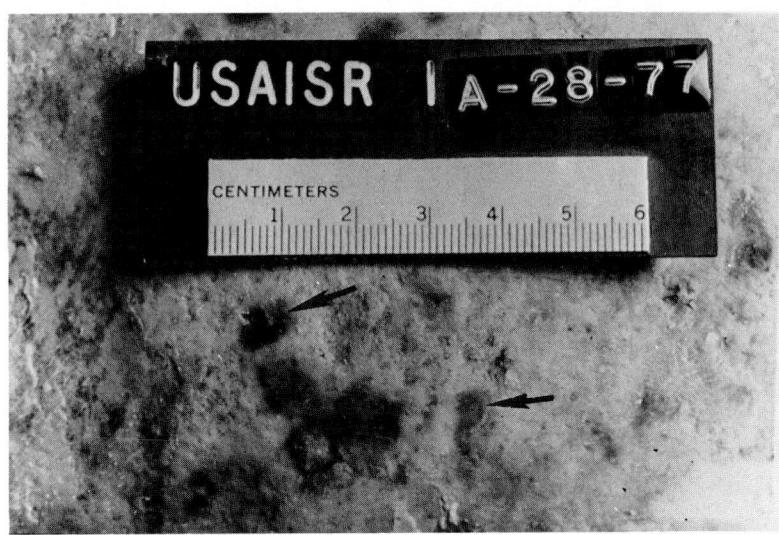

Fig. 6–4.   The entire burn wound must be examined closely for evidence of focal invasive infection (arrows).

the small arteries and veins, arterioles, capillaries, and venules. Thrombi may be found in these involved blood vessels; but, when present, they resemble blood clots (Fig. 6–6). Fibrin and/or platelet thrombi are rare. The presence of the latter may suggest a concomitant disseminated intravascular coagulopathy. Perineural bacillary cuffing and lymphatic involvement have also been reported in invasive *P. aeruginosa* infections.[10]

An acute inflammatory process associated with the classical, rapidly progressing invasive *P. aeruginosa* infection of the burn wound is often scanty to absent. In fact, if the presence of polymorphonuclear leukocytes were considered a prime element for the diagnosis of ecthyma gangrenosum, most cases would remain undiagnosed. The green pus usually associated with burn wound infections by *P. aeruginosa* may represent a favorable sign suggesting colonization of eschar or focal, superficial, well-controlled infection instead of invasive infection.[10] A recent review of *Pseudomonas aeruginosa* infections in burn patients considers these topics in more detail.[11]

Pseudomonas aeruginosa INFECTION WITHOUT VASCULAR INVOLVEMENT

The "pathognomonic" vascular or perivascular pattern has been described and emphasized in *Pseudomonas* infection for decades[2,3,12] and is certainly a useful diagnostic feature when present. In reality, it is often not

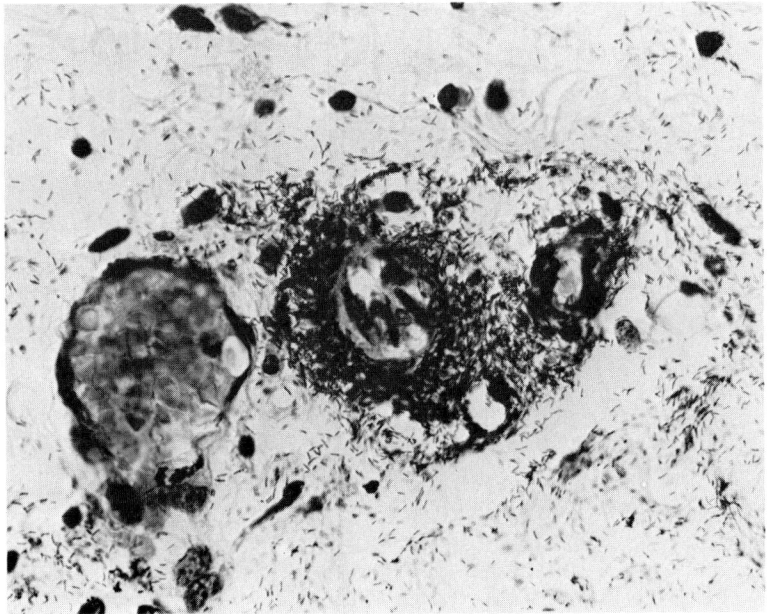

Fig. 6–5. A prominent perivascular infiltrate of gram-negative bacilli is characteristic of infections caused by *Pseudomonas aeruginosa*. Many bacteria are seen at the periphery of this arteriole. (Brown and Bren stain, × 540)

found in the one or two sections that a pathologist may receive as a biopsy specimen, even though invasive *Pseudomonas* infection is present. Therefore, the absence of perivascular distribution of bacteria does not rule out invasive infection by *P. aeruginosa*.

ECTHYMA GANGRENOSUM

"Metastatic" cutaneous lesions of *P. aeruginosa* (ecthyma gangrenosum) appear in concert with, or shortly after, "invasive" lesions of the burn wound (Fig. 6–7). These lesions involve burned and unburned skin, including skin graft donor sites. Characteristically, four progressive clinical stages are noted in ecthyma gangrenosum.[13] The earliest form (stage 1) is a very nonspecific small area of edema which resembles a histamine reaction (inconspicuous in burned skin). Shortly thereafter, this area of edema turns dark brown (stage 2) due to prominent vascular dilatation and congestion and multiple microscopic areas of hemorrhage (Fig. 6–8) which develop

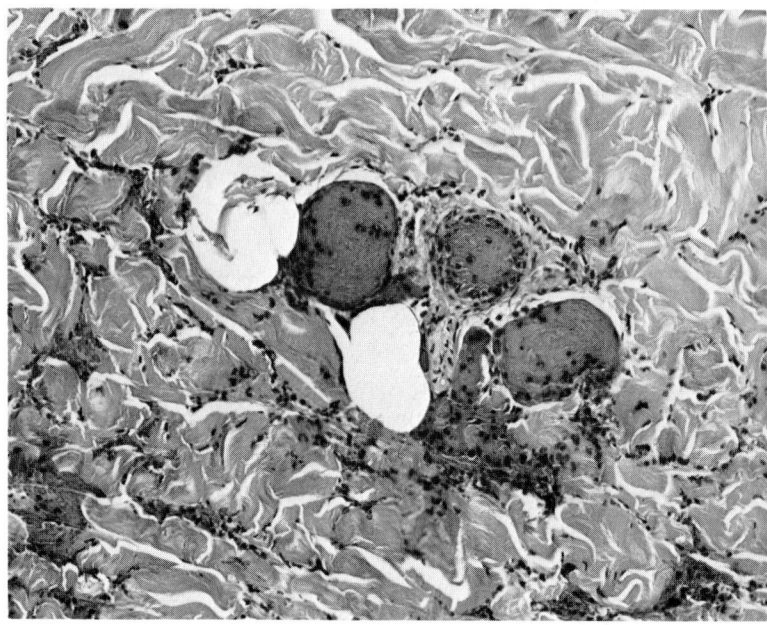

Fig. 6–6.   Small blood vessels in infected burn wounds usually contain blood clots rather than well developed fibrin thrombi. [Hematoxylin and eosin (H & E), × 135]

following vascular invasion and destruction by *P. aeruginosa.* In stage 3, the lesion becomes vesicular. This vesicular change is usually intraepidermal or subepidermal and leads to necrosis of the epidermis (Fig. 6–9A,B). The final and most characteristic phase (stage 4) occurs when the vesicle bursts, leaving a central grayish-brown ulcer (Fig. 6–7) surrounded by a pale zone which in turn is surrounded by a dark brown or purple ring. The pale rim surrounding the central necrosis represents avascular tissue, while the outer dark brown rim is due to vascular dilatation and congestion.[13]

Ecthyma gangrenosum has been emphasized throughout the literature, perhaps because it is so readily apparent to both the clinician and the pathologist. Morphologically, it seems to be identical to the metastatic *P. aeruginosa* lesions that may be encountered in viscera as well as in the burn wound.

Teplitz et al.[2,3] investigated in burned and unburned septic rats the concept that *P. aeruginosa* bacteremia could lead to hematogenous burn wound infection. Burn wounds were apparently hematogenously seeded only when high intravenous doses of *P. aeruginosa* were injected into burned rats. The quantity of bacterial inoculum required to produce these lesions in the rat exceeded the number of microorganisms found in the blood of patients that have *P. aeruginosa* septicemia. As Teplitz implied, however, since such a phenomenon occurs in the rat, it must be considered as an alternative mechanism for human burn wound infection. Logically, it would seem unlikely that blood-borne pseudomonads would selectively infect unburned skin (ecthyma gangrenosum lesions) and simultaneously avoid areas of thermally injured skin.

SELECTION OF BIOPSY SPECIMENS

Although we classically view ecthyma gangrenosum and invasive *Pseudomonas* infection as rapidly progressive, infectious lesions with violaceous, serpentine appearances, one must realize that some invasive or metastatic lesions are bland and nonsuppurative and may initially have a quite innocent appearance. Subtle changes such as a dull, tannish, or pink hue within the burn wound or surrounding unburned skin may be the only indication of serious infection.[5] Hence, the physician may have to resort to repetitive burn wound biopsies of small areas with minimal changes when the clinical pictures suggest early septicemia.

The number of microorganisms invading viable tissue below the burn wound eschar is highly variable. Isolated areas of the burn wound may be

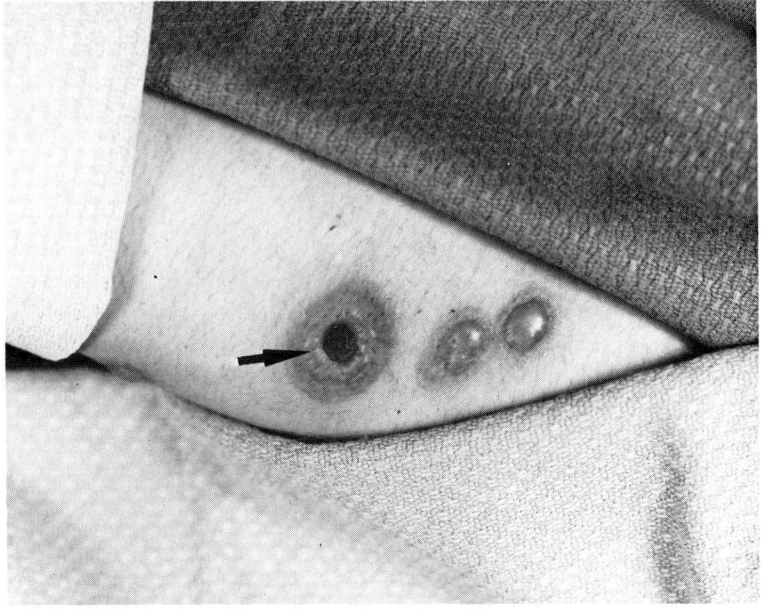

Fig. 6–7. Metastatic cutaneous lesions caused by *Pseudomonas aeruginosa* (ecthyma gangrenosum). The advanced lesion is ulcerated (arrow).

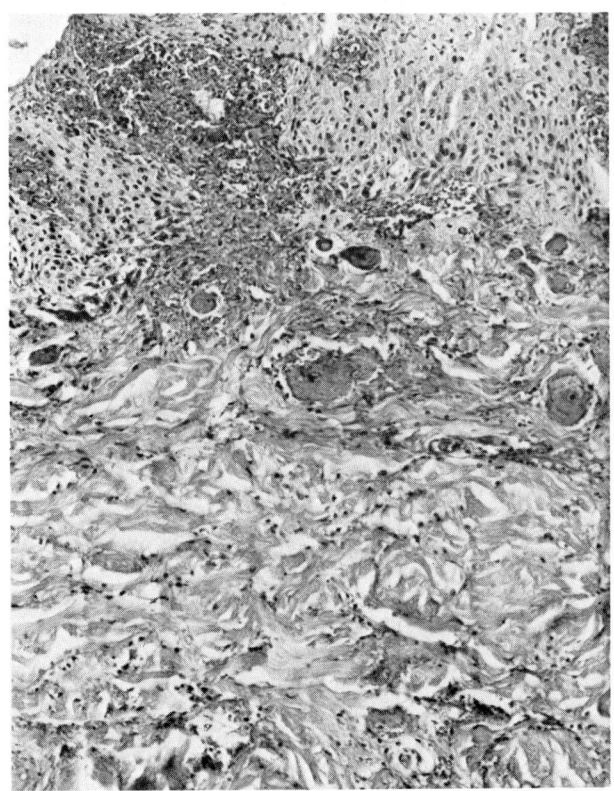

Fig. 6–8. Extensive vasodilation, congestion, and hemorrhage are characteristic of acute ecthyma gangrenosum. (H & E, × 135)

massively invaded by *P. aeruginosa,* while contiguous areas have no evidence of invasive infection. This observation underscores the necessity for selection of representative biopsy and postmortem specimens. It is often useful to evaluate retrospectively all biopsies, along with subsequent clinical information and microbiological data. Patients with significant gram-negative infections often manifest hypothermia, leukopenia, ileus, and rapid, shallow breathing. Prompt recognition of invasive *Pseudomonas* infection of the burn wound and subsequent aggressive therapy has led to the survival of several patients with ecthyma gangrenosum.[14,15]

ATYPICAL LESIONS PRODUCED BY
Pseudomonas aeruginosa

The classical lesions produced in tissue by *P. aeruginosa* are necrotizing and often contain very few acute inflammatory cells. However, the appearance of suppurative foci with numerous polymorphonuclear leukocytes, small numbers of histiocytes, and mycotic-like "grains" and "granules" is a most uncommon but potentially confusing presentation.

We have recently recognized in a burn patient "granules" in a suppurative pyelonephritic lesion which simulated actinomycosis (Fig. 6–10). Large numbers of *P. aeruginosa* and a few colonies of *E. coli* were isolated from renal cultures. Lesions of this type may easily be mistaken for those caused by a wide variety of microorganisms that are known to induce the so-called "botryoid" reaction. These granules are basophilic with hematoxylin and eosin stains and form irregular aggregates of crystals with a suggestion of pallisading at the edge of each aggregate. Gram's tissue stain demonstrates numerous bacilli within the center of each granule. It has been suggested that botryomycotic lesions are the result of chronic stimulation or irritation due to persistence of the organisms. The granulomatous nature of the inflammation supports this contention.[16]

The second type of structure is termed a "grain." It is a 40–60 μ-diameter spherule with a thin, PAS-positive, weakly acid-fast, gram-negative capsule surrounding several coccobacillary structures (Fig. 6–11). These "grains" resemble the endosphores of *Coccidioides immitis.* However, in *P. aeruginosa* "grains," the capsule is quite thin and the internal coccobacillary structures are smaller and more elongated than the endospores seen in tissues infected with sporangia of *Coccidioides immitis.* Similar structures have been identified in a burned guinea pig inoculated with *P. aeruginosa,* strain 1244, and have been noted in other experimental *Pseudomonas* infections.[17] Specific fluorescent antibody reaction has identified the internal structures of the "grain" as *P. aeruginosa.*[18]

PLEOMORPHIC FORMS OF
Pseudomonas aeruginosa

Extreme pleomorphism of bacillary forms of *P. aeruginosa* has been observed in burn wound eschar and in invaded viable tissue. As demonstrated in Figure 6–12, a striking filamentous form of *P. aeruginosa* may resemble superficially *Actinomyces* sp. or *Nocardia* sp. No granules or suppurative foci are identified in lesions of this type. The filamentous forms do not branch and are negative for acid-fast stains. Other than the unusual filamentous bacterial forms, the morphologic features were identical to those of invasive *P. aeruginosa* infection. Quantitative cultures of such lesions yield large numbers (10^8/g tissue) of *P. aeruginosa.* We speculate that this filamentous form of *P. aeruginosa* was induced by mafenide therapy.

Similar, and even more pleomorphic, filamentous forms have been identified in *Proteus vulgaris* exposed to penicillin. The more pleomorphic shapes were noted with higher doses of penicillin which prevented division

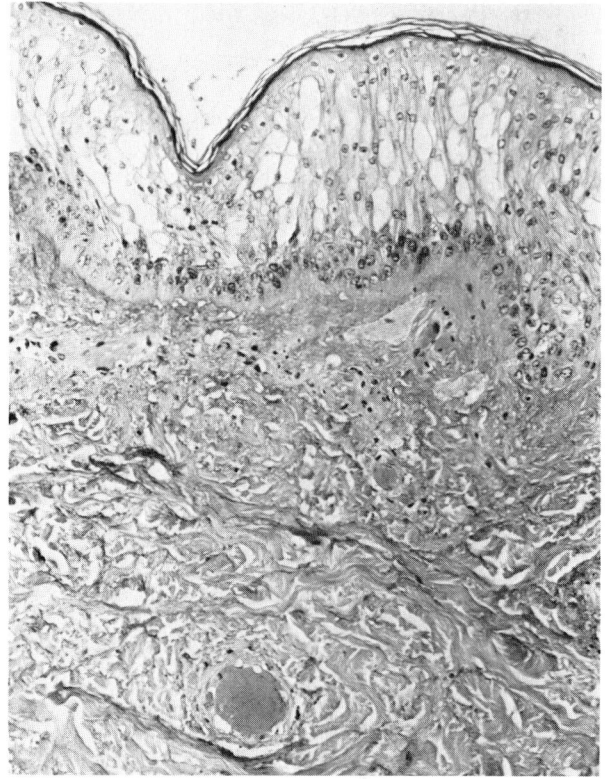

Fig. 6–9A.

of the bacteria. Similar changes were noted in *Salmonella typhi* and *Pseudomonas aeruginosa*. The filamentous forms in the latter were sometimes branched.[19]

Pseudomonas aeruginosa, GREEN URINE SYNDROME

"Green urine syndrome" is a clinical entity which has been associated with *Pseudomonas* septicemia in burn patients. Careful biochemical analysis of urine has demonstrated that the pigment responsible for the green color is verdoglobin, an intermediary product of hemoglobin metabolism. Verdoglobin may accumulate as a result of inhibition of hemoglobin catabolism in reticuloendothelial cells.[20] Verdoglobin possesses a characteristic fluorescent activity in ultraviolet light and thus can be detected by fluorescent tests of the urine. In clinical studies of 300 patients, Stone et al.[20] demonstrated verdoglobinuria in patients with *P. aeruginosa* burn wound infection. Those patients had third degree burns involving more than 18

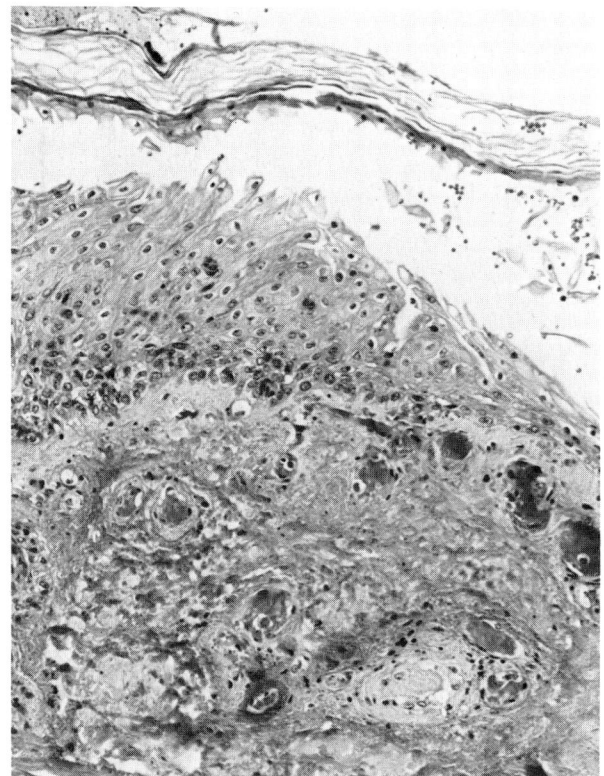

Fig. 6–9B Vesiculation (A) and epidermal necrosis (B) lead to ulceration in advanced metastatic cutaneous lesions caused by *Pseudomonas aeruginosa*. (H & E, × 135)

percent of their total body surface.[21] Microorganism fractionation of a strain of *P. aeruginosa* known to produce this syndrome demonstrated that extracts from the "slime layer" were responsible for the production of verdoglobinuria. Removal of this hemolysin from the slime layer delayed detection of verdoglobin in the urine.[20]

OTHER GRAM-NEGATIVE BACTERIAL INVASIVE INFECTIONS

Quantitative bacterial counts of the burn wound have demonstrated a variety of other gram-negative microorganisms. In our unit, these isolates have included *Enterobacter cloacae, Klebsiella pneumoniae,* and

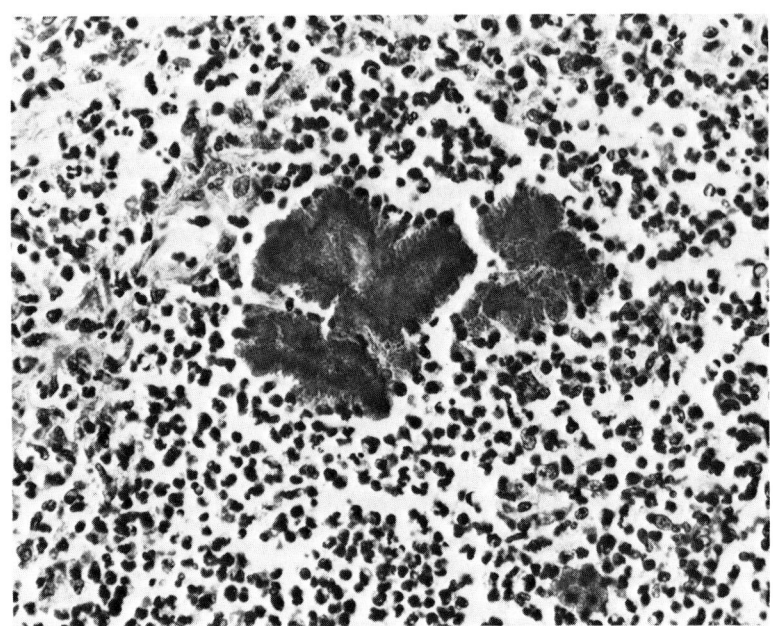

Fig. 6–10. Granules may be form in areas of suppuration associated with *Pseudomonas aeruginosa* infection. (H & E, × 342)

Eschericha coli. The classical perivascular distribution and vascular invasive characteristics of *P. aeruginosa* are not found in these other gram-negative bacterial infections. Diagnosis of burn wound invasion is made by the identification of necrosis extending beyond the burn wound and the presence of bacteria in viable tissue.

QUANTITATIVE WOUND CULTURES

There are scattered reports of the utilization of quantitative cultures of the burn wound without histologic examination for the diagnosis of invasive bacterial infection (See Chapter 2). This is probably a serious error. It is not unusual to find cases in which large number of bacteria ($>10^5$/g tissue) were cultured from the burn wound; but histologically, the organisms were confined to eschar. This is particularly true of *Staphylococcus* sp., which is almost invariably seen as a colonizing (Fig. 6–2) rather than as an invasive organism. Also, with the widespread use of topical antibiotics, focal invasive infection may be found even though the eschar may be seen microscopically to contain few bacteria.

STAPHYLOCOCCUS SPECIES

In the late 1940s and early 1950s, *Staphylococcus aureus* was the predominant life-threatening microbe affecting burn patients. Antibiotic suppression of this microorganisms has led to emergence of *P. aeruginosa* as the predominant bacterial microorganism associated with fatal infections in the burn patient.

As stated earlier, *Staphylococcus* sp. is predominantly found in the superficial eschar and seldom invades the burn wound. However, when a suppurative infection occurs at the insertion site of a central venous catheter, *Staphylococcus aureus* is frequently the causative microorganism. Routine histologic examination of soft tissue near the sites of central venous catheter placement at postmortem examination may demonstrate a suppurative infectious process and abundant gram-positive cocci. Cultures of these areas often demonstrate that the cocci are coagulase-positive *Staphylococcus*. These staphylococcal-infected cutdown sites and the resulting septic thromboemboli probably contribute to many of the staphylococcal bacteremias in burn patients (See Chapter 9).

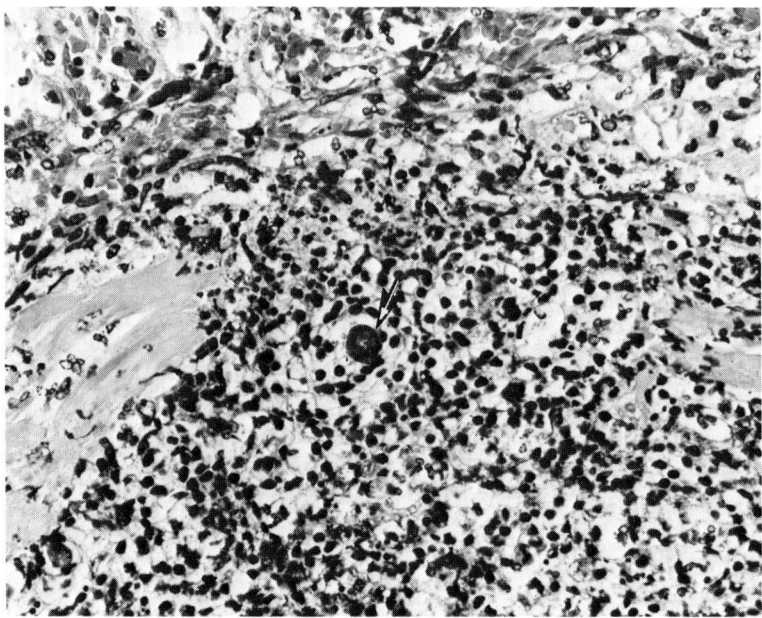

Fig. 6–11. Rounded aggregates of "grains" (arrow) of *Pseudomonas aeruginosa* may be seen in areas of intense pyogranulomatous inflammation caused by this organism. (H & E, × 342)

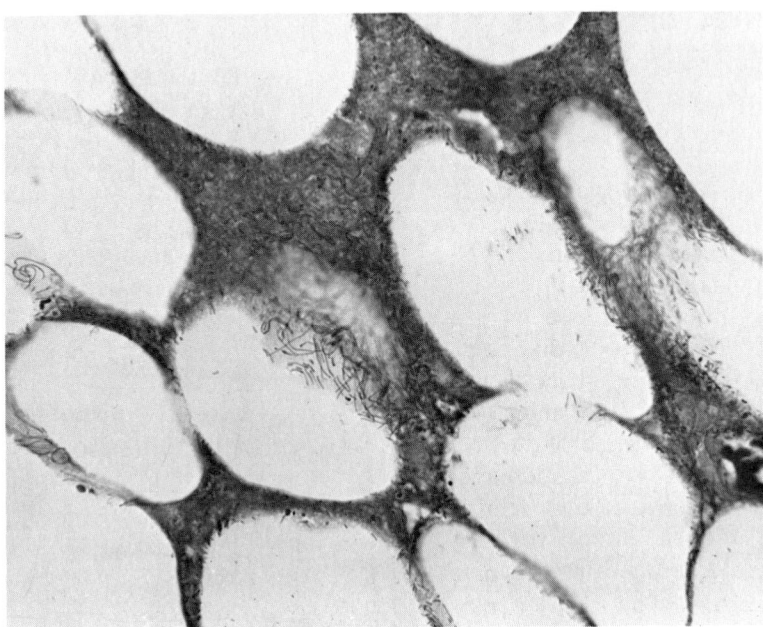

Fig. 6–12.   A filamentous form of *Pseudomonas aeruginosa* may be found in burn wound eschar or in infected viable tissue. (Brown and Bren, × 540)

STREPTOCOCCUS SPECIES

Group A hemolytic *Streptococcus* was a significant pathogen in the burn patient up to the early 1940s. Cellulitis related to streptococcal infection in the burn wound is currently quite unusual with the use of penicillin and current topical antibiotic agents. Quantitative cultures of burn wound specimens indicate that significant streptococcal infections may occur with as few as 300 microorganisms/g tissue.[22] This makes histologic diagnosis of streptococcal invasive infection very difficult, because these low levels of bacteria are practically impossible to detect histologically in biopsy specimens even when appropriate special stains are used. This sharply contrasts with *P. aeruginosa,* which usually requires 10^5 organisms/g tissue before a significant infection results. Robson et al.[22] relate this disparity in pathogenicity to the different modes of host defense; complement acts against gram-negative bacteria while lysosomes destroy gram-positive bacteria. They further hypothesize that beta-hemolytic streptococci protect themselves by activating plasmin to break down fibrin and by having decreased susceptibility to lysosomal enzymes in comparison to other microorganisms.

"GAS GANGRENE" AND CLOSTRIDIAL INFECTIONS

There have been two recent reports of clostridial infections in burn patients.[23,24,25] These are considered rare but dangerous infections because of the potent proteolytic exotoxins released by the *Clostridium* species, which are anaerobic, gram-positive, spore-forming bacteria. These exotoxins destroy blood and tissue cells. *Clostridium* sp. are common soil contaminants and may be cultured routinely from both animal and human feces.[26] Deep massive burn wounds involving muscle are particularly susceptible to clostridial infections.[25] Crepitation of tissue may not be appreciable as described classically.

Deep clostridial infections usually have extensive muscle necrosis related to the release of exotoxin. Individual muscle fibers undergo coagulative necrosis and fragmentation, and cross-striations are lost (Fig. 6–13). Gas-filled spaces widely separate sarcolemmal sheaths from the individual muscle fibers. The quantity of exudate varies. The intensity of the acute inflammatory infiltrate is generally inversely proportional to the severity of clostridial infection.[25,27] Gram's tissue stain demonstrates many large, gram-positive bacilli (Fig. 6–14). Mixed bacterial flora are commonly iso-

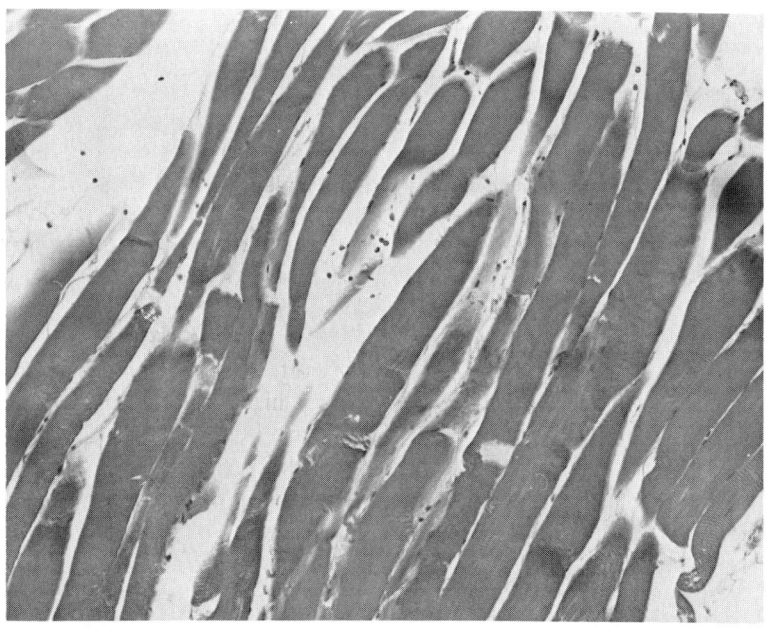

Fig. 6–13. Muscle necrosis associated with a clostridial infection. (H & E, × 342)

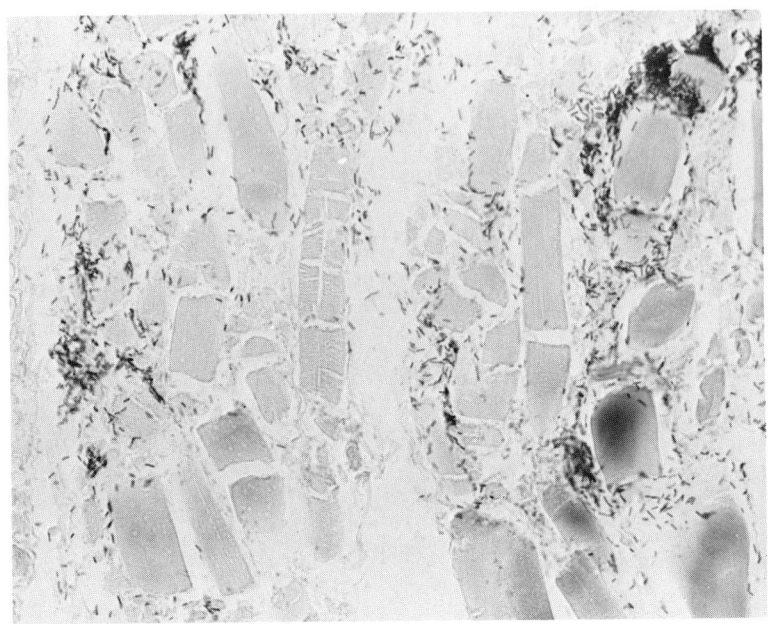

Fig. 6–14. Large gram-positive bacilli in a zone of coagulation necrosis caused by clostridial infection. (Brown and Bren, × 342)

lated from "clostridial" infections, and these isolates often include several clostridial species as well as a variety of gram-positive cocci and gram-negative aerobic and falcultative anaerobic bacillary microorganisms.[25] Significant clostridial infections may be preceded by other bacterial infections, e.g., *Proteus vulgaris*.[27]

Currently, prompt and aggressive debridement of nonviable tissue in the burn wound and timely performance of escharotomy or fasciotomy have practically eliminated this lethal infection from the list of complications found in the burn patient.

Crepitation in soft tissue is not pathognomonic for clostridial infection. Other microorganisms producing gaseous changes in tissue include *Bacterioides* sp., anaerobic *Streptococcus*, *Proteus*, Enterococcus, *E. coli*, *Enterobacter*, *Staphylococcus*, and *Pseudomonas*.[26]

DETERMINING THE ORIGIN OF SEPTICEMIA

When a blood culture from a burn patient is found to contain *P. aeruginosa* or *Klebsiella pneumoniae*, it is all too tempting to ascribe the bacterial source to "burn wound sepsis" in the absence of pulmonic

Table 6—1 *Sources of Microorganisms in Burn Patients with Septicemia[a]*

Gram-Positive Cocci	Gram-Positive or Negative Microorganisms	Gram-Negative Bacilli
Intravenous cannulae	Necrotizing laryngitis	Burn wound "invasion"
Tracheostomy stoma	Necrotizing tracheo-bronchitis or bronchopneumonia	Cystitis
Thrombophlebitis		Prostatitis
? Ischemic entero-colitis	Inhalation injury	Manipulation of the burn wound
Infectious endocarditis/valvulitis	Empyema	Ischemic enterocolitis
Manipulation of burn wound	Carious teeth with gingival infection/abscess	Pyelonephritis, ascending Perionitis
	Skeletal fractures	

[a] These represent the most characteristic sites. Obviously, it is possible that any source may give rise to gram-positive and/or gram-negative microorganisms.

infiltrate or other obvious sources of infection. Similarly, isolation of *Staphylococcus aureus* in the blood may suggest septic thrombophlebitis, cannula infection, burn wound infection or specimen contamination. In a review of several hundred burn autopsies, a broad spectrum of possible infectious sites that may contribute to septicemia have been identified. These are summarized briefly in Table 6–1.

MICROBIAL SYNERGISM IN INFECTIOUS DISEASE

Experiments have shown that there is a synergistic effect on mortality when mice are infected with murine cytomegalovirus, *Pseudomonas aeruginosa*, *Staphylococcus aureus*, and *Candida albicans*.[29] It is possible that similar mixed infections affect mortality in the burn patient (see Chapter 8).

CHANGING PATTERNS OF INFECTION

The changing patterns of infections are certainly a problem in the current treatment of burn patients as well as other victims of severe forms of trauma.[28] Briefly stated, the current and future problems in infectious disease faced by physicians include

1. Management of secondary infections due to antibiotic therapy
2. Synergistic action of mixed bacterial or viral–bacterial infections
3. Increasing number of infections by "L" forms and other atypical bacterial forms
4. Anaerobic bacterial infections
5. Mycotic infections
6. Viral infections
7. Impairment of host defenses (primary and secondary to burns).

References

1. Rabin, E. R., Graber, C. D., Vogel, E. H., Jr., Finkelstein, R. A., and Tumbusch, W. T.: Fatal *Pseudomonas* infection in burned patients. A clinical, bacteriologic and anatomic study. New Engl. J. Med. 265:1225–1231, 1961.
2. Teplitz, C., Davis, D., Mason, A. D., Jr., and Moncrief, J. A.: *Pseudomonas* burn wound sepsis: I. Pathogenesis of experimental *Pseudomonas* burn wound sepsis. J. Surg. Res. 4:200–216, 1964.
3. Teplitz, C., Walker, H. L., Raulston, G. L., Mason, A. D., and Moncrief, J. A.: *Pseudomonas* burn wound sepsis: II. Hematogenous infection at junction of burn wound and unburned hypodermis. J. Surg. Res. 4:217–222, 1964.
4. Pruitt, B. A., Jr., and McManus, A. T.: Opportunistic infections in severely burned patients. Am. J. Med. 76S:146–154, 1984.
5. Teplitz, C.: Pathology of burns. *In* The Treatment of Burns. Chapter 2, 2nd. ed. C. P. Artz and J. A. Moncrief, eds. Philadelphia, W. B Saunders Co., 1969, pp 22–88.
6. Lindberg, R. B., Moncrief, J. A., and Mason, A. D., Jr.: Control of experimental and clinical burn wound sepsis by topical application of sulfamylon compounds. Ann. N. Y. Acad. Sci. 150:950–960, 1968.
7. Woolfrey, B. F., Fox, J. M., and Quall, C. O.: An evaluation of burn wound quantitative microbiology. I. Quantitative eschar cultures. Am. J. Clin. Pathol. 75:532–537, 1981.
8. Lindberg, R. B., and Latta, R. L.: Phage typing of *Pseudomonas aeruginosa*: Clinical and epidemiologic considerations. J. Infect. Dis. 130:33–42, 1974.
9. Sasaki, T. M., Welch, G. W., Herndon, D. N., Kaplan, J. Z., Lindberg, R. B., and Pruitt, B. A., Jr.: Burn wound manipulation—induced bacteremia. J Trauma 19:46–48, 1979.
10. Teplitz, C.: The pathology of burns and the fundamentals of burn wound sepsis. *In* Burns. A Team Approach. C. P. Artz, J. A. Moncrief, and B. A. Pruitt, Jr., eds. Philadelphia, W. B. Saunders Co., 1979, pp 45–94.
11. Pruitt, B. A., Jr., Lindberg, R. B., McManus, W. F., Mason, A. D., Jr.: Current approach to prevention and treatment of *Pseudomonas aeruginosa* infections in burned patients. Rev. Infect. Dis. 55:889–897, 1983.

12. Fraenkel, E.: Uber allgemeininfektionendurch den *Bacillus pyocyaneus.* Virchows Arch. F. Path. Anat. 183:405–440, 1906.
13. Dorff, G. J., Geimer, N. F., Rosenthal, D. R., and Rytel, M. W.: *Pseudomonas* septicemia. Ilustrated evolution of its skin lesion. Arch. Intern. Med. 128:591–595, 1971.
14. Loebl, E. C., Marvin, J. A., Heck, E. L., Curreri, P. W., and Baxter, C. R.: The use of quantitative biopsy cultures in bacteriologic monitoring of burn patients. J. Surg. Res. 16:1–5, 1974.
15. Solem, L. D., Zashe, D., and Strate, R. G.: Ecthyma gangrenosum. Survival with individualized antibiotic therapy. Arch. Surg. 114:580–583, 1979.
16. Kimmelstiel, P., and Easley, C. A., Jr.: Experimental botryomycosis. Am. J. Pathol. 16:95–102, 1940.
17. Koeda, T., Abe, N., and Konno, S.: Experimental *Pseudomonas* granuloma in the skin of guinea pigs. Nat. Inst. Anim. Health Quart. 14:9–16, 1974.
18. Sawada, T., Koeda, T., Muromatsu, K., and Sazawa, H.: Specific fluorescent antibody staining of the grain in skin lesion of guinea pig infected with *Pseudomonas aeruginosa.* Jap. J. Vet. Sci. 37:621–625, 1975.
19. Fleming, A., Voureka, A., Kramer, I. R. H., and Hughes, W. H.: The morphology and motility of *Proteus vulgaris* and other organisms cultured in the presence of penicillin. J. Gen. Microbiol. 4:257–269, 1950.
20. Stone, H. H., Martin, J. D., Jr., Kolb, L.: The mechanism and treatment of verdoglobinuria in *Pseudomonas* sepsis. Surg. Forum 15:48–49, 1964.
21. Stone, H. H.: Review of *Pseudomonas* sepsis in thermal burns; verdoglobin determination and gentamicin therapy. Ann. Surg. 163:297–305, 1966.
22. Robson, M. C., and Heggers, J. P.: Surgical infection. II. The b-hemolytic streptococcus. J. Surg. Res. 9:289–292, 1969.
23. Davies, D. M., Brown, J. M., Bennett, J. P., et al.: Survival after major burn complicated by gas gangrene, acute renal failure, and toxic myocarditis. Br. Med. J. 1:718–719, 1979.
24. Davies, D. M.: Gas gangrene as a complication of burns. Scand. J. Plast. Reconstr. Surg. 13:73–75, 1979.
25. Monafo, W. W., Jr., Brentano, L., Gravens, D. L., Kampson, R., and Mayer, C. A.: Gas gangrene and mixed-clostridial infections of muscle complicating deep thermal burns. Arch. Surg. 92:212–221, 1966.
26. Hart, G. B., Lamb, R. C., Strauss M. B.: Gas gangrene: I. A collective review. J. Trauma. 23:991–1000, 1983.
27. Anderson, W. A. D., and Kissane, J. M.: *In* Pathology, Vol. 1, 7th. ed. St. Louis, The C. V. Mosley Co., 1977, p 376.
28. Altemeier, W. A.: The significance of infection in trauma. Bull. Am. Col. Surg. 57:7–16, 1972.
29. Hamilton, J. R., Overall, J. C. Jr., Glasgow, L. A.: Synergistic effect on mortality in mice with murine cytomegalovirus and *Pseudomonas aeruginosa, Staphylococcus aureus,* or *Candida albicans.* Infect Immun. 14:982–989, 1976.

Chapter 7

INFECTIOUS COMPLICATIONS— FUNGAL

With the introduction of topical antibiotic therapy, the incidence of mycotic infection has increased 10-fold in the burn patient.[1,2] The most significant mycotic pathogens of the burn wound include Zygomycetes, *Aspergillus*, *Candida* and *Fusarium*.[1-9]

A wide variety of factors are known to alter host resistance to mycotic organisms. Third degree burns develop mycotic colonization 3.5 times more commonly than second degree burns.[2] Broad spectrum antibiotics are known to predispose to mycotic infections.[10] Coexistent bacterial infections have been shown experimentally to depress mycotic growth;[11,12] however, a coexistent viral infection appears to have a synergistic action.[13]

Mycotic spread generally involves contiguous skin and soft tissue (muscle). Visceral fungal infections are uncommon in burn patients and consist only of anecdotal case reports.[3,5,6]

Culturing mycotic organisms from surgical or autopsy specimens has sometimes been quite difficult, even when there is substantial clinical and histologic evidence of significant mycotic infection. In addition to the actual submission of tissue for culture, direct culture of the wound with a cotton swab applicator may provide an additional mode for recovery of mycotic organisms. However, because of slow and variable culture growth characteristics, it is important to recognize that histologic evaluation is probably the fastest and single most reliable mode of examination to (1) differentiate between bacterial and mycotic infection, (2) establish the presence of "invasive" infection as opposed to colonization, and (3) provide a morphologic diagnosis to the surgeon in a timely manner (usually less than 4 hr) (see Chapter 3) which will facilitate early treatment.

With experience, the pathologist may, in many cases, accurately diagnose burn wound candidiasis, aspergillosis, and zygcomycosis; however, definitive identification is accomplished only by culture results. From

Fig. 7–1. Two colonies of *Candida* species in superficial layers of eschar. Pseudohyphae may be seen in eschar, but seldom invade viable tissue. (Gomori's methenamine silver stain, × 135)

the practical point of view, the identity of a fungus in a surgical specimen is of secondary importance. The vital determination to be made is whether or not the fungus is present as a colonizing (Fig. 7–1) or invasive organism (Fig. 7–2).

CANDIDA SPECIES

Candida sp. are the most frequently observed nonbacterial organisms with the burn wound. They appear as colonies of yeasts and pseudohyphae in superficial eschar, usually within 1–2 weeks postburn (Fig. 7–1). *Candida* sp. seldom invade viable tissue, but this fact should not lull the clinician into a false sense of security, as invasive candidal infections have occurred in burn patients,[14–19] and have been successfully treated with amphotericin when diagnosed early in the disease course.[20,21] The literature is replete with examples of disseminated candidiasis in debilitated or immunologically compromised patients. In addition to the burn wound, *Candida* sp. may gain

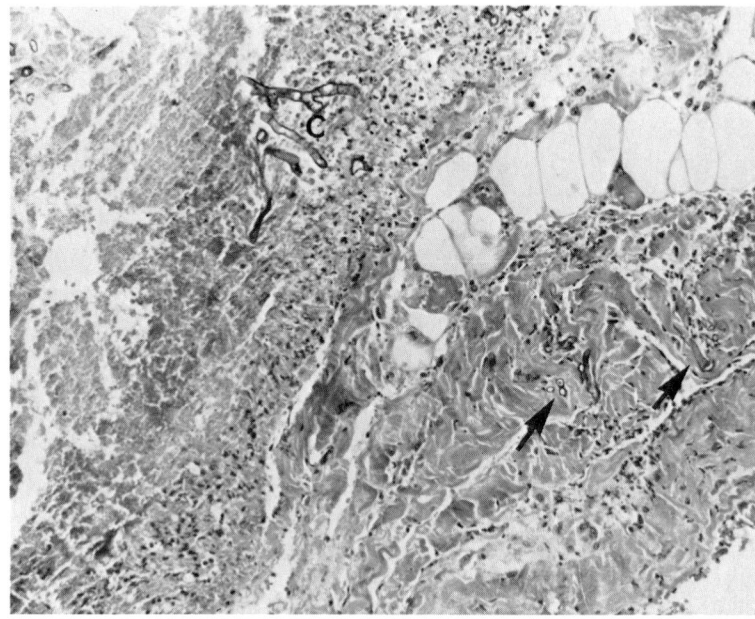

Fig. 7–2. Burn wound with colonizing (C) and invasive (arrows) fungal hyphae. (H & E, × 135)

entrance through a variety of portals in the thermally injured patient, including central venous catheters, oropharynx, or upper gastrointestinal tract.[18] Candidal passage through gastrointestinal mucosa may occur even in the absence of necrosis (persorption).[9]

Candida sp. are a relatively frequent isolate from the burn wound, but only sparsely colonize burn patients with smaller wounds.[22] Because Candida sp. are commonly isolated simultaneously from the mouth or feces, these sites may be the reservoir of Candida in these patients. With severe burns, however, the incidence and degree of colonization of the burn wound is increased; and candidal septicemia becomes apparent.[20,23] Candida sp. rarely invade viable tissue below the burn wound in a manner that we might detect histologically; therefore, the burn wound biopsy for histological examination is not useful. Spebar and Lindberg suggested that blood cultures be utilized as evidence of candidal invasion and have data that appears to support this contention. However, one should also seriously consider the gastrointestinal tract (oral thrush, monilial esophagitis or enterocolitis) as the actual source of septicemia while the burn wound may be passively colonized.

ZYGOMYCETES

In contrast to the ubiquitous *Candida* sp., the Zygomycetes are infrequently found in the burn wound; but, when present, they are more likely to be significant wound invaders.[1,3,4,7,8,24] The significance of invasive zygomycotic infections lies in its ability to cause serious local disease and rapidly spread to adjacent viable tissue (Fig. 7–2). Therefore, its natural history necessitates aggressive therapy by the physician, including wide excision of the affected burn wound, and at times even amputation. Bruck et al.[24] emphasized the importance of histologic examination of biopsy specimens and the difficulty experienced in culturing these microorganisms from tissue specimen. In his study of 30 patients with invasive mycotic infection, nine patients (30 percent) died from fungal invasion and 7 of the 15 survivors required amputation to control the infection. Visceral involvement by Zygomycetes via hematogenous spread in the burn patient is rare but can occur.

Clinically, in invasive zygomycotic infections the burn wound or adjacent skin or both undergo a dark brown to black discoloration indicative of hemorrhagic ischemic necrosis (Fig. 7–3). Histologically, one classically

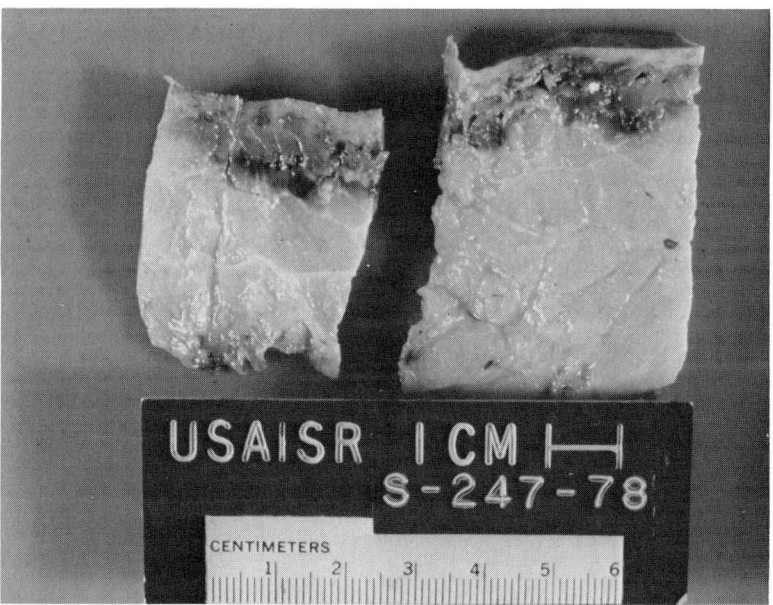

Fig. 7–3. Excised burn wound with characteristic hemorrhage in subcutaneous fat. Invasive phycomycotic infection was diagnosed microscopically.

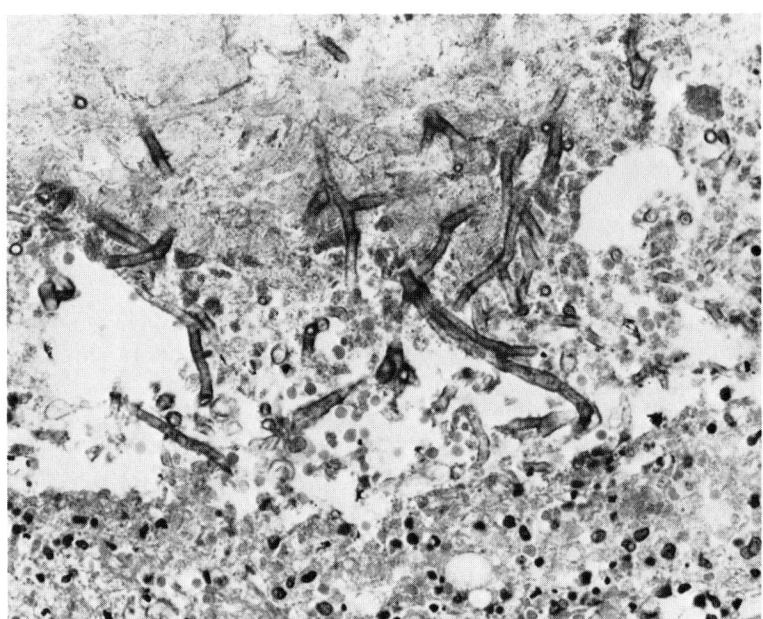

Fig. 7–4. Broad nonseptate hyphae with variable angles of branching are characteristic of *Zygomycetes*. (H & E, × 342)

sees nonseptate, broad hyphae with nonparallel sides. Branching occurs at variable angles (Fig. 7–4). Zygomycetes are most clearly apparent on slides stained with hematoxylin and eosin stains. PAS and silver stains tend to demonstrate Zygomycetes less prominently. Vascular invasion is commonly seen in association with vascular thrombosis (Fig. 7–5). The resultant necrosis and hemorrhage within soft tissue further provide an optimal milieu in the tissue surrounding the original burn wound for additional zygomycotic growth and explain the rapid spread of this infection. Exceedingly low cultural isolation of Zygomycetes is somewhat improved by the use of tissue slices instead of finely ground tissue samples. Histologic indications for invasive fungal infection are similar to those for invasive bacterial infection: (1) microorganisms at the border of, or involving, viable tissue and (2) extension of necrosis beyond (laterally or deeper than) the burn wound. The dissemination of mycotic infection must be interrupted immediately by surgical intervention and antimycotic agents administered if systemic spread has occured or is considered likely.

Although the burned skin is the most common site of involvement by zygomycosis, the pathologist should be aware of rhinocerebral zygomycosis, especially in the burn patient with diabetic ketoacidosis. Unilateral proptosis should lead to retrobulbar fat biopsy, culture, and debridement of the nasal and oral cavities, and roentgenographic studies of the sinuses.[25]

Aspergillus SPECIES

Aspergillus sp. are more commonly found in burn wounds than Zygomycetes, but currently are seen less often as invasive pathogens. Stone et al.[26] have recently reported 18 burn patients with *Aspergillus* infections of their burn wounds. Such infections were often associated with air conditioning filters heavily colonized with *Aspergillus* sp. *Aspergillus* burn wound infections resulted in a high mortality rate (78 percent), could not be controlled with topical or parenteral antimycotic therapy, and were best treated by amputation or deep excision of affected areas.[26]

Classically, *Aspergillus* sp. has slender, septate hyphae which branch dichotomously at approximately 45° angles. The fungal hyphae extend into adjacent viable tissue and may also invade blood vessels in a manner similar to Zygomycetes (Figs. 7–6 and 7–7). *Aspergillus* sp. found on the burn wound may occasionally have pleomorphic elements similar to those which

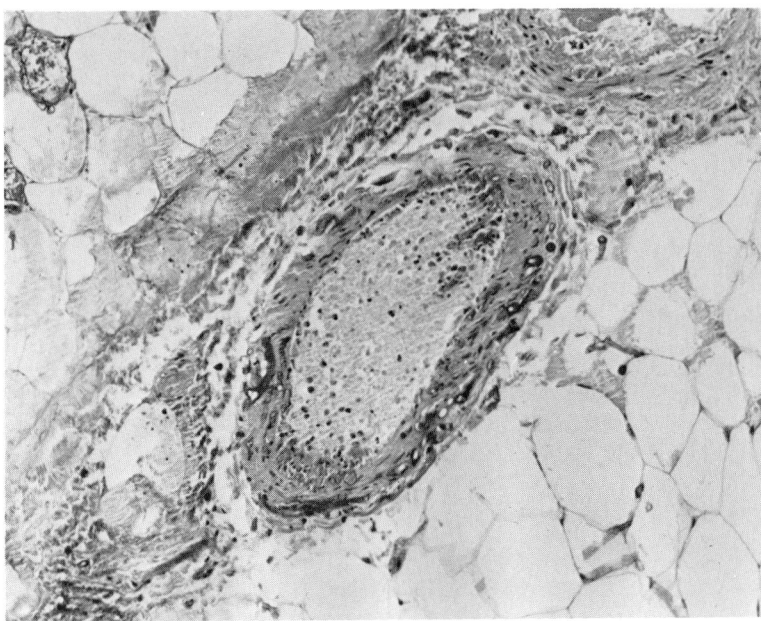

Fig. 7–5. Fungal hyphae within the wall of a small artery in a burn wound. (H & E, × 135)

Fig. 7–6. "Fruiting bodies" of *Aspergillus* species may be seen in massively colonized burn wounds. (PAS, × 54)

have been identified in granulomatous inflammatory processes of the paranasal sinuses and periorbital tissues.[27]

Infections of the burn wound by *Aspergillus niger* have also been noted. This microorganism is readily recognized by dark black fungal growth on the burn wound surface (See Chapter 3, Fig. 3–7A). Because of this pigmentation, the mycotic organism can usually be identified at a very early stage (prior to invasion of the burn wound). In other mycotic infections, the dark brown to black color is due to hemorrhagic necrosis of tissue and is an indicator that invasive infection has already occurred. The darkly pigmented superficial colonies of *Aspergillus niger* are occasionally submitted as biopsies of the burn wound suspected of having invasive infection,[28] but histologic examination of the biopsy enables one to differentiate between the two conditions.

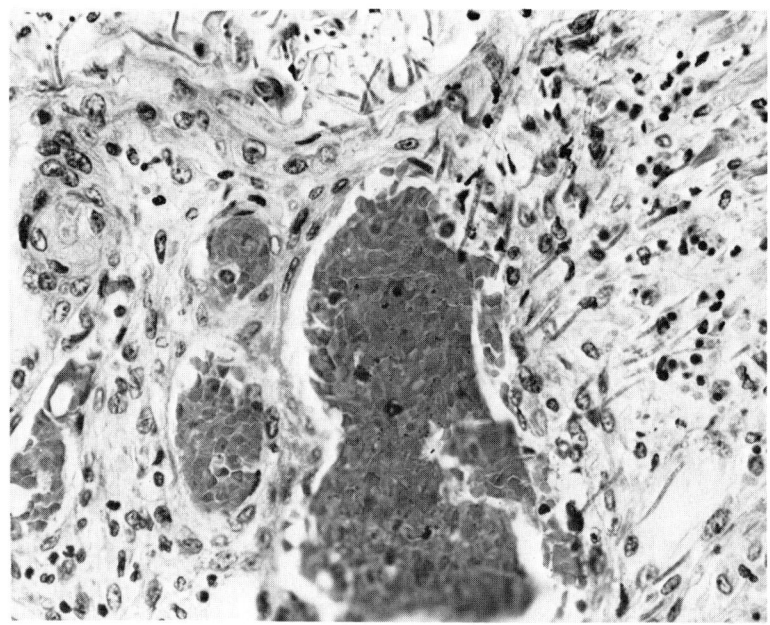

Fig. 7–7. Invasive *Aspergillus* infection in a burn wound. Hyphae often penetrate small blood vessels as shown here. (PAS, × 135)

OTHER MYCOTIC INFECTIONS OF THE BURN WOUND

A variety of other mycotic organisms have been identified histologically and culturally in the burn wound, including *Geotrichum, Fusarium, Helminthosporium, Alternaria, Penicillium,* and *Cephalosporium*.[9,29,30] Significant infection by these mycotic organisms is rare. To date, no specific histologic pattern of tissue invasion has been identified for these potentially pathogenic fungi.

—————————————————————————— **References**

1. Nash, G., Foley, F. D., Goodwin, N. M., Jr., Bruck, H. M., Greenawald, K. A., and Pruitt, B. A., Jr.: Fungal burn wound infection. J.A.M.A. 215:1664–1666, 1971.
2. Bruck, H. M., Nash, G., Stein, S. M., and Lindberg, R. B.: Studies on the

occurrence and significance of yeasts and fungi in the burn wound. Ann. Surg. 176:108–110, 1972.

3. Rabin, E. R., Lundberg, G. D., and Mitchell, E. T.: Mucormycosis in severely burned patients. Report of two cases with extensive destruction of the face and nasal cavity. New Engl. J. Med. 264:1286–1289, 1961.

4. Foley, F. D., and Shuck, J. M.: Burn-wound infection with *Phycomycetes* requiring amputation of hand. J.A.M.A. 203:596, 1968.

5. Nash, G., Foley, F. D., and Pruitt, B. A., Jr.: *Candida* burn-wound invasion. A cause of systemic candidiasis. Arch. Pathol. 90:75–78, 1970.

6. Abramowsky, C. R., Quinn, D., Bradford, W. D., and Conant, N. F.: Systemic infection by *Fusarium* in a burned child. The emergence of a saprophyte strain. J. Pediatr. 84:561–564, 1974.

7. Salisbury, R. E., Silverstein, P., and Goodwin, M. N., Jr.: Upper extremity fungal invasions secondary to large burns. Plastic & Reconstr. Surgery 54:654–659, 1974.

8. Majeski, J. A., and MacMillian, B. G.: Fatal systemic mycotic infections in the burned child. J. Trauma 17:320–322, 1977.

9. Wheeler, M. S., McGinnis, M. R., Schell, W. A., and Walker, D. H.: *Fusarium* infection in burned patients. Am. J. Clin. Pathol. 75:304–311, 1981.

10. Seelig, M. S.: The role of antibiotics in the pathogenesis of *Candida* infections. Am. J. Med. 40:887–917, 1966.

11. Hummel, R. P., Oestreicher, E. J., Maley, M. P., and MacMillian, B. G.: Inhibition of *Candida albicans* by *Escherichia coli* in vitro and in the germfree mouse. J. Surg. Res. 15:53–58, 1973.

12. Hummel, R. P., Maley, M. P., Miskell, P. W., and Altemeier, W. A.: Suppression of *Candida albicans* by *Escherichia coli*. J. Trauma 15:413–418, 1975.

13. Hamilton, J. R., Overall, J. C., Jr., and Glasgow, L. A.: Synergistic effect on mortality in mice with murine cytomegalovirus and *Pseudomonas aeruginosa*, *Staphylococcus aureus* or *Candida albicans* infections. Infect. Immun. 14:982–989, 1976.

14. Wilson, A. W.: Treatment of *Candida* septicemia in a severely scalded child with amphotericin B. Post Grad. Med. J. 46:623–625, 1970.

15. Law, E. J., Kim, O. J., Stieritz, D. D., and MacMillan, B. G.: Experience with systemic candidiasis in the burned patient. J. Trauma 12:543–552, 1972.

16. MacMillan, B. G., Law, E. J., and Holder, I. A.: Experience with *Candida* infections in the burn patient. Arch. Surg. 104:509–514, 1972.

17. Rayner, C. R. W.: Disseminated candidiasis in a severely burned patient—Case report. Plast. Reconstr. Surg. 51:461–463, 1973.

18. Codish, S. D., and Tobias, J. S.: Managing systemic mycosis in the compromised host. J.A.M.A. 235:2132–2134, 1976.

19. Stone, H. H., Kolb, L. D., Currie, C. A., E., Geheber, C. E., and Cuzell, J. Z.: *Candida* sepsis: Pathogenesis and principles of treatment. Ann. Surg. 179:697–710, 1974.

20. Law, E. J., Holder, I. A., and MacMillan, B. G.: *Candida paropsilosis* fungemia in burn patients: report of three cases. Burns. 10:203–206, 1984.

21. Gauto, A., Law, E. J., Holder, J. A., and MacMillan, B. G.: Experience with

amphotericin B in the treatment of systemic candidiasis in burn patients. Am. J. Surg. 133:174–178, 1977.

22. Kidson, A., and Lowbury, E. J. L.: *Candida* infection of burns. Burns. 6:228–230, 1980.
23. Spebar, M. J., and Lindberg, R. B.: Fungal infection of the burn wound. Am. J. Surg. 138:879–882, 1979.
24. Bruck, H. M., Nash, G., Foley, F. D., and Pruitt, B. A., Jr.: Opportunistic fungal infection of the burn wound with *Phycomycetes* and *Aspergillus*. Arch. Surg. 102:476–482, 1971.
25. Pruitt, B. A., Jr.: The burn patient: II. Later care and complications of thermal injury. Current Probl. Surg. 16:1–95, 1979.
26. Stone, H. H., Cuzzell, J. Z., Kolb, L. D., Moskowitz, M. S., and McGowan, J. E., Jr.: Aspergillus infection of the burn wound. J. Trauma 19:765–767, 1979.
27. Emmons, C. W., Binford, C. H., and Utz, J. P.: *In* Medical Mycology, 2nd. ed. Philadelphia, Lea and Febiger, 1970, pp 266–267.
28. Panke, T. W., McManus, A. T., and Spebar, M. J.: *Aspergillus niger* infection of the burn wound: Gross appearance simulating ecthyma gangrenosa. Am. J. Clin. Pathol. 72:230–232, 1979.
29. Baker, T. J., and Peterson, J. E.: An apparently proteolytic fungus isolated from a burn patient. Plast. Reconstr. Surg. 24:209–213, 1959.
30. Lindberg, R. B., Contreras, A. A., and Smith, H. O. D., Jr.: The role of fungi in burn wound infection: Observation of biopsy and autopsy tissues from seriously burned soldiers. *In* Annual Research Report. US Army Institute Surgical Research, Brooke Army Medical Center, Fort Sam Houston, Tex, FY 1974.

Chapter 8 ────────────────────────────────────

INFECTIOUS
COMPLICATIONS—
VIRAL

Although significant viral infections in burn patients are uncommon, limited herpetic infections may be encountered in surgical or autopsy specimens. These infections usually affect morbidity but not mortality in the burn patient, and probably reflect the compromised host defense that may exist. Herpesvirus, cytomegalovirus, vaccinia, and varicella are the only recognized viral pathogens in burn patients and are listed in order of decreasing frequency.

HERPESVIRUS

Scattered endemics of *Herpes simplex* infection appear to affect burn patients. Clinically, herpesvirus has a predilection for partial thickness burns, especially those of the face where the nasolabial areas are commonly involved. The lesions are small and measure 2–4 mm[1] (Fig. 8–1). Similar lesions may occur in unburned skin. The herpetic vesicle is fragile and easily ruptured. Subsequent secondary bacterial and mycotic infection is common and may present as shallow crusted erosions at the burn wound margin. Light microscopic examination of the skin biopsy is usually adequate for morphologic diagnosis (Fig. 8–2). Selected scrapings of the vesicle or crusted lesions may be submitted for cytologic examination. Electron microscopic examination (Fig. 8–3) and viral cultures are necessary for a conclusive diagnosis.

At postmortem examination, herpesvirus infection may be disseminated or confined to a single organ.[1,2] The most common sites include the laryngotracheobronchial tree, pulmonary alveoli, and the esophagus (Figs. 8–4A,B). Necrotizing liver and adrenal lesions in infants and children due to

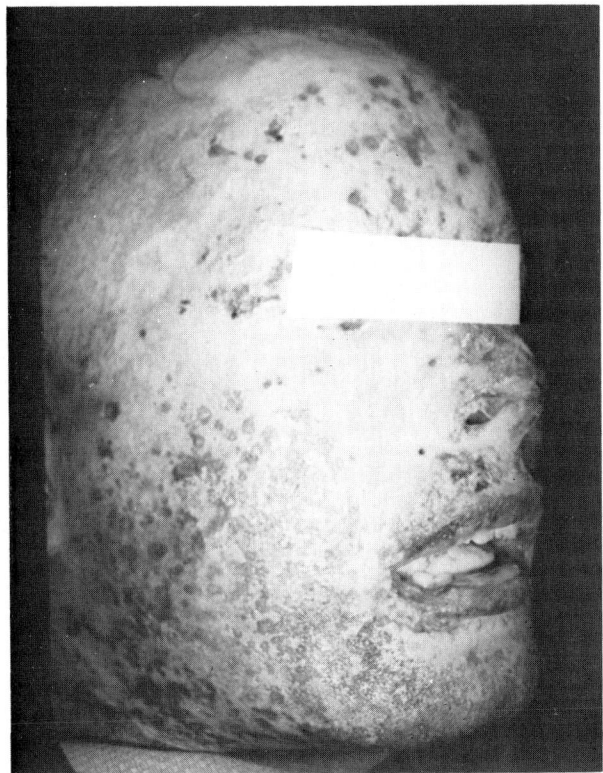

Fig. 8–1. Cutaneous herpesvirus infection in a burn patient. Lesion may occur in burned (partial thickness) or unburned skin.

Herpes simplex are uncommon in adults.[1] Herpetic involvement of the respiratory system appears to be relatively common.[2] Nash et al. reported a series of 15 cases with pulmonary herpetic lesions in burn patients. Nine of these cases were identified by a retrospective review of cases from a 4-year period. In another study, cytologic examination of sputum specimens of fatally burned patients demonstrated characteristic herpesvirus inclusions in 14 percent of patients starting on the eighth postburn day.[3]

Histologically, the herpetic inclusions in burn patients are similar to classic descriptions regardless of the type of cell infected. Large, eosinophilic, intranuclear inclusions (Cowdry type A) are surrounded by a thin pale clear area and accentuated by peripheral margination of chromatin (Fig. 8–5). Sometimes a greatly enlarged homogenous, eosinophilic nucleus with no distinct "halo" is seen (Fig. 8–6). In the skin, the diagnostic cells may be found in an area of second degree burn where residual squamous

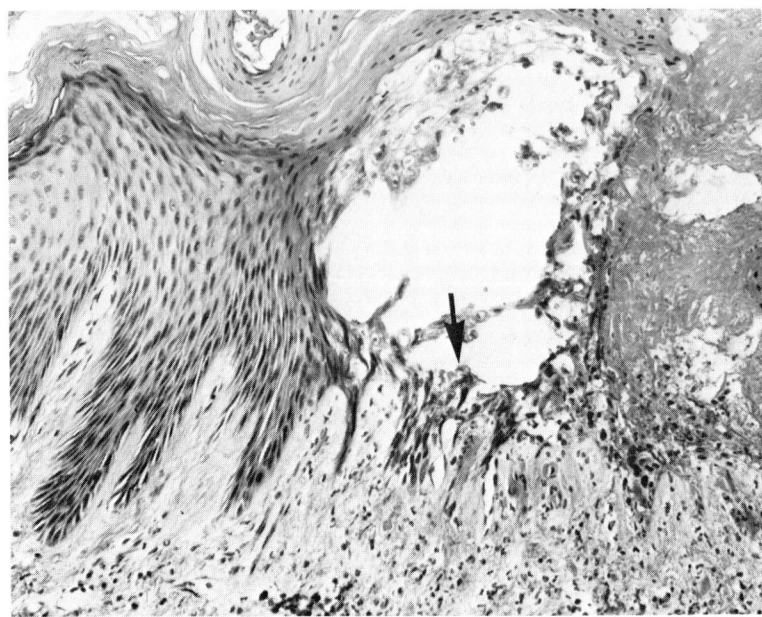

Fig. 8–2. A biopsy specimen of a suspected cutaneous herpetic lesion. High magnification of the swollen cells (arrow) will often reveal characteristic intranuclear inclusions (see Fig. 8–8). (H & E, × 135)

epithelium surrounds intradermal adnexal structures (Fig. 8–7). Most often, the diagnostic inclusions are difficult to identify. Suggestive changes include a small erosion with prominent mycotic colonization (usually *Candida* sp. or *Aspergillus* sp.) in a partial thickness cutaneous burn (Fig. 8–8). Others have mentioned secondary bacterial infections that appear to complicate viral lesions and suggest such lesions as sources for *Pseudomonas* septicemia.[1]

The predilection of viral infections for second degree burns and donor sites is unexplained. It has been hypothesized that such lesions may be related to the availability of nucleoprotein moieties in regenerating epidermal cells for viral replication.[1] Similar extensive viral eruptions may occur in children with eczema. Simplistically, the immune defenses are impaired in the thermally injured patient, and the protective skin barrier is interrupted. Both these processes permit a variety of microbiologic growth including viral infections.

In the respiratory tract, erosive mucosal lesions should be examined for herpetic lesions. Histologically, one may identify typical Cowdry type A intranuclear inclusions. The other characteristic changes include syncytial giant cells and ballooning degeneration of epithelial cells. Bacterial and

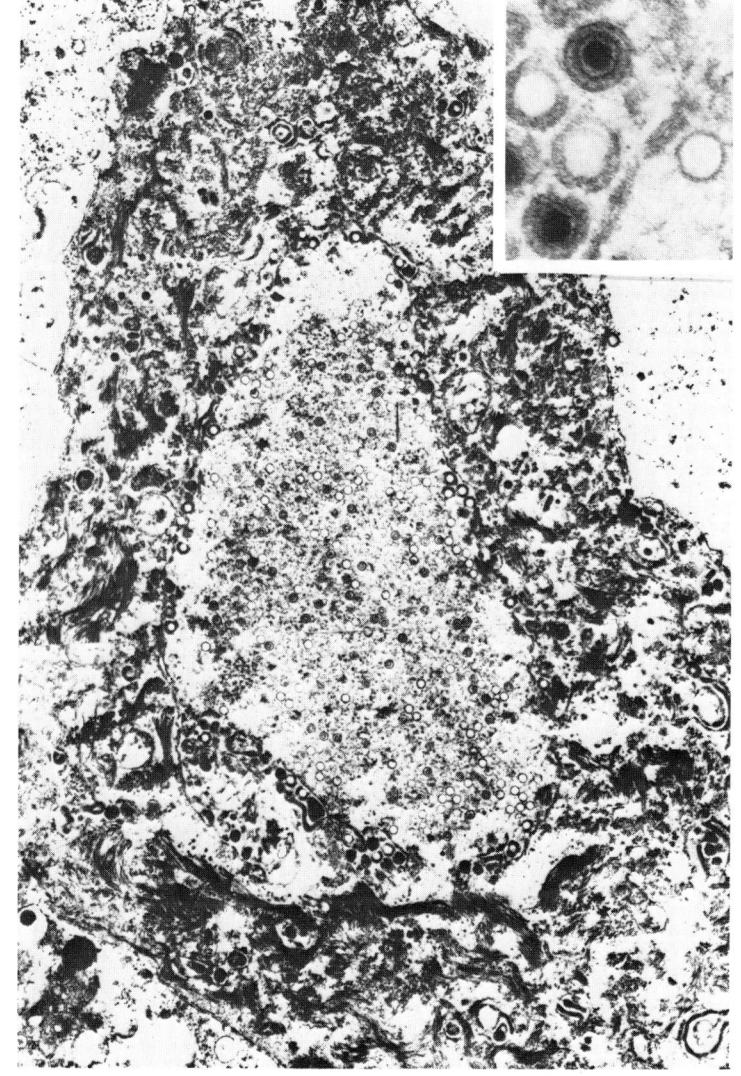

Fig. 8–3. Electron photomicrograph of a degenerating epithelial cell with multiple intranuclear herpesvirus particles. Inset shows detail of virus particles. (cell × 11,900, inset × 68,850)

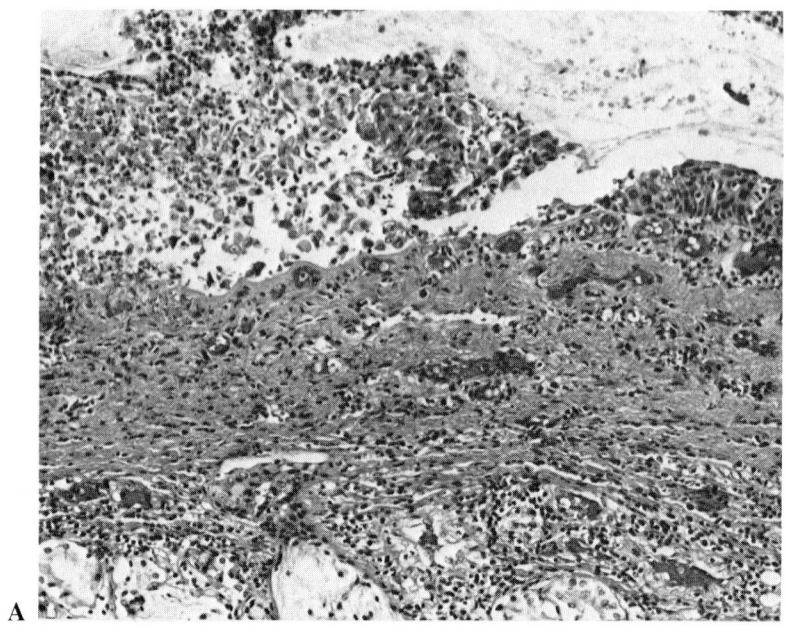

A

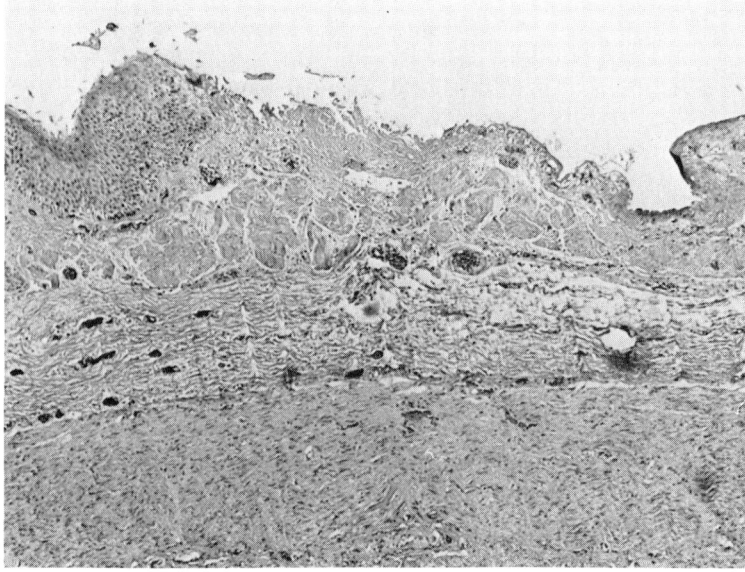

B

Fig. 8–4. Focal ulcerative lesions of the trachea (**A**) or esophagus (**B**) should be examined for evidence of herpesvirus infection. As in the skin, these lesions may be colonized by yeasts. (H & E, × 135)

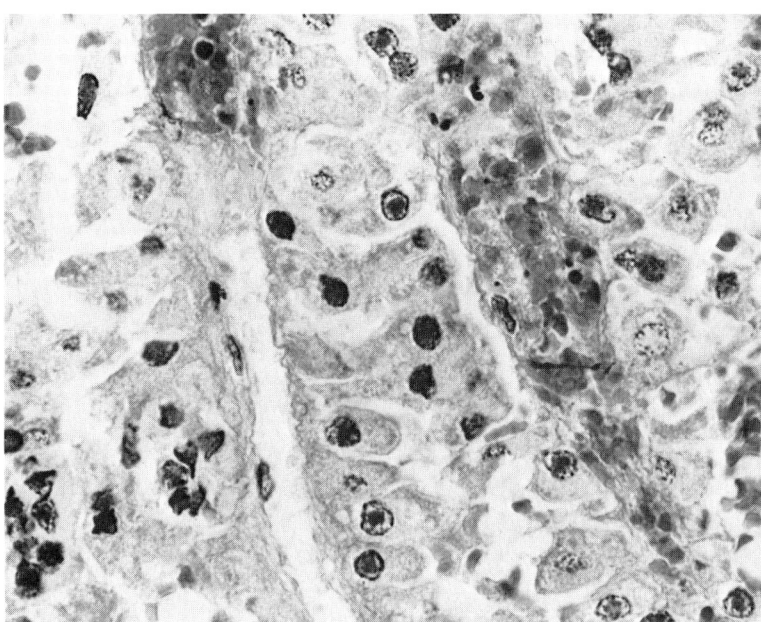

Fig. 8–5. Herpesvirus inclusions in cells from the adrenal cortex of a burn patient. (H & E, × 380)

fungal superinfections often obscure the classical viral nature of these lesions.[2] Pulmonary alveolar lesions also contain Cowdry type A inclusions within areas of necrotic debris and intraalveolar proteinaceous exudate with or without hemorrhage.

In addition to involvement of skin, lung and upper gastrointestinal tract, systemic herpetic infections may also involve liver, spleen, bone marrow, and adrenals. Clinical deterioration in the presence of cutaneous herpesvirus infection suggests the possibility of disseminated systemic herpes infection. A liver biopsy can be diagnostic[4] in such patients but may be contraindicated if a coagulopathy is present or if extensive burns provide only contaminated surfaces through which a percutaneous liver biopsy would have to be performed.

Routine, periodic, serologic studies indicates that herpes simplex occurs at a much higher incidence (20 percent) than one might expect.[5] It typically occurs in the more severely burned patient. Viral cultures of urine, upper respiratory tract, and suspicious cutaneous lesions are very useful; and electron microscopy may be diagnostic. Histologic examination of skin lesions may give a prompt diagnosis.[6]

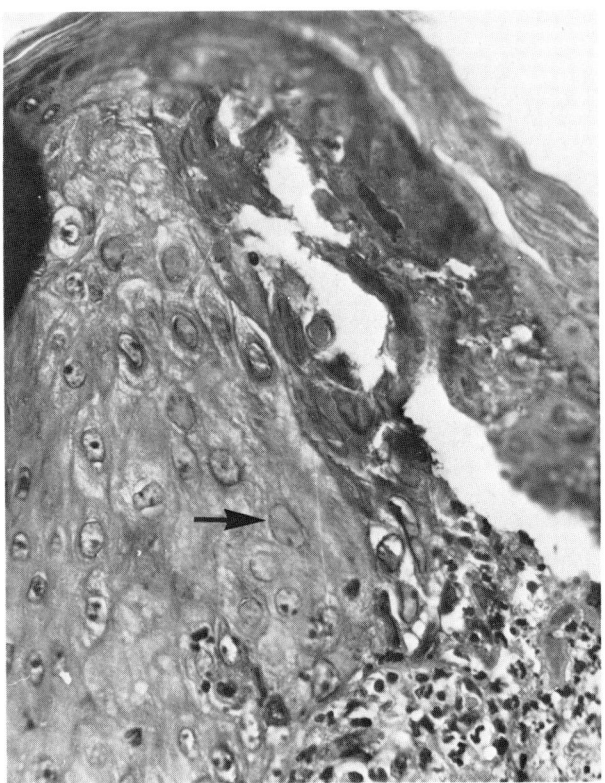

Fig. 8–6. Enlarged nuclei of esophageal mucosal cells contain homogenous eosinophilic herpetic inclusions (arrow). (H & E, × 270)

CYTOMEGALOVIRUS

Disseminated cytomegalic inclusion disease is primarily a disease of newborns, but may also occur in adults.[7] Widely disseminated, severe cytomegalovirus (CMV) infections in the burn patient have been uncommon.[8] It is quite likely that some cases are overlooked. Macasaet et al.[9] reported an incidence of 6.8 percent recovery of virus (culture techniques) from 502 unselected autopsies. Classical viral inclusions in tissue sections were identified in less than half of those cases with positive viral cultures. Seeman and Konigova investigated 74 severely burned patients (greater than 20 percent third degree burns) for complement-fixing antibodies directed against CMV.[10] Fifty-four of these 74 patients (73 percent) were seropositive for CMV. Seven patients with high antibody titers developed a clinical

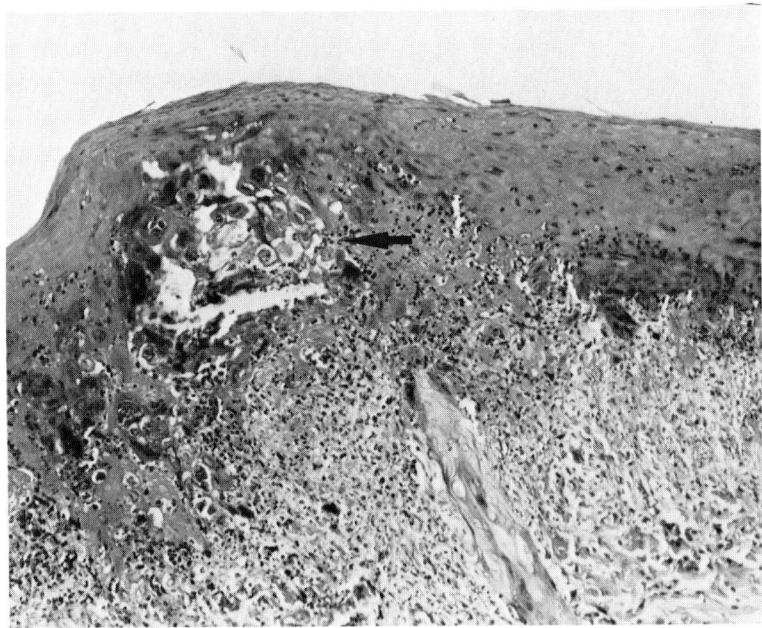

Fig. 8—7. Herpesvirus may infect residual or regenerating (arrow) epithelium in partial thickness burns. (H & E, × 135)

picture of hepatitis. Linneman and MacMillan[11] demonstrated a 33 percent incidence of cytomegalovirus (CMV) disease by complement fixation tests and/or viral cultures in burned children. Unexplained fever in this pediatric burn population was sometimes demonstrated to be secondary to CMV infection. The above data obviously stress the necessity for viral cultures and serologic studies to identify cases with CMV.

Several factors increased the likelihood of developing a cytomegalovirus infection and include: extensive third degree burns, multiple blood transfusions, and prolonged hospitalization. This epidemiologic data helps select the population of burn patients at risk. The literature on renal transplant patients is replete with data on CMV infections and hypothesizes that any two of the following manifestations suggest active CMV infection in a patient with positive serology: unexplained fever, myalgia, arthralgia, pneumonitis, hepatomegaly, splenomegaly, leukopenia, lymphocytosis, atypical lymphocytosis, and abnormal liver function tests.[12] Although atypical lymphocytes are a typical feature of CMV infections, they may not be present.[13] In renal transplant patients, almost all patients who developed a primary CMV infection were symptomatic, while those patients with reactivated (secondary) CMV were mostly asymptomatic. The above data obviously stress the necessity for viral cultures and serological studies to

Fig. 8–8. Skin biopsy with herpesvirus inclusions (arrow) and colonizing yeasts (Y). (H & E, × 342)

identify cases with CMV. In thermal injury, the only reported case of CMV had widely disseminated visceral lesions.[8] Histologically, the inclusions of CMV are both intracytoplasmic and intranuclear. The intranuclear inclusions measure up to 15 microns and may be eosinophilic to basophilic (Fig. 8–9). The intracytoplasmic inclusions are basophilic, measure 2–4 and are less common than the intranuclear inclusions.[14]

VACCINIA

A single case of vaccinia virus infection has been reported in a 3-year-old boy who sustained a 25 percent second degree burn. On the 16th postburn day, the patient developed small vesicles and erythematous papules with ulcerated white centers. Cultures yielded vaccinia virus. The patient had been vaccinated 7 days prior to the burn.[15]

Another burn case is reported of a 4-month old baby whose father had previously been vaccinated for smallpox. The child developed vesicular lesions on the previously burned hand 1 week after admission to the hospital. Live vaccinia virus was isolated in the child's lesions.[5]

VARICELLA

Foley et al.[1] mentioned five childhood cases of varicella infection in burn patients. Three of these patients had varicella infection prior to hospitalization, and two developed infections after admission to the same ward. Donor sites and areas of healing, second degree burn were primarily affected. A similar epidemic of varicella infection in a pediatric burn ward has been reported.[16] Undoubtedly, many cases of varicella-herpes zoster are not recognized because they are asymptomatic and identified only with serological studies or viral cultures.[5]

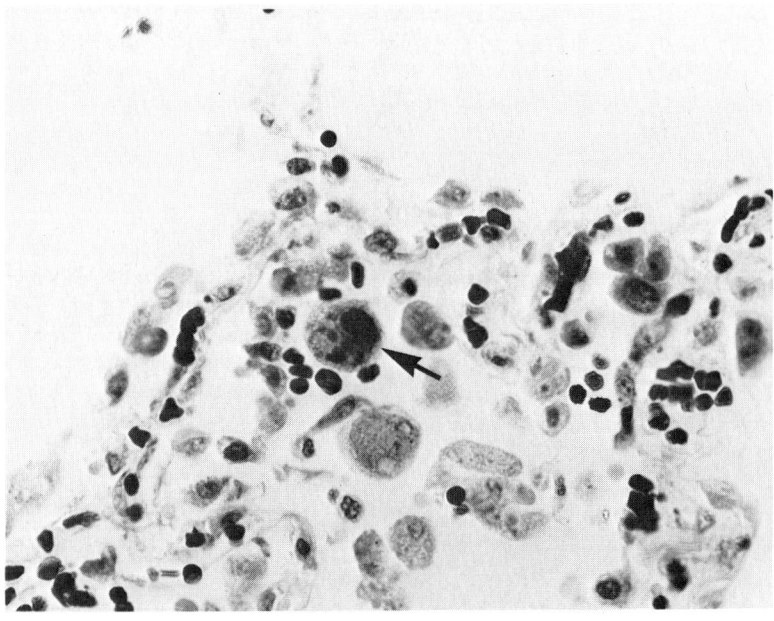

Fig. 8–9. Intranuclear cytomegalovirus inclusion (arrow) in an alveolar macrophage. (H & E, × 342)

References

1. Foley, F. D., Greenawald, K. A., Nash, G., and Pruitt, B. A., Jr.: Herpesvirus infection in burned patients. N. Engl. J. Med. 282:652–656, 1970.
2. Nash, G., and Foley, F. D.: Herpetic infection of the middle and lower respiratory tract. Am. J. Clin. Pathol. 54:857–863, 1970.
3. Cooney, W. D., Dzuira, B., Harper, R., and Nash, G.: The cytology of sputum from thermally injured patients. Acta Cytol. 16:433–437, 1972.
4. Aronson, M. D., Phillips, C. F., Gump, D. W., Albertini, R. J., and Phillips, C. A.: Vidarabine therapy for severe herpesvirus infection. An unusual syndrome of chronic varicella and chronic immunologic deficiency. J.A.M.A. 235:1339–1342, 1976.
5. Matthews, S. C. W., Lewick, P. L., Coombes, E. J., Ely, D. W., Pead, P. J., and Saeed, A. A.: Viral infections in a group of burned patients. Burns 6:55–60, 1979.
6. Pruitt, B. A., Jr.: Complications of thermal injury. Clin. Plast. Surg. 1:667–691, 1974.
7. Wong, T., and Warner, N.E.: Cytomegalic inclusion disease in adults. Report of 14 cases with review of the literature. Arch. Pathol. 74:403–422, 1962.
8. Nash, G., Asch, M. J., Foley, F. D., and Pruitt, B. A., Jr.: Disseminated cytomegalic inclusion disease in a burned adult. J.A.M.A. 214:587–588, 1970.
9. Macasaet, F. F., Holley, K. E., Smith, T. F., and Keys, T. F.: Cytomegalovirus studies of autopsy tissue. II. Incidence of inclusion bodies and related pathologic data. Am. J. Clin. Pathol. 63:859–865, 1975.
10. Seeman, J., and Königova, R.: Cytomegalovirus infection in severely burned patients. Acta Chir. Plast. 18:142–151, 1976.
11. Linnemann, C. C., and MacMillan, B. G.: Viral infections in pediatric burn patients. Am. J. Dis. Child. 135:750–753, 1981.
12. Suwansirikul, S., Rao, N., Dowling, J. N., and Ho, M.: Primary and secondary cytomegalovirus infection. Clinical manifestations after renal transplantation. Arch. Intern. Med. 137:1026–1029, 1977.
13. Deepe, G. S., Jr., MacMillan, B. G., and Linnemann, C. C., Jr.: Unexplained fever in burn patients due to cytomegalovirus infection. J.A.M.A. 248:2299–2301, 1982.
14. Bodey, G. P., Wertlake, P. T., Douglas, G., and Levin, R. H.: Cytomegalic inclusion disease in patients with acute leukemia. Ann. Intern. Med. 62:899–906, 1965.
15. Torstenson, O. L., Meyer, W. I., and Quie, P. G.: Burn-wound infection with viruses. (Abstract) N. Engl. J. Med.: 282:1272–1273, 1970.
16. Weintraub, W. H., and Lilly, J. R.: Chickenpox in the burned child: A recent epidemic in a pediatric burn unit. ABA Sixth Annual Meeting, Cincinnati, Oh, April 4, 1974.

Chapter 9

CARDIOVASCULAR SYSTEM

The purpose of this chapter is to review the cardiovascular complications of burn injury. Most of the lesions in this system are related to cannula trauma and sepsis. However, other less common but fatal cardiovascular complications will also be discussed.

COMPLICATIONS OF INTRAVASCULAR CANNULAE

Because of the critical fluid requirements of severely burned patients, venous cannulae are widely used in the early stages of treatment. Bland (noninfected) as well as septic thrombi develop frequently in the burn population, usually in sites of central venous cannulation.[1,2] The early thrombi are initiated when the cannula physically injures venous endothelium and exposes basement membrane and/or collagen. This causes platelet aggregation and results in a nidus of a thrombus.

Microscopically, this *bland* (noninfected) thrombus is composed of fibrin, platelets, and a few leukocytes (Fig. 9–1, A,B). If the physical irritation of the cannula persists, the thrombus develops a more mixed composition as erythrocytes are incorporated into the thrombus. A septicemic episode, or extension of infection from the skin at the cannula insertion site, may lead to bacterial colonization of the thrombus (Fig. 9–2). A *septic* thrombus creates a focus for continuous septicemia until it is removed or otherwise therapeutically controlled. If the thrombus is large, postmortem examination may demonstrate bacteria within its center (Fig. 9–3), even in patients who have had negative blood cultures in the immediate antemortem period. Bacteria in this location present serious diagnostic and therapeutic implications for the clinician.

Several factors are thought to contribute to the development of

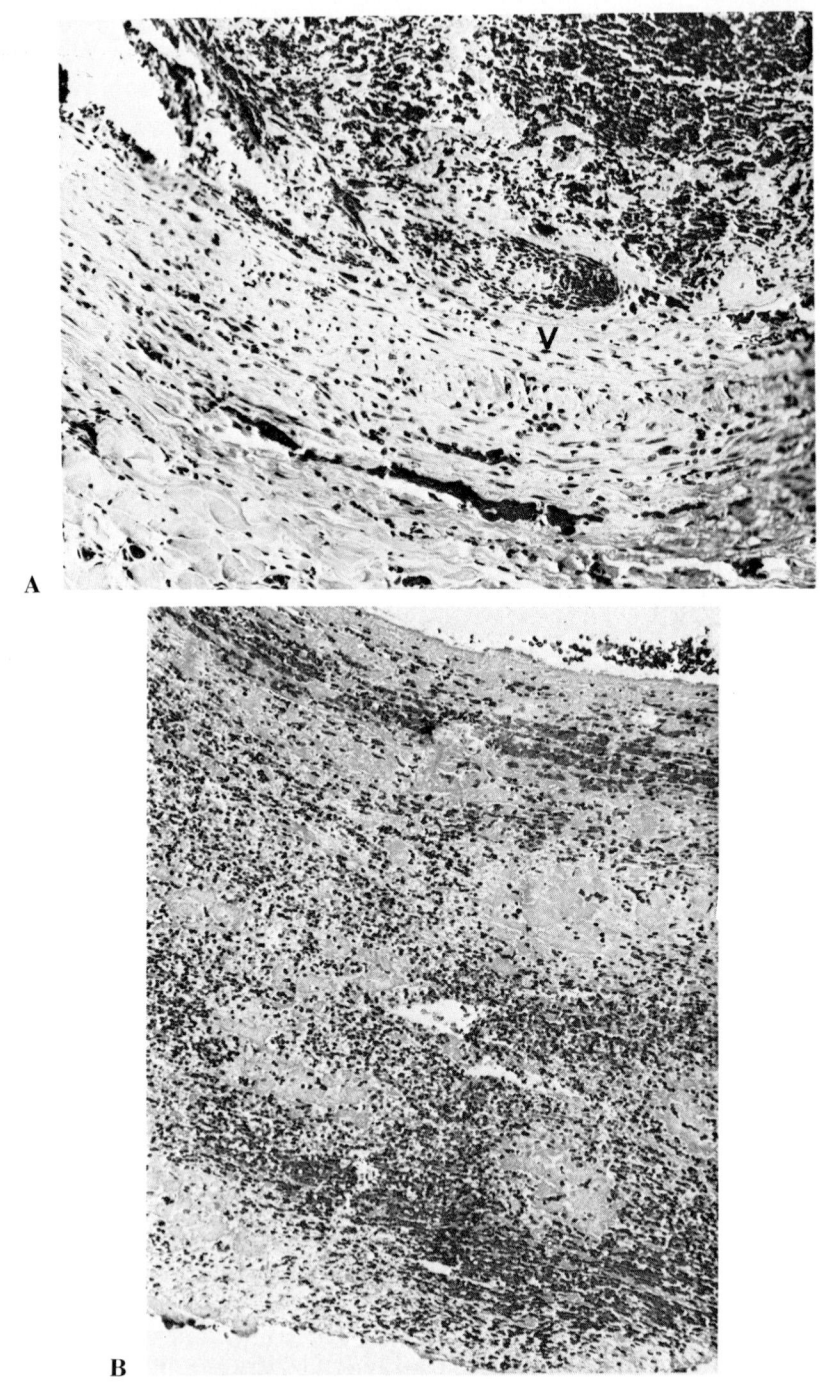

Fig. 9–1. (A): A sterile acute thrombus composed mainly of loosely arranged fibrin, platelets, and erythrocytes attached to the wall of the cephalic vein. (V). (H & E × 126) (B): More advanced acute sterile thrombus with a mixed composition of fibrin, platelets, and erythrocytes. (H & E × 126)

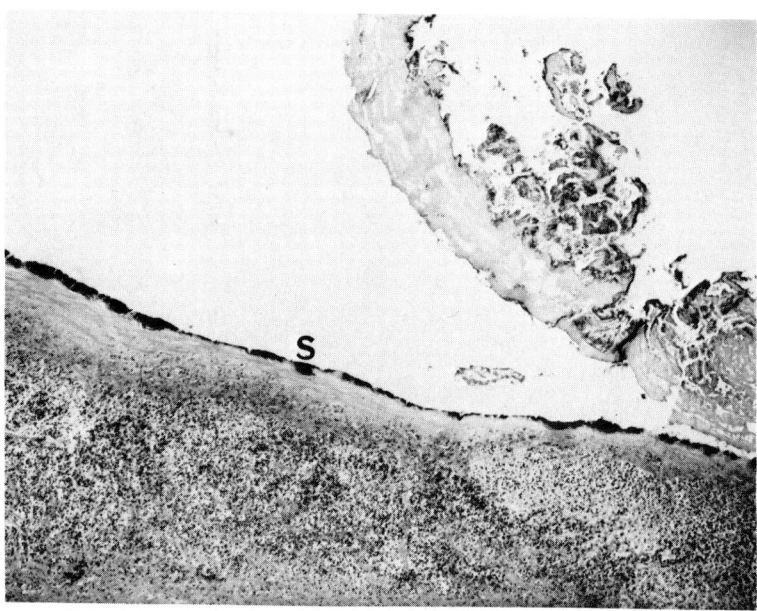

Fig. 9–2. Superficial staphylococcal colonization of a venous thrombus from the superior vena cava. Dark superficial zone represents colonies of *Staphylococcus* (S). (H & E × 504)

thrombophlebitis in patients with intravenous catheters. These include: pH and composition of infusate, duration of infusion, size and composition of the cannula, presence of bacteria, venous blood flow characteristics, presence of a hypercoagulable state, skin preparation, techniques of venipuncture, and presence of burn in the area of catheterization.[3] Thrombophlebitis, especially involving central veins, is a common event (37 percent of 139 consecutive autopsies). This incidence has been recently reduced by careful, aseptic insertion of such cannulae and by limiting the time that the cannula remains in any one site to 72 hr or less.[4,5]

In the discussion of invasive infection (Chapter 6), we stated that gram-negative bacilli were the most frequent invaders of the burn wound. Gram-positive cocci are found in the superficial eschar and on normal skin of burn patients; however, it is this organism, *Staphylococcus aureus*, that most often extends along the cannula into blood vessels and colonizes the thrombus. Not infrequently, microabscesses are seen at postmortem examination of routine sections of tissue taken from the insertion site of central venous cannulae. In over 90 percent of recent cases of infectious thrombophlebitis at the Institute of Surgical Research, *Staphylococcus* sp. was identified as the infectious agent in these thrombi. Septic thrombi

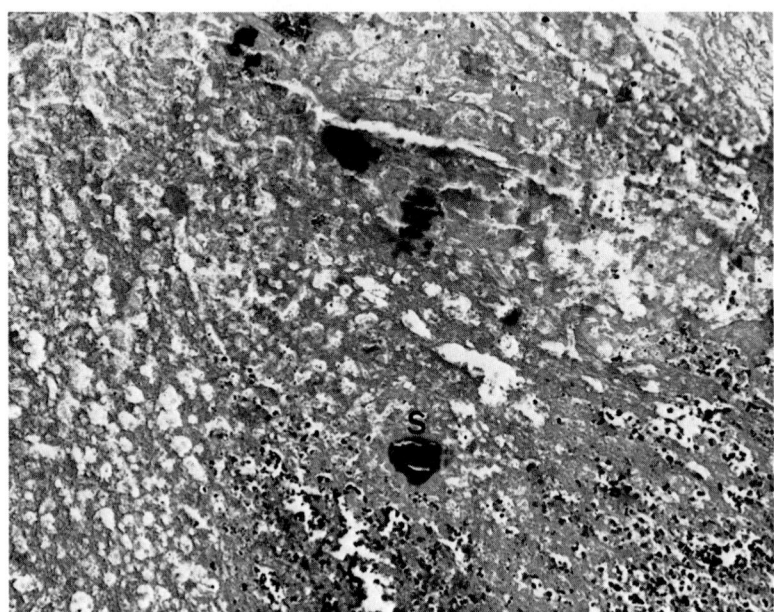

Fig. 9–3. Septic thrombi often contain deep profuse growths of *Staphylococcus* (S). Blood cultures are often negative unless the bacteria are exposed to the surface of the thrombus. (H & E × 144)

involving cardiac valves and endocardium have a similar high incidence of staphylococcal infection. Gram-negative bacterial infections, especially *Providencia stuartii*, played a more important role in septic venous thrombi in earlier[5,6] years; fungal infections (especially *Candida* sp. and also *Aspergillus* sp., *Fusarium* sp.)[5] were occasionally identified.

Pruitt et al.[7] reported that essentially any cannulated blood vessel (saphenous, iliac, femoral, jugular, or small superficial arm vein) could be affected by septic thrombophlebitis. With central venous cannulation, infectious thrombophlebitis of the subclavian, brachiocephalic, and superior vena cava (Fig. 9–4) is a relatively frequent event and is much more difficult to detect than thrombophlebitis in a peripheral vein. Because of the difficulty of diagnosing and treating septic thrombophlebitis when it is associated with a central venous cannula, peripheral veins should be used whenever possible.[4] Peripheral veins are more easily examined and excised if septic thrombophlebitis develops.

Arteries have less frequent complications related to cannulation; however, the use of the *balloon-tipped pulmonary arterial cannula* has led to pulmonary artery thrombosis,[8] perforation of the pulmonary artery,[9] septic endocarditis,[10] and fatal pulmonary hemorrhage.[11]

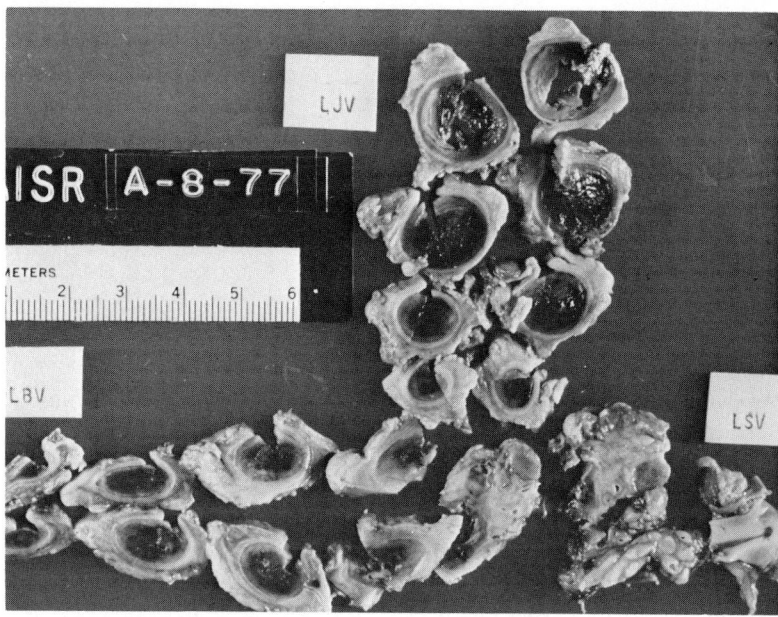

Fig. 9—4. Cannulation may lead to septic thrombosis of the jugular (LJV), subclavian (LSV) or brachiocephalic vessels (LBV).

VERRUCOUS AND INFECTIOUS ENDOCARDITIS/VALVULITIS

When found in the heart, septic thrombi most commonly involve the cardiac valves (Fig. 9–5), but have also been noted on the endocardial (nonvalvular) surface. Endocardial septic thrombi were most frequently noted in the right atrium. The predominance of lesions on the right side of the heart at our facility has been noted since 1968, when central venous cannulae were first utilized; this right-sided predominance of lesions has been recently reported by another burn center.[10] Both bland (or "verrucous") thrombi and septic thrombi are found on the cardiac valvular and endocardial surfaces. In a retrospective study of patients with infectious cardiac valvular and endocardial lesions, Sasaki et al.[6] investigated the position of the cannulae by examining chest roentgenograms. They noted that most of the patients with infectious valvular or endocardial lesions had a central venous and/or Swan-Ganz cannula extending into the heart prior to death. Underlying valvular or endocardial disease is known to predispose to thrombosis; however, no evidence of preexisting scarring or significant atherosclerosis was seen in the right heart of the patients at autopsy.

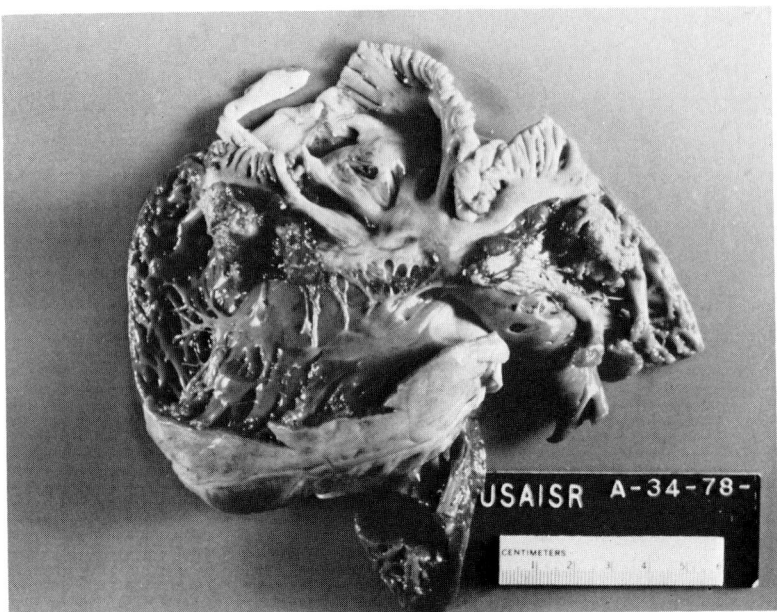

Fig. 9–5. Septic thrombi involving the cardiac valves should be cultured and examined histologically for staphylococcal bacteria.

Occasionally, left-sided cardiac thrombi were associated with only mild atherosclerosis of the valves. The data strongly implicate use of central venous cannulae in the development of the valvular and endocardial lesions.[6] Similar cardiac valvular lesions have been also documented in patients with other disease processes who have had indwelling cardiac catheters.[12,13]

Pseudomonas aeruginosa is a rare cause of endocarditis in burn patients. Rabin et al.[14] found that in 13 cases of endocarditis (126 burn fatalities) the lesions were due to *P. aeruginosa* in two, and that in one case the lesions were due to both *P. aeruginosa* and *Staphylococcus*. The remaining 10 cases were caused by *Staphylococcus*. He found that *Staphylococcus* septicemia was present at some time in all 13 cases and suggested that the endocardial lesions might well have been initially caused by *Staphylococcus* and later become superinfected with *P. aeruginosa*.

The absence of histologic and microbiologic evidence of microorganisms should be interpreted with caution. The diagnosis of "verrucous endocarditis" ordinarily implies a noninfectious state, and such a designation is given to cardiac valvular thrombi at postmortem examination when no microorganisms are found. However, recent data suggest that infectious endocarditis may be present despite the lack of apparent organisms.[15]

Systemic antibiotics may eliminate or inhibit the microorganisms' ability to grow. This is complicated by the fastidious growth of some bacteria (*Hemophilus* sp.). Hence, premortem blood cultures, possibly associated with evidence of necrosis of the cardiac valve at postmortem examination, may be the best supportive evidence of infectious valvulitis.

MYOCARDIAL ABSCESSES

Small abscesses of the myocardium are seen in approximately 8 percent of burn patients at autopsy (Fig. 9–6). Histologically, the foci have characteristic liquefactive necrosis and only a mild acute inflammatory infiltrate which extends into the surrounding tissue. Special stains usually demonstrate *Staphylococcus* (Fig. 9–7) within the lesion which are characteristically found in patients with staphylococcal septicemia. Only rarely are other microorganisms identified in these myocardial abscesses in our experience. Srivastava and MacMillan reported eight patients with myocardial abscesses.[16] In addition to *Staphylococcus aureus*, gram-negative infections also caused these abscesses. They also noted infectious epicarditis in three patients, and pericarditis in five patients. Some of these lesions occurred in patients who developed septic thrombophlebitis following the use of intra-

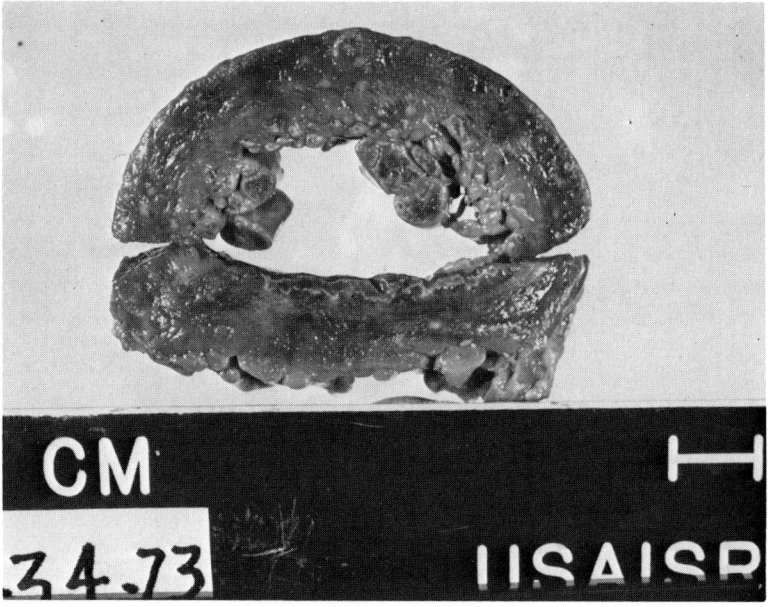

Fig. 9–6. Pale rounded lesions represent myocardial abscesses.

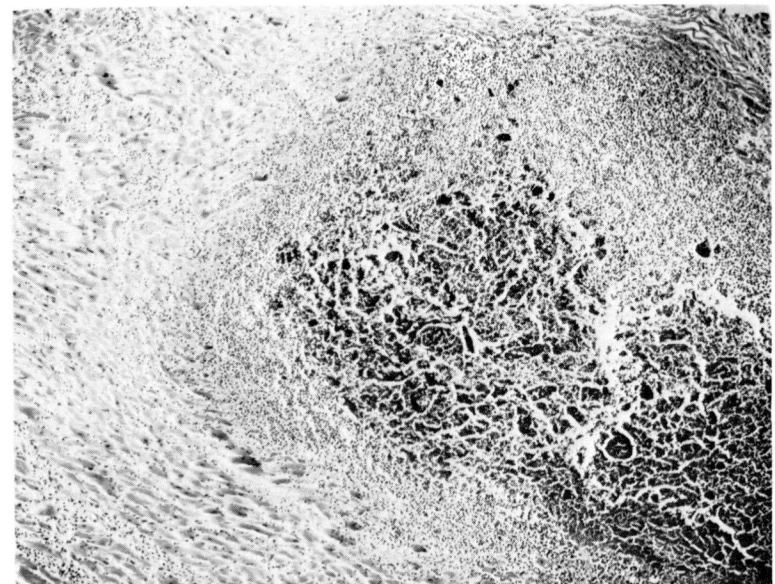

Fig. 9–7.   Staphylococcal myocardial abscess with adjacent myocarditis.

venous catheters. Others had no apparent preceding infectious cause. The cardiac lesions were often not appreciated, and mortality in such patients was high.

The source of the microorganisms is usually an infected thrombus within the cardiovascular system. Occasionally the source may be staphylococcal bronchopneumonia, necrotizing laryngitis, or tracheobronchitis. The origin of infection is occasionally not evident at autopsy. However, identification of such intramyocardial abscesses should stimulate the pathologist to perform a more thorough search of the cardiovascular system, particularly the cannulated veins, for staphylococcal septic thrombi.

SUBENDOCARDIAL HEMORRHAGE

Sevitt reported variably sized subendocardial hemorrhage on the left ventricular side of the interventricular septum along the high pressure outflow tract.[17] Its close relationship to the Purkinje fibers suggested functional importance. These lesions appeared to be agonal and were not an immediate postburn lesion.

MYOCARDIAL INFARCTS

A recent study demonstrated that 4.1 percent of patients (12 of 293 burn-related deaths) had a myocardial infarct at postmortem examination (Fig. 9–8).[18] Eight of the cases were not detected clinically. In two cases, infarction was clinically suspected prior to death, and in two other of the 12 cases, infarction was definitely diagnosed premortem. Age was certainly not a consistent limiting factor, because 25 percent of the cases occurred in patients under 30 years of age. The only common factors in the group were that all patients had large burns (usually greater than 45 percent of the total body surface) and a prolonged in-hospital course. Only 42 percent of the cases had severe coronary atherosclerosis which caused the myocardial infarct (Table 9–1).

History of prior heart disease was found in only three of 12 patients. These included myocardial infarction, angina without clinical evidence of myocardial infarction (at postmortem examination, an old myocardial infarct was seen), and "uncomplicated" hypertensive cardiovascular disease.

This study demonstrated that burn patients with myocardial infarcts appeared to separate into two clearly defined groups. The *first* group consisted of six older patients (mean age of 59.5 years) who had severe coronary vascular atherosclerosis or hypertension. Postmortem examination

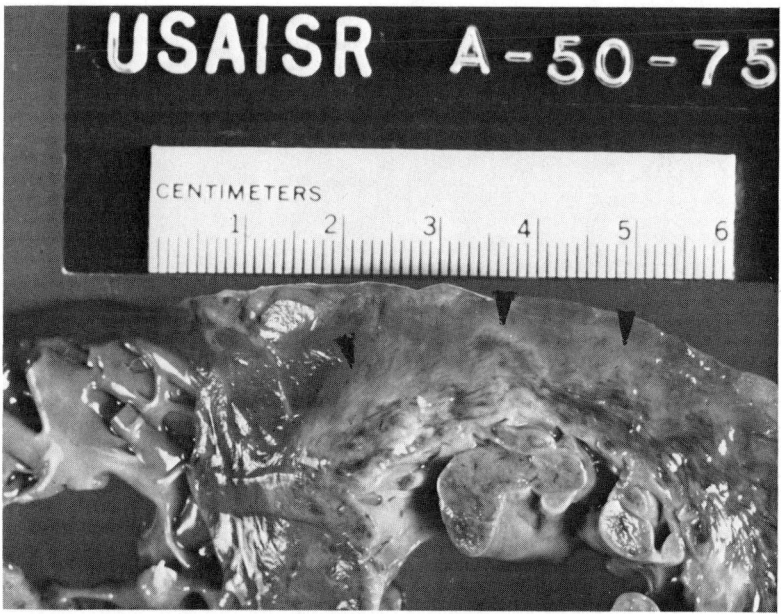

Fig. 9–8. Acute myocardial infarction in a burn patient who died 10 days after suffering a 50 percent total body surface burn. The zone of infarction (arrows) was found mainly in the left ventricular wall.

indicated that these patients developed ischemic heart disease after the first week of hospitalization, but the condition was undiagnosed clinically except for the two cases mentioned above. This group of patients did not have multisystemic disease complicating their treatment and therefore represented a group of patients who were potentially treatable.

The *second* group consisted of six younger patients (mean age of 29.2 years) who usually developed myocardial infarcts late in the burn course. They lacked significant atherosclerotic coronary vascular disease. The myocardial infarcts in these patients occurred in the milieu of multisystem failure which included serious infections of the burn wound and/or lung, septicemia, and renal or hepatic failure. The causes of myocardial infarct in these younger patients were diverse. One patient had a staphylococcal mediastinal infection which extended into the base of the heart and presumably caused the associated extensive septal infarct. Other uncommon etiologies included infarcts related to disseminated intravascular coagulopathy in two patients and subendocardial infarcts probably related to severe ischemia in two patients. Even if myocardial infarction were recognized clinically in these patients, it is unlikely that treatment would be effective. Ischemia has been proposed as a cause of a form of concentric myocardial infarction.[19] which was similar to the subendocardial infarction seen in the two burn patients. A recent report has associated subendocardial infarction with lightning injury.[20]

Table 9–1 *Conditions Associated with Myocardial Infarcts in Burn Patients*

No. of Cases	Ages	Pathogenesis
5	43, 58, 65, 72, 78	Severe (focal, multifocal) atherosclerotic coronary vascular disease.
2	40, 55	Severe fibrin thrombi in small blood vessels as part of disseminated intravascular coagulopathy
2	8, 26	Subendocardial hemorrhage, ? due to severe ischemia
1	12	Abscess at base of left ventricle, ? predisposing to severe septal infarct
1	41	Hypertension
1	34	Unknown

Disseminated intravascular coagulopathy causing a myocardial infarct in two burn patients presented in this study is certainly not unique. Cole presented a case of myocardial infarction in a 31-year-old woman with only mild to moderate coronary atherosclerosis.[21] She had many widely distributed microvascular thrombi, including some in the heart, in addition to a thrombus in the coronary artery. The latter thrombus was striking in that it was incompatible with the very mild degree of atherosclerosis.

Sevitt[22,23] has also reported myocardial infarcts in trauma patients with only mild to moderate, nonstenosing atherosclerosis insufficient to explain thrombogenesis and ischemic heart disease. He has postulated that a hypercoagulable state most likely explains the presence of these coronary thrombi as well as the occasional case of renal artery thrombosis.[22]

Most myocardial infarcts in burn patients are unrecognized clinically because of the multitudinous problems affecting the often obtunded or semicomatose patient. Acute hypotension in combination with a normal body temperature is more likely to be treated as septicemia due to *P. aeruginosa* in the burn patient than as a myocardial infarct. Cardiogenic shock related to a myocardial infarct may go unrecognized. The severely debilitated patient is unable to report the clinically suggestive cardiac symptoms. Furthermore, some patients may be so young (8, 12, 26 years old in the reported group) that myocardial infarction would not be prominent in the differential diagnosis.

We are aware of an earlier study of acute myocardial infarction in 11 thermally injured patients.[24] Although the number of cases is small in both reports, our findings are in general agreement. Additionally, that report emphasized the need to evaluate carefully all cardiac arrhythmias occurring in burn patients, especially those seen in patients with severe thermal injury.

Part of the evaluation of acute myocardial infarct includes measurement of creatine kinase (CK) and specifically the "MB" fraction. Great care should be exercised in interpreting CK, especially in patients with electrical burns, because CK-MB fraction may be elevated whenever there is extensive skeletal muscle involvement. Ancillary changes in the EKG are necessary for the diagnosis.

HYPERTENSION IN BURNED CHILDREN

Hypertension in burn victims is seen mostly in young children, predominately males who have extensive (greater than 40 percent total body surface) burn. Some investigators have reported an increased mortality rate, while others have not noted this effect. Elevated catecholamine levels have been observed in burned children with and without hypertension. Lowrey[28] has noted a high incidence (46 percent) of encephalopathy in burned hypertensive children.[28–31]

At postmortem examination, the cardiomegaly of hypertensive children is found to be due to cardiac hypertrophy. Microscopically, foci of interstitial edema, mononuclear infiltrates, and necrosis are also seen. Microorganisms are not seen in lesions of this type.[32] These changes are nonspecific and may be related to any of a variety of metabolic derangements as well as anemia.

COMPLICATIONS OF INTRAVENOUS HYPERALIMENTATION WITH FAT EMULSIONS

Recent demonstration of the beneficial effects of fat emulsion has led to the use of lipid in the intravenous hyperalimentation fluid[33] to prevent or correct deficiencies of essential fatty acids. To date, we as well as others[34] have been unable to document complications due to fat emulsion therapy. However, Hessov et al.[35] have reported three patients who apparently expired from excessive intravenous administration of fat emulsion [20 percent Intralipid (Cutters)]. In these latter cases, "mild white, shiny clotted lumps" were identified in the small blood vessels of the heart. In two of these cases, similar small vessels of the lungs, cerebrum, and kidneys were also involved. These "white lumps" in the small blood vessels were seen histologically as amorphous accumulations of Sudan Black III-positive material. The material contained no fatty cells or hematopoietic cells, and thus was easily differentiated from fat emboli associated with multiple fractures. Freund et al.[36] reported a similar case with lesions related to intravenous administration of fat emulsions in a cachectic man dying from sepsis. Such cases appear to be uncommon.[34] and may be related to the more concentrated lipid emulsion used in Europe.[35,37]

MYOCARDIAL DEPRESSION IN THE BURN PATIENT

Experimental data have demonstrated serum factor(s) from burn patients that cause myocardial depression in animals.[38–42] The precise identity and source of the factor(s) is unknown, although origin of such factors in the burn wound is highly probable. Such factors appear to alter both the function of human heart muscle and cardiac output.[41] In a burned rat model, Rubanyi et al. have found recent evidence that endogenous nickel may play a role in coronary vasospasm and myocardial depression in acute burn shock.[43]

The burn patient's inability to maintain adequate circulation[41] (based on arterial pressure and cardiac output) has often been ascribed to left ventricular failure induced by the serum myocardial depressant factor. The echocardiographic studies of Dorethy[44] in patients with postburn or septic shock did not support this concept. The primary findings in his patients were a decreased left ventricular intravascular volume. The only echocardiographic evidence of myocardial depression was found in two patients with myocardial abscesses and septic shock.[44]

--- **References**

1. O'Neill, J. A., Jr., Pruitt, B. A., Jr., and Foley, F. D., et al.: Suppurative thrombophlebitis: A lethal complication of intravenous therapy. J. Trauma 8:256–267, 1968.
2. Stein, J. M., and Pruitt, B. A., Jr.: Suppurative thrombophlebitis. A lethal iatrogenic disease. N. Eng. J. Med. 282:1452–1455, 1970.
3. Welch, G. W., McKeel, D. W., Jr., Silverstein, P., et al.: The role of catheter composition in the development of thrombophlebitis. Surg. Gynec. Obstet. 138:421–424, 1974.
4. Warden, G. D., Wilmore, D. W., and Pruitt, B. A., Jr.: Central venous thrombosis: A hazard of medical progress. J. Trauma 13:620–626, 1973.
5. Pruitt, B. A. Jr., McManus, W. F., Kim, S. H., and Treat, R. C.: Diagnosis and treatment of cannula-related intravenous sepsis in burn patients. Ann. Surg. 191:546–553, 1980.
6. Sasaki, T. M., Panke, T. W., Dorethy, J. F., Lindberg, R. B., and Pruitt, B. A., Jr.: The relationship of central venous catheter position to acute right-sided endocarditis in severe thermal injury. J. Trauma. 19:740–743, 1979.
7. Pruitt, B. A., Jr., Stein, J. M., Foley, F. D., Moncrief, J. A., and O'Neill, J. A., Jr.: Intravenous therapy in burn patients. Suppurative thrombophlebitis and other life-threatening complications. Arch. Surg. 100:399–404, 1970.
8. Goodman, D. J., Rider, A. K., Billingham, M. E., and Schroeder, J. S.: Thromboemboli complications with the indwelling balloon-tipped pulmonary arterial catheter. N. Eng. J. Med. 291:777, 1974.
9. Chum, M. H., and Ellestad, M. H.: Perforation of the pulmonary artery by a Swan-Ganz catheter. N. Eng. J. Med. 284:1041–1042, 1971.
10. Ehrie, M., Morgan, A. P., Moore, F. D., and O'Connor, N. E.: Endocarditis with the indwelling balloon-tipped pulmonary artery catheter in burn patients. J. Trauma 18:664–666, 1978.
11. Golden, M. S., Pinder, T., Jr., Anderson, W. T., and Cheitlin, M. D.: Fatal pulmonary hemorrhage complicating use of a flow-directed balloon-tipped catheter in a patient receiving anticoagulant therapy. Am. J. Cardiol. 32:865–867, 1973.
12. Kahn, J. V. D. B., Fowler, N. O., and Doerger, P.: Right heart catheter lesions: Any significance? Am. J. Clin. Pathol. 82:137–147, 1984.
13. Liepman, M. K., Jones, P. G., and Kauffman, C. A.: Endocarditis as a complication of indwelling atrial catheters in leukemic patients. Cancer 54:804–807, 1984.

14. Rabin, E. R., Graber, C. D., Vogel, E. H., Jr., Finkelstein, R. A., and Tumbusch, W. T.: Fatal *Pseudomonas* infection in burned patients. A clinical, bacteriologic and anatomic study. N. Eng. J. Med. 265:1225–1231, 1961.

15. Pesanti, E. L., and Smith, I. M.: Infective endocarditis with negative blood cultures. An analysis of 52 cases. Am. J. Med. 66:43–50, 1979.

16. Srivastava, R. K., and MacMillan, B. G.: Cardiac infection in acute burn patients. Burns 6:48–54, 1979.

17. Sevitt, S.: A review of the complications of burns, their origin and importance for illness and death. J. Trauma 19:358–369, 1979.

18. Panke, T. W., McLeod, C. G., Jr., and Langlinais, P. C.: Myocardial infarction: An uncommon antemortem diagnosis in the thermally injured patient. (Unpublished data)

19. Gotlieb, A., Masse, S., Allard, J., Dobell, A., and Huang, S.: Concentric hemorrhagic necrosis of the myocardium. A morphological and clinical study. Human Pathol. 8:27–37, 1977.

20. Brian, A. W., McManus, W. F., Goodwin, C. W., and Pruitt, B. A., Jr.: Lightning injury with survival in five patients. J.A.M.A. 253:243–245, 1985.

21. Cole, F. M.: Myocardial infarction after burns. Brit. Med. J. 2:1575–1576, 1963.

22. Sevitt, S.: Coronary thrombosis following injury and burns. Med. Sci. Law 13:185–191, 1973.

23. Sevitt, S.: *In* Reactions to Injury and Burns and Their Clinical Importance, 1st. ed. Philadelphia, J. B. Lippincott Co., 1974, p 187.

24. Andes, W. A., and Hunt, J. L.: Myocardial infarction in the thermally injured patient. Amer. Burn Assoc., 6th Ann. Meeting, Cincinnati, Oh, April, 1974.

25. Nanji, A. A., and Filipenko, J. D.: Non-myocardial source of CK-MB in a patient with electrical burn injury. Burns 10:372–373, 1984.

26. Seigel, A. J., Silverman, L. M., and Holman, B. L.: Peripheral source of MB band of creatine kinase in alcoholic rhabdomyolysis. J.A.M.A. 244:580–582, 1980.

27. Lott, J. A., Speicher, C. E., and Ayers, L. W.: Typhoid fever toxemia with associated destruction of skeletal muscle. Clin. Chem. 26:1361–1362, 1980.

28. Lowry, G. H.: Hypertension in children with burns. J. Trauma 7:140–144, 1967.

29. Falkner, B., Roven, S., DeClement, F. A., and Bendlin, A.: Hypertension in children with burns. J. Trauma 8:213–217, 1978.

30. Douglas, B. S., and Broadfoot, M. J.: Hypertension in burnt children. Aust. N. Z. J. Surg. 42:194–196, 1972.

31. Popp, M. B., Friedberg, D. L., and MacMillian, B. G.: Clinical characteristics of hypertension in burned children. Ann. Surg. 191:473–478, 1980.

32. Joshi, V. V.: Effects of burns on the heart. A clinicopathological study in children. J. Amer. Med. Assoc. 211:2130–2134, 1970.

33. Wilmore, D. W., Moylan, J. A., Helmkamp, G. A., and Pruitt, B. A., Jr.: Clinical evaluation of a 10 percent intravenous fat emulsion for parenteral nutrition in the thermally injured patient. Ann. Surg. 178:503–511, 1973.

34. Lee, H. A.: Parenteral Nutrition in Acute Metabolic Illness. New York, Academic Press, Inc., 1974, pp 419–420.

35. Hessov, I., Melsen, F., and Haug, A.: Postmortem findings in three patients treated with intravenous fat emulsions. Arch. Surg. 114:66–68, 1979.

36. Freund, U., Krausz, Y., Levis, I. S., and Eliakam, M.: Iatrogenic lipidosis

following prolonged intravenous hyperalimentation. Amer. J. Clin. Nutr. 28:1156–1160, 1975.

37. Pruitt, B. A., Jr.: Editorial: Intravenous fat emulsions. Arch. Surg. 114:68, 1979.

38. Raffa, J., and Trunkey, D. D.: Myocardial depression in acute thermal injury. J. Trauma 18:90–93, 1978.

39. Hakim, A. A.: Effects of digitoxin on "acute burn serum inhibitor" depressed heart papillary muscles. Pharm. Acta Helv. 50:178–184, 1975.

40. Fozzard, H. A.: Myocardial injury in burn shock. Ann. Surg. 154:113–119, 1961.

41. Okamoto, A., Kaye, M., Coleman, T. B., and Glaviano, V. V.: Hemodynamic and metabolic alterations of the heart in burn shock. Circ. Shock. 1:243–250, 1974.

42. Temples, T. E., Burns, A. H., Nance, F. C., and Miller, H. I.: Effect of burn shock on myocardial function in guinea pigs. Circ. Shock. 14:81–92, 1984.

43. Rubanyi, G., Szabo, K., Balogh, I., Bakos, M., Gergely, A., and Kovach, A. G. B.: Endogenous nickel release as a possible cause of coronary vasoconstriction and myocardial injury in acute burn of rats. Cir. Shock. 10:361–370, 1983.

44. Dorethy, J. F.: Echocardiographic evaluation of left ventricular performance in the severely burned military population. *In* Annual Progress Report. US Army Institute of Surgical Research, BAMC, Fort Sam Houston, Tex, pp 433–445, Fy 1978.

————————————————————————————

Chapter 10

RESPIRATORY
SYSTEM

Although the upper respiratory tract and lung of the burn victim are susceptible to many serious pathological changes and complications, the typical findings at autopsy are often nonspecific and may be identical to those seen in routine (nonburn) autopsy populations. Certain specific syndromes typically occur in burn patients, and even though it is often impossible to assign significance to all these lesions as to the cause of death, these conditions should be considered in all burn autopsies regardless of the clinical history (Table 10–1).

INHALATION INJURY

With presently employed, improved techniques in early resuscitation, burn patients now escape shock and renal failure, which were once the major causes of early death. The incidence of significant pulmonary injury has been estimated to be between 15 and 22 percent in burn patient populations.[1,2] In spite of the remarkable reparative capacity of the respiratory system, the mortality rate in such cases has been estimated to be between 48 and 86 percent.[3] In patients with serious burns and inhalation injury, the actual mortality rate greatly exceeds that expected for patients that have cutaneous burns only.[4]

Clinically, patients with a significant inhalation injury have (1) a history of thermal injury within a closed space; (2) wheezing, hoarseness and dyspnea; (3) facial and oropharyngeal burns; and (4) carbonaceous sputum.[5] A recent article indicates that a history of unconsciousness is probably the best historical indicator of inhalation injury.[6] Inhalation injury is confirmed by noting prominent erythema and ulcerations of the tracheobronchial mucosa with carbonaceous particles (soot) by broncoscopy. Abnormal retention of Xenon 133 in lung scans[7] indicates pulmonary parenchymal

Table 10–1 *Pulmonary Complications in Burn Patients*

Inhalation injury

  Smoke inhalation
  Steam burn
  Acute laryngeal edema

Tracheostomy

  Hemorrhage
  Stomal and tracheobronchial infection
  Tracheomalacia
  Tracheoesophageal fistula

Bronchiectasis

Pneumonia

  Airborne bronchopneumonia
  Hematogenous pneumonia
  Melioidosis pneumonitis
  Aspiration pneumonia

Hyaline membrane disease

Acute pulmonary edema

Pulmonary hemorrhage

Thromboembolism

Empyema

Atelectasis

Chest wall trauma and thermal injury

---

damage. Measurement of arterial blood gases (provided cardiovascular function is normal) and early carboxyhemoglobin determinations are useful secondary aids in diagnosing inhalation injury. (See Chapter 16 for carbon monoxide poisoning).

In patients with inhalation injury, interstitial pulmonary edema is always a prominent feature reflecting the extent of parenchymal injury. Head[8] analyzed the alveolocapillary interface with a fluorescene-labeled albumen complex and assayed bronchial aspirates for this complex and for surfactant content. Combination of fibrooptic bronchoscopy observations with the quantity of "leaked" fluorescene–albumen was predictive of which patients with inhalation injuries would develop significant, early respiratory failure that often lasted 7–10 days. Surfactant levels, usually normal during the first week, were predictive of which patients might develop significant pulmonary problems in the second postburn week when the levels were low. Levels of surfactant may be dependent on adequate dietary palmitate for the synthesis of dipalmitoye lecithin.[8] In an experimental model of smoke inhalation (wood and kerosene), it was found that the smoke affected the surfactant level by inactivating surfactant activity rather than by decreasing its production.[9]

Upper airway and lung inhalation injuries are usually secondary to inhalation of gaseous or particulate products of incomplete combustion and rarely are due to heat unless steam has been inhaled.[10] Patients with inhalation injury due to steam more rapidly develop pulmonary edema and congestion. As expected, black soot is not identified, and the tracheobronchial mucosa undergoes rapid dissolution complicated by bronchopneumonia, which develops as early as the third or fourth postburn day.[11] Natural and synthetic building and insulation materials, furniture, carpeting, clothing, bedding, and fuels all contain substances that, when partially combusted, have direct injurious effects on the lung and upper airways. These toxic substances include: acetic acid, acetic anhydride, oxides of nitrogen, acrolein, sulphur dioxide, formaldehyde, and cyanide.[12,13]

The more soluble chemicals (ammonia, sulfur dioxide, chlorine, and hydrogen chloride) tend to involve the upper respiratory tract, while insoluble chemicals (aldehydes, phosgene, and nitrogen oxides) tend to destroy the lower respiratory tract. High concentrations and/or prolonged exposures of any of these chemicals will cause extensive damage.[14,15] The exact action of most of these substances is not known; however, some apparently cause immediate loss of function or destruction of respiratory cilia and a dose-related necrosis of respiratory mucosa. Others appear to decrease pulmonary surfactant activity.[9] The resultant lack of clearing and subsequent formation of cellular mucinous casts lead, in many cases, to small airway obstruction. Additionally, carbon-filled macrophages may cause delayed damage by settling into smaller airways and subsequently releasing hydrolases and other enzymes. This delayed process may explain why some victims of inhalation injury have a temporary "clear period" of 1–2 days and then develop acute respiratory distress. It is very interesting that unburned firemen treated for severe smoke inhalation most often recover within 24 hour.[16] Long-term physiologic impairment of the respiratory system appears minimal in nonsmoking firemen, even when their employment exceeds 25 year. Firemen with a significant smoking history do have mild obstructive changes in their small airways. One fireman who did have a severe exposure (42 percent carboxyhemoglobin level) did still have mild impairment 2.5 year later.[17]

The pulmonary lesions seen at autopsy depend on the survival time post injury. In patients that survive a few hours, the only significant pulmonary findings may be singed nasal hairs, carbonaceous debris in the upper airways

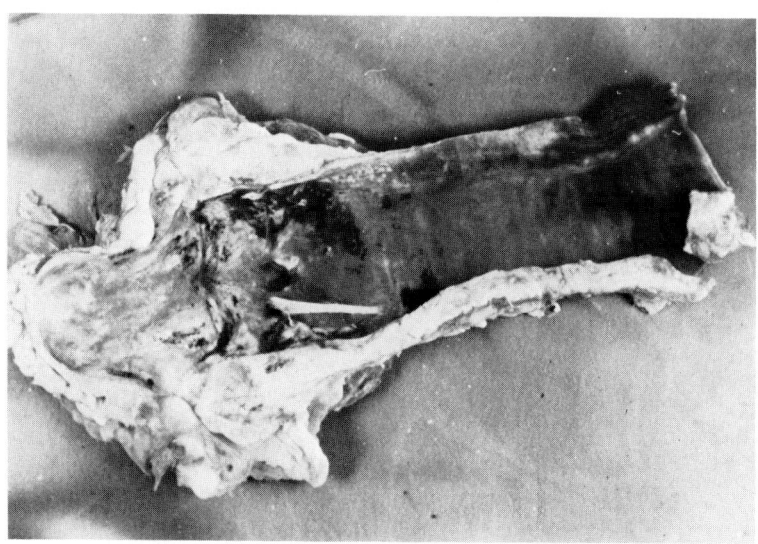

Fig. 10—1. Carbonaceous debris adherent to edematous tracheal mucosa of a burn victim who died 2 days following injury.

(Fig. 10–1), and pulmonary edema. Asphyxiation is often the cause of death in these patients. In patients dying within 1–2 days following inhalation injury, the tracheal, bronchial, and bronchiolar mucosa may show acute necrosis, sloughing and mild suppuration. Small airways may be obstructed by sloughed mucosal cells and exudate (Fig. 10–2). Occasionally, bronchioles will contain sloughed tall, ciliated epithelium, indicating that necrotic mucosa from large airways has settled into distal bronchioles (Fig. 10–3). Large macrophages containing carbonaceous debris are seen in alveolar spaces and in small airways. Mucosal epithelium may show evidence of regeneration after approximately 48 hr. In animal experiments, flattened squamous epithelium is often seen extending outward from the necks of mucosal glands in an apparent attempt to cover the denuded airways (Fig. 10–4) in a manner comparable to follicular squamous epithelium covering a second degree burn wound surface.[18]

Patients who survive the first few days following a severe inhalation injury seldom show the uncomplicated necrotizing laryngotracheobronchitis described above. Bacterial pneumonias, primarily gram-negative, account for most of the fatal cases of inhalation injury.[1,2,5] Tracheostomy and mechanical suction can cause deep tracheal mucosal ulcerations and contribute to serious soft tissue infections.[19] Burn wound sepsis and infected cannulated veins can lead to hematogenous pneumonia in burn patients.

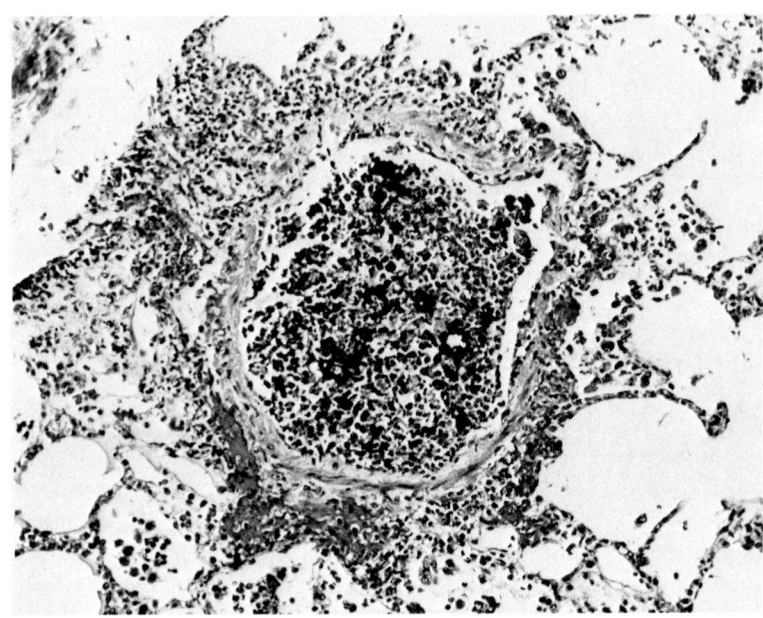

Fig. 10–2. Small airway obstructed with necrotic cellular debris, carbon, and polymorphs. Note that respiratory mucosa is absent. (H & E, × 144)

When advanced, these and the other complications discussed below may completely obscure the typical lesions of inhalation injury.

INFECTIOUS COMPLICATIONS OF
TRACHEOSTOMY

Tracheostomies in severely burned patients may be associated with several complications (Table 10–2). Infections of the soft tissues, especially if the tracheostomy passes through burned skin, often extend from the stoma. If the tracheostomy is performed just inferior to the thyroid, staphylococcal infection of the tracheostomy stoma may extend into the thyroid and cause staphylococcal abscesses.

A severe necrotizing tracheobronchitis may also be seen, often secondary to *Staphylococcus* sp. or gram-negative bacilli.[20] *Candida* sp. may also populate the necrotic debris in the tracheostomy and tracheobronchial tree but is usually noninvasive. Severe, necrotizing inflammatory processes involving the trachea may result in the formation of a pseudomembrane composed of necrotic debris, fragmented respiratory epithelium, inflammatory cells, erythrocytes, and colonies of bacteria.

Another site for staphylococcal or gram-negative necrotizing infection is the larynx. This infection is related to nasotracheal or endotracheal intubation which causes pressure necrosis of the vocal cords.

In the previous discussion of sources for staphylococcal septicemia, we mentioned the above sites and wish to stress again these potential sources of serious systemic infections.

COMPLICATIONS OF TRACHEAL INTUBATION

Earlier publications have defined these entities in detail.[21–23] The most common lesion in the upper respiratory tract in a series of burn patients studied by Teplitz[24] was mucosal erosion caused by the tip of the tracheostomy tube. Great care in the selection of the appropriately sized, pliable endotracheal and tracheostomy tubes combined with the use of more selective criteria for the performance of tracheostomies have reduced the incidence of these complications. However, severely burned patients with

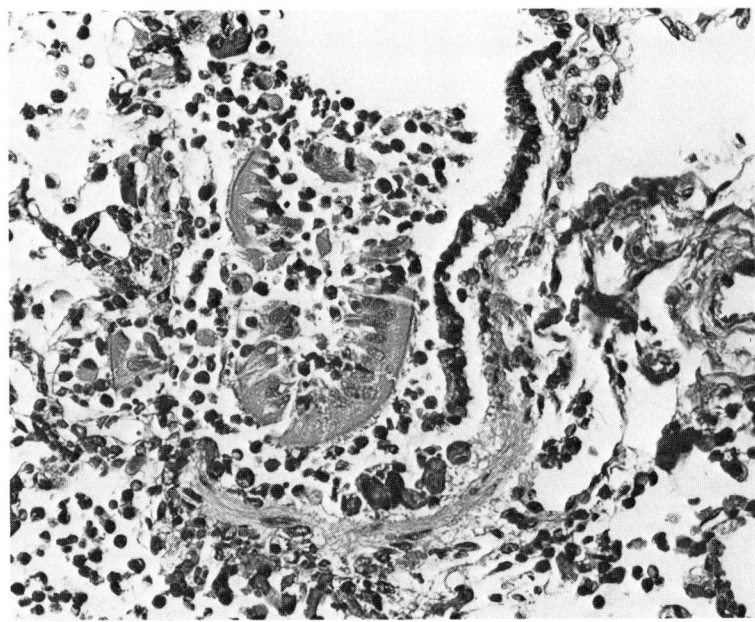

Fig. 10–3. Distal airways and occasionally alveolar spaces may contain casts of necrotic tall respiratory epithelium that have sloughed and settled into the lung following upper airway inhalation injury. This finding is usually seen in patients who die within 1–2 days following inhalation injury. ($\times$ 315)

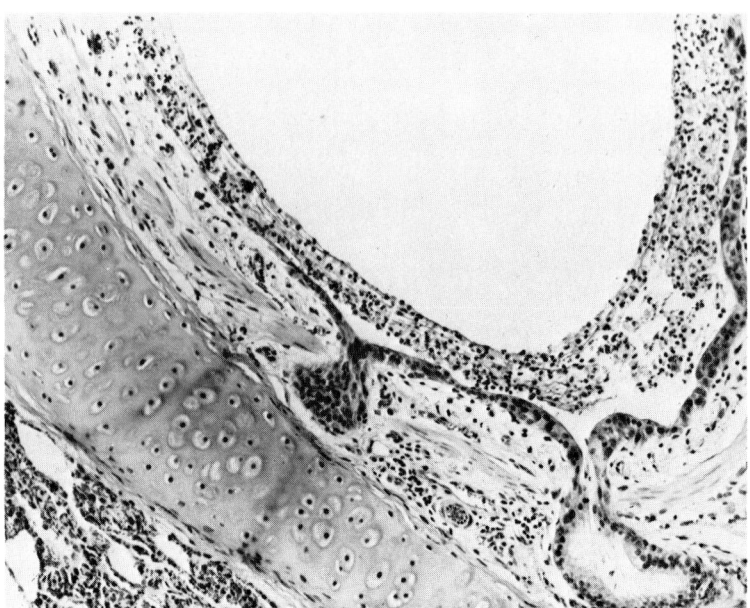

Fig. 10–4. At about 48 hr following inhalation injury, evidence of mucosal repair may be seen. This section shows extension of flattened squamous epithelium from the neck of a submucosal gland. (H & E, × 270)

recurrent complications of the respiratory and other systems often require prolonged assisted ventilation.

There are several direct and indirect complications of tracheal intubation which include mucosal edema and subsequent acute ulcers of the *vocal cords*. Later fibrous stenosis or fibrous webbing may occur between the vocal cords, and vocal cord polyps may also develop. When the endotracheal tube is replaced in a patient by a tracheostomy tube, there is a much higher incidence of *subglottic stenosis*,[25–27] and this problem is particularly magnified if microbiologic antisepsis around the tracheostomy stoma is not meticulously maintained. Bacterial overgrowth of the stoma leads to outright infection and tissue necrosis, and the healing phase of this process is often associated with disproportionate fibrosis and severe *laryngotracheal stenosis*.[26,27] Laryngeal stenosis also occurs if the cuff is inflated too high (at the level of the cricoid cartilage).[25]

Laryngeal and tracheal stenosis is usually apparent at the time of extubation or within the next 1–2 weeks. Occasionally, onset of respiratory difficulty does not occur until months to years later, and the inciting cause, laryngotracheal stenosis, may not be immediately apparent.[27]

With the advent of soft pliable cuffs, the high incidence of tracheoma-

Table 10–2 *Complications of Tracheostomy in the Burn Patient*

Tracheal erosions and/or tracheitis

Tracheomalacia at site of tracheostomy

Primary bacterial infection or secondary colonization of necrotic tissue

  Tracheostomy stoma
  Surrounding soft tissue *e.g.*, thyroid
  Tracheobronchial
  Larynx (tracheal intubation)

Pneumothorax

Bleeding

  mild
  severe with extensive pulmonary consolidation

Subcutaneous emphysema

Occlusion of one bronchus by tracheal tube (low placement)

Tube dislodged

Vocal cord paralysis—necrosis

Tracheopleural fistula

Tracheoesophageal fistula

Subglottic stenosis

---

[a] Adapted from Pruitt, B. A., Jr., Flemma, R. J., DiVincenti, F. C., Foley, F. D., and Mason, A. D., Jr.: Pulmonary complications in burn patients. J. Thor. Cardiovasc. Surg. 59, 7–18, 1970.

lacia[21,22,24] has greatly diminished.[27] With prolonged intubation of the severely burned patient, there is the threat of soft tissue pressure necrosis with exposure of the cartilaginous rings. The rings of cartilage may themselved be eroded and/or a necrotizing bacterial cellulitis may develop around the rings and further accelerate their dissolution. This leads to marked dilatation of the trachea and transmural erosions, which predispose to the fistula formation mentioned below. All such lesions serve as sources of microbiologic dissemination and may result in fatal pneumonia or septicemia.

Occasionally prolonged tracheal intubation leads to the development of a *tracheoesophageal fistula* in the burn patient.[23,28,29] Simultaneous use of a nasogastric tube, severe tracheal inhalation injury, septicema, and hypotension may all predispose to this lesion. *Tracheoarterial fistula* may also occur[23] and may result in a fatal exsanguination.[30]

The mere presence of a tracheostomy stoma, often extending through a burn site, offers a new portal of entry for microorganisms. The patient's endotracheal or tracheostomy tube will inhibit his ability to cough and thwart the mucociliary process, resulting in pooling of secretions containing bacteria and toxins in the tracheobronchial tree. Contamination by oral secretions and gastric aspiration are additional problems that threaten to accelerate all of the above processes.[27]

BRONCHIECTASIS AS A COMPLICATION OF INHALATION INJURY

Three cases of bronchiectasis have been described in patients with severe inhalation injury. One case had lesions involving the right lung which were noted on bronchography 4 months after injury.[31] Another case had bilateral extensive lesions involving all lobes of both lungs, and the patient expired 5.5 months after thermal injury with severe bronchiectasis, bronchopneumonia, and renal failure (Fig. 10–5).[32] Bronchiectasis was not diagnosed antemortem in this patient. The third case occurred in a 5-year-old who sustained an inhalation injury in a tenement fire. She was discharged from the hospital 3 months post burn but required readmission for recurrent pneumonia. Bronchography demonstrated stenosis of the bronchus to the lower lobe of the left lung with bronchiectasis distally.[33]

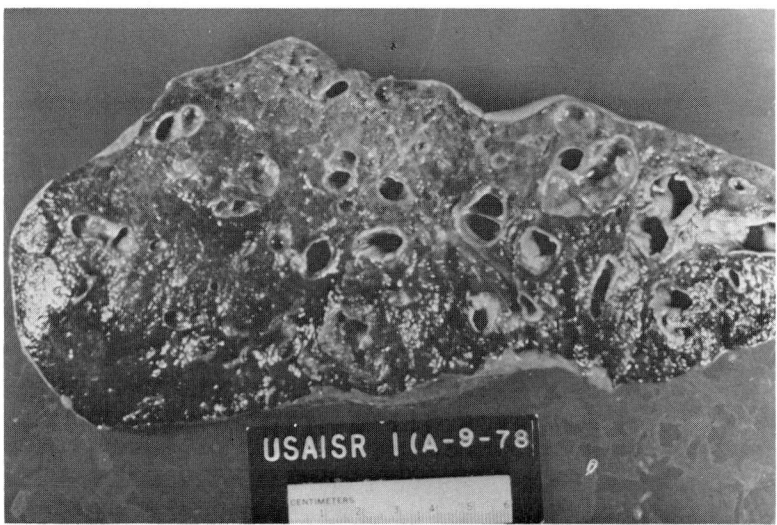

Fig. 10–5. Severe bronchiectasis in a burn patient who expired 5.5 months after thermal injury.

Because of the paucity of cases, the pathogenesis is not known with certainty. The case presented by Donnellen et al. demonstrates that fibrosis and torsion with subsequent stenosis of a bronchus is one predisposing condition for bronchiectasis in the burn patient with inhalation injury. However, findings of bronchial distortion and stenosis were not observed in the other two cases.[31,32] In severe inhalation injury, extensive bronchial and bronchiolar mucosal necrosis occurs. Subsequent bacterial colonization of the necrotic tissue leads to infection of the bronchial wall. In some patients, this process continues and leads to chronic inflammation and scarring consequent to bronchiolar smooth muscle injury. Loss of muscular tone leads to further bronchial dilation, which renders the lung even more susceptible to infection and continuation of this process.

CHRONIC COMPLICATIONS OF INHALATION INJURY

Mucosal "polyps" of the tracheobronchial tree have been infrequently reported 1–3 months after inhalation injury.[34,35] They are described as multiple, broad-based, erythematous and friable structures that focally or extensively involve the trachea and bronchi. Histologically, they are composed of granulation tissue in the submucosa. The overlying respiratory mucosa may undergo squamous metaplasia. Endobronchial polyps appear to resolve spontaneously in 6–8 months. They are probably more common than reported because they may be asymptomatic complications of tracheobronchial injury from smoke inhalation.

Severe inhalation injury may occasionally lead to *bronchiolitis obliterans*.[36,37] Initial respiratory symptoms begin as early as 2 months post burn and lead to death in 1–2.5 years after burn injury. Several factors cause broncheolitis, including toxic chemicals and vapors, certain viral infections, collagen vascular disease such as rheumatoid arthritis, association with interstitial pneumonitis, lymphoma, myasthenia gravis, and idiopathic causes. Histologically, progressive peribronchial, subepithelial fibrosis leads to a narrowed bronchial lumen and panacinar emphysema. Therapy for broncheolitis obliterans, often instituted later in the course of disease, appears ineffective.

PNEUMONIA CAUSED BY
Pseudomonas aeruginosa

Hematogenous pulmonary lesions caused by *P. aeruginosa* are often subpleural, darkly hemorrhagic, and infarct-like. Histologically, massive alveolar necrosis and large numbers of gram-negative rod-shaped bacteria are seen. These bacteria in early lesions proliferate in alveolar capillaries and later appear in massive numbers in perivascular tissues. Peribronchiolar connective tissue may contain many bacteria; but in the early stages, there is very little exudate compared to other types of gram-negative hematogenous pneumonia or airborne bronchopneumonia.[19] Acute hemorrhage and fibrin deposition are also prominent features of pulmonary *Pseudomonas* infection.

The classical pattern of capillary and arterial necrosis is seen only in the early stages of hematogenous *Pseudomonas* pneumonia. The lack of acute inflammatory cells in contrast to the extensive areas of coagulative necrosis often gives a morphologic appearance of a markedly accelerated pulmonic lesion comparable to ecthyma gangrenosum of the skin. The occasional finding of lesions with large numbers of polymorphonuclear leukocytes in *Pseudomonas* pneumonia suggests that the infection is less aggressive. The difference in invasive rates is very likely related to the variable pathogenicity of different strains of *P. aeruginosa,* as well as to the host's ability to resist infections (Chapter 6). A form of *P. aeruginosa* pneumonia which has an extensive hemorrhagic, confluent, infectious process may resemble pneumococcal, lobar pneumonia in some cases. Only the finding of perivascular gram-negative bacillary organisms and microbiologic cultures confirm the diagnosis of *P. aeruginosa* pneumonia. Teplitz et al.[38] considered the extensive hemorrhage in such cases to be due to diffuse tissue invasion and necrosis rather than to bacterial "vasculitis." However, extensive tissue necrosis and hemorrhage may obliterate the microvasculature and efface any evidence of a prior bacterial "vasculitis."

AIRBORNE BRONCHOPNEUMONIA VS HEMATOGENOUS PNEUMONIA

Bronchopneumonia (airborne pneumonia) is a frequent complication in severely-burned patients. Grossly, it may be difficult to distinguish bronchopneumonia from hematogenous pneumonia. Clinically, airborne pneumonia generally occurs relatively early in the postburn course, whereas hematogenous pneumonia tends to occur later.[39] Morphologically, the inflammatory process in airborne bronchopneumonia is centered around bronchioles and bronchi (Fig. 10–6). In contrast, hematogenous pneumonia has a perivascular distribution (Fig. 10–7). In the advanced forms of either of these inflammatory processes, it is often difficult to make exact morphologic distinctions.

STAPHYLOCOCCAL PNEUMONIA

The morphological pattern of staphylococcal pneumonia suggests that it may develop from either hematogenous or airborne sources (Table 10–3). In the early stages, the tissue appears firm, red-blue, edematous, and glisten-

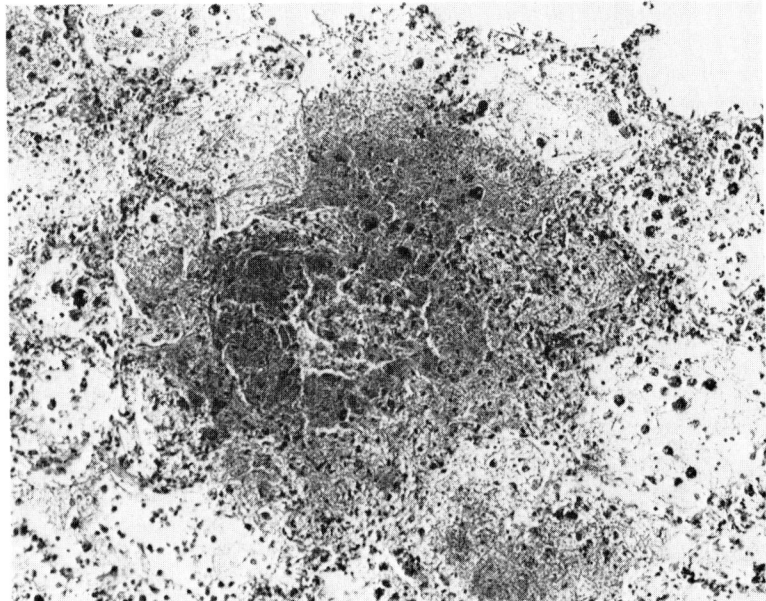

Fig. 10–6. The finding of heavy bacterial growth in and around airways with extension of necrosis and inflammation into lung parenchyma is suggestive of an airborne source. (H & E, × 126)

ing. Later, lesions in most cases are focal and multifocal gray-yellow patches of consolidated tissue, some parts of which have evolved into abscesses. Microscopically, this inflammatory process is most commonly suppurative and necrotizing. Colonies of coccoid bacteria are readily apparent within the center of the abscesses. Only coagulase-positive *Staphylococcus aureus* has been cultured from lesions of this appearance.

Occasionally, small septic thromboemboli and pulmonary infarcts are seen in cases of staphylococcal pneumonia. These septic emboli most commonly arise from vegetations on the tricuspid valve, but may also arise from septic mural thrombi in cannulated veins.

GRAM-NEGATIVE BACTERIAL PNEUMONIA

Gram-negative bacillary microorganisms other than *P. aeruginosa* are less frequent causes of all forms of fatal pneumonia in the burn patient. Examples of these microorganisms are *Klebsiella pneumoniae, Enterobacter cloacae, Proteus* sp., and *E. coli*. Morphologically, these infections have

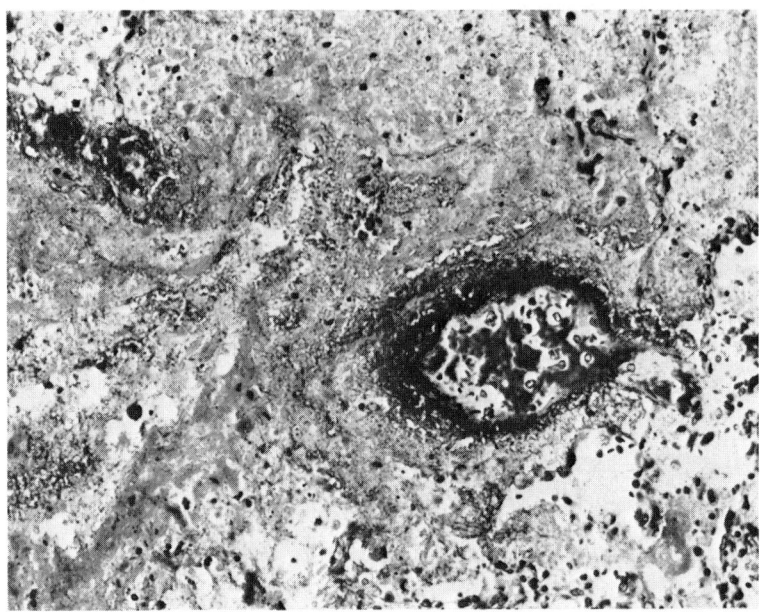

Fig. 10–7. In uncomplicated hematogenous pneumonia, heavy bacterial growth and early inflammation usually involve blood vessels. (H & E × 189)

a pattern that could be either hematogenous or airborne. Possible sources from which pulmonary infections may arise are listed in Table 10–3.

Klebsiella pneumoniae is characterized by consolidated red-gray lesions which exude a typical sticky exudate. Microscopically, the process is necrotizing and has a dense eosinophilic, probably proteinaceous fluid containing acute inflammatory cells. Abscesses may form later in the course of the disease. The other gram-negative microorganisms associated with a pneumonic process have red-gray consolidated areas but usually lack the characteristic viscid, sticky exudate.

MYCOTIC PNEUMONIA

Mycotic dissemination to the lung causing significant pulmonary disease in burn patients is rare. *Candida* sp. are frequently seen histologically in sputum, but this microorganism is rarely a serious pathogen. Most phycomycotic infections in burn patients are confined to the burn wound and adjacent unburned skin. These microorganisms may spread rapidly and

Table 10–3 *Sources for Staphylococcal and Gram-Negative Pneumonia*

Hematogenous

  Staphylococcal septicemia
  Cardiac valvular septic vegetations
  Venous septic thrombi
  Abscesses around cannulae
  Visceral abscesses
  Soft tissue infection—injection site abscess
  Gram-negative bacterial septicemia
    Burn wound
    Urinary tract infection

Tracheobronchial infection secondary to

  Inhalation injury
  Tracheostomy
  Tracheal intubation

Pulmonary parenchymal injury secondary to

  Inhalation injury of alveoli
  Aspiration pneumonia, secondarily infected

extensively to deep subcutaneous tissue, fascia, and muscle, but the lungs and other organs are usually not affected.

We are aware of only one recent case of pulmonary infection by *Aspergillus* sp. in a burn patient. The microorganisms were obviously invading pulmonary parenchyma, but an associated bacterial pneumonia was also present.

PULMONARY MELIOIDOSIS

Melioidosis, an infectious disease caused by the gram-negative bacillus *Pseudomonas pseudomallei,* is most frequently recognized in burn patients as a respiratory infection which may range in severity from a mild bronchitis to an acute septicemia with overwhelming pneumonia. It also occurs in a chronic localized granulomatous form which may involve the lungs, lymph nodes, bones, or skin. The chronic form of this disease may resemble tuberculosis or other chronic bacterial diseases histologically, clinically, and radiographically.[40] Numerous case reports of acute melioidosis, several of which involved US soldiers who suffered burn injuries in Viet Nam, have been published.[40-45] These and several other cases have been presented, along with an excellent histologic description of the lesions, in a review article.[46]

The characteristic lesion in the acute form of melioidosis is a focal suppurative process which may be found in almost any organ. Large histiocytes may be numerous in the acute form of melioidosis. The lesions are usually multiple, and when the acute disease persists for several days, small abscesses develop (see Chapter 6 for complete histologic description). Clinically, when pulmonary symptoms predominate, patients with acute melioidosis have large confluent masses of bronchopneumonia and hemorrhage. Bacterial stains will usually reveal the causative organisms within the suppurative lesions.

ASPIRATION PNEUMONIA

This condition is seen both as an uncommon primary cause of death and, more frequently, as an insignificant terminal event in burn patients. The diagnosis is easily made when food particles are detected on gross or microscopic examination of the lung (Fig. 10–8). A purely chemical pneu-

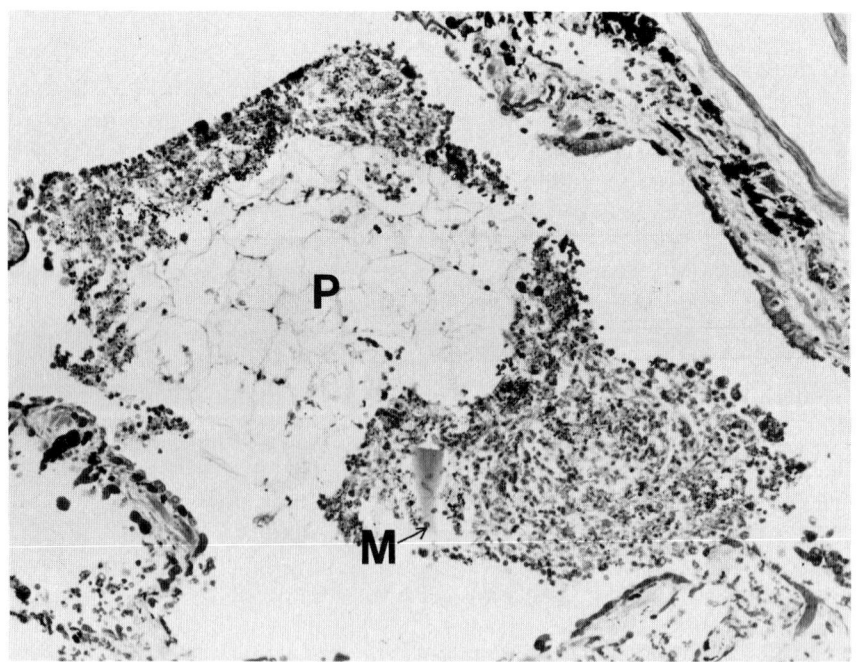

Fig. 10–8. Plant material (P) and muscle fiber (M) in a small airway of a burn patient with aspiration pneumonia. (H & E, × 144)

monia may result when highly acidic gastric juices, devoid of food particles or bacteria, are aspirated.[47,48] This form of aspiration pneumonia characteristically has widespread hemorrhage and edema, and could possibly be mistaken for certain forms of acute inhalation injury.

The pathologist is often faced with the difficult, if not impossible, task of deciding the significance of aspiration pneumonia when it is seen in concert with other complications in the lung. The cases in which there is a short interval between aspiration and death are especially challenging diagnostic problems. The mass of lung parenchyma affected and the clinical history are the most valuable guides in evaluating such cases.

PULMONARY EMBOLISM

Although there are conflicting studies of the incidence of pulmonary embolism in burn patients, this clinical entity currently is seldom seen at autopsy as a primary cause of death. During a 2-year period in which 111 autopsies on burn patients were conducted, pulmonary emboli were found in 10 cases. In none of these cases was pulmonary embolism the major cause of death.[49] Our experience contrasts with that of others[50,51] who have reported much higher incidences and associated mortality. We have no exact explanation for the present low incidence of pulmonary thromboembolism, but differences in patient populations, time period studied, frequency of use of central venous catheters, other complications, and the criteria for clinical and pathological diagnosis may account for some of the discrepancies in the reported incidence.

HYALINE MEMBRANE DISEASE

The clinical term "respirator lung syndrome" is not synonymous with the morphologic diagnosis "hyaline membrane disease."[31] Hyaline membrane formation is generally regarded as a nonspecific response of the lung to a variety of injurious substances and infections.[52] Hyaline membranes are often seen in patients who have neither been placed on a respirator nor treated with oxygen (Fig. 10–9). They are commonly found in burn patients who have had pulmonary edema, pleural effusions, viral infections, and bacterial pneumonias.[31]

Other nonspecific changes which accompany hyaline membrane disease consist of a reactive hyperplasia of type II pneumocytes with prominent interstitial edema and fibroblastic proliferation.

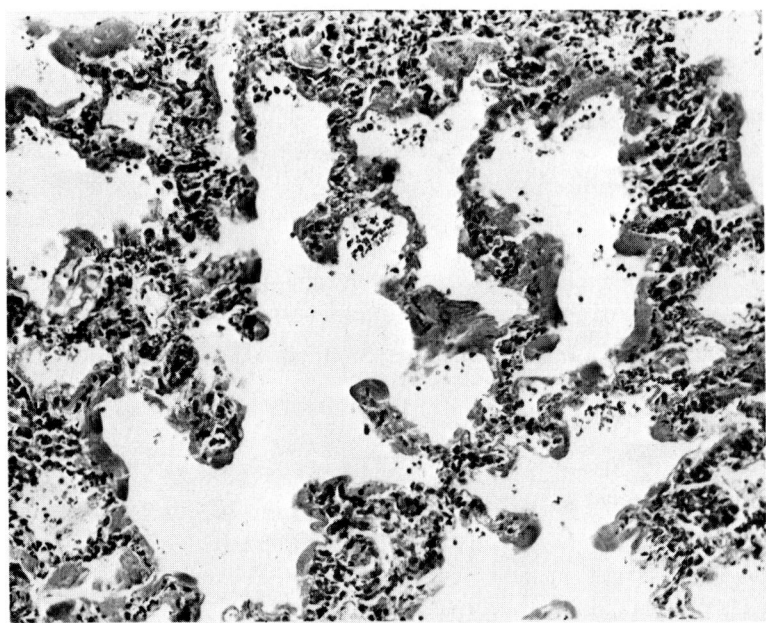

Fig. 10–9. Hyaline membrane formation is commonly found in burn patients following pulmonary edema, effusions, viral or bacterial infections, and inhalation injury. (H & E × 108)

PULMONARY HEMORRHAGE

Focal to diffuse acute parenchymal hemorrhages are commonly seen within the lungs of severely burned patients. Diverse etiologies may be associated with this finding (Table 10–4). Two of the more common and characteristic patterns of hemorrhages are those associated with inhalation injury or certain necrotizing infections such as *P. aeruginosa*. The latter are distinguished by the presence of a necrotizing pattern and numerous bacilli centered around bronchioles or blood vessels ("*Pseudomonas* vasculitis"). Other types of hemorrhage (e.g., secondary to vigorous tracheobronchial suctioning) may be entirely nonspecific and require careful clinicopathologic correlation to determine their cause. Accurate and complete postmortem examinations should include a thorough review of clinical information with the clinician(s). Only in this way can the specific pathologic finding (pulmonary hemorrhage) be linked with the correct causative factor (e.g., tracheobronchial suctioning). Such careful clinicopathologic correlations provide meaningful data for accurate assessment of therapeutic regimens.

Table 10–4 *Clinical Entities Associated with Pulmonary Hemorrhage in the Burn Patient*

Pulmonary hypertension

    Congestive heart failure
    Head trauma

Necrotizing infections

    Pseudomonas aeruginosa
    Mycotic (especially *Phycomycetes*)—rare
    Viral disease

"Inhalation injury"

Disseminated intravascular coagulopathy

Pulmonary thromboembolus with hemorrhagic infarct

Gastric acid aspiration

Fat emboli following traumatic skeletal fractures

Iatrogenic
    Vigorous tracheobronchial suctioning of secretions
    Inadequate hemostasis during tracheostomy procedure

EMPYEMA

Pneumothorax in burn patients with cutaneous burns of the thorax presents a difficult problem. The burn wound is invariably colonized by bacteria, and a thoracostomy significantly increases the incidence of empyema.[29] Both gram-positive cocci and gram-negative bacilli have been identified in these inflammatory processes. Empyema complicating pneumonia is uncommon.

SPUTUM, CYTOLOGIC EXAMINATION

Cytologic examinations of sputum specimens have yielded additional interesting data.[53] As mentioned in Chapter 8, cytologic changes consistent with herpes hominis were identified in 14 percent of fatally burned patients. They appeared from 8 to 28 days (average 17 days) post burn. Sputa containing *Candida* sp. were usually noted on the first day post burn. Candidal infection is seldom seen clinically or at postmortem exam despite the ubiquitous nature of the organism. These facts support Spencer's contention that only repeated isolation of this fungus suggests a diagnosis of invasive infection.[54]

Sputum cytologic examination to determine the presence of "inhalation injury" is useful only when specimens are studied in the first 48 hour post burn. Beyond that point, complicating infections may produce the same nonspecific but striking atypicality of respiratory epithelium that is seen with an inhalation injury.[53]

ANTECEDENT LUNG DISEASE

Teplitz[55] has discussed several factors predating thermal injury which are obviously of prognostic and clinical importance, but they have not been evaluated in any series of burn patients. These include a history of heavy smoking, chronic bronchitis, and emphysema. Many of these factors are not significant in the young burn patient, but must be carefully considered in the elderly patient.

TRAUMA TO THORACIC WALL

A variety of acute thoracic injuries are discussed by Sevitt.[56] These include traumatic damage to the chest with fractures of the ribs and concomitant contusions and lacerations of the lungs. Hemothorax and tension pneumothorax are frequent complications in these cases.

Deep burns of the thoracic wall have been seen in patients with electrical injuries. Costal chondritis or osteomyelitis may occur as a result of electrical injury itself or may be related to necrosis and desiccation of overlying soft tissue which exposes the underlying cartilage and bone.[39]

INTRAALVEOLAR MACROPHAGE ACCUMULATIONS

In up to 10 percent of burn patients, large numbers of macrophages appear to focally and diffusely consolidate pulmonary parenchyma.[57] The macrophages almost always contain pigment which only occasionally stain positively for hemosiderin. Usually this pigment is black and appears to be anthracotic pigment or is yellowish brown and special stains suggest lipofuscin. Electron microscopy on one such case[58] indicates that these cells are indeed macrophages and contain numerous lysosomes, a prominent Golgi apparatus and occasional phagocytized foreign debris. Teplitz[55] found that 80 percent of burn patients with such large accumulations of macrophages were burned in enclosed spaces or had were involved in an explosion. In addition, 40 percent of such patients had clinical respiratory signs or symptoms in the early or mid-postburn hospital course. Increased pulmonary macrophages have been noted in cigarette smokers; one must wonder what contribution this has had to the number of pulmonary macrophages in burn patients.

References

1. Achauer, B. M., Allyn, P. A., and Furnas, D. W.: Pulmonary complications: The major threat to the burn patient. Ann. Surg. 17:311–319, 1973.
2. Stone, H. H., and Martin, J. D.: Pulmonary injury associated with thermal burns. Surg. Gynec. Obstet. 129:1242–1246, 1969.
3. Levine, B. A., Petroff, P. A., Slade, C. L., and Pruitt, B. A., Jr.: Prospective trials of dexamethasone and aerosolized gentamicin in the treatment of inhalation injury in the burned patient. J. Trauma 18:188–193, 1978.
4. Agee, R. N., Long, J. M., and Hunt, J. L.: Use of ^{133}Xenon in early diagnosis of inhalation injury. J. Trauma 16:218–224, 1976.
5. DiVincenti, F. C., Pruitt, B. A., Jr., and Reckler, J. M.: Inhalation injuries. J. Trauma 11:109–117, 1971.
6. Moylan, J. A.: Inhalation injury. J.B.C.R. 3:51–53, 1982.
7. Moylan, J. A., Jr., Wilmore, D. W., Mouton, D. E., and Pruitt, B. A., Jr.: Early diagnosis of inhalation injury using 133-Xenon lung scan. Ann. Surg. 176:477–484, 1972.
8. Head, J. M.: Inhalation injury in burns. Am. J. Surg. 139:508–512, 1980.
9. Nieman, G. F., Clark, W. R., Jr., and Wax, S. D.: The effect of smoke inhalation on pulmonary surfactant. Ann. Surg. 191:171–181, 1980.
10. Moritz, A. R., Henriques, F. C., Jr., and McLean, R.: The effects of inhaled heat on the air passages and lungs: an experimental investigation. Amer. J. Pathol. 21:311–331, 1945.
11. Chu, C.: New concepts of pulmonary burn injury. J. Trauma 21:958–961, 1981.
12. Phillips, A. W.: Pulmonary complications in burned patients. Bahama International Conference on Burns, Dorrance & Co, Philadelphia, 1964.
13. Welch, G. W., Lull, R. J., Petroff, P. A., Hander, E. W., McLeod, C. G., Jr, and Clayton, W. H.: The use of steroids in inhalation injury. Surg. Gynec. Obstet. 145:539–544, 1977.
14. Crapo, R. O.: Smoke-inhalation injuries. J.A.M.A. 246:1694–1696, 1981.
15. Cahalane, M., and Demling, R. H.: Early respiratory abnormalities from smoke inhalation. J.A.M.A. 251:771–773, 1984.
16. Teplitz, C.: Pathology of burns and fundamentals of burn wound sepsis. *In* Artz, C. P., and Moncrief, J. A. The Treatment of Burns, 2nd. ed. Philadelphia, W. B. Saunders Co., 1969, p 53.
17. Loke, J., Farmer, W., Matthay, R. A., Putman, C. E., and Smith, G. J. W.:

Acute and chronic effects of fire fighting on pulmonary function. Chest 77:369–373, 1980.

18. Walker, H. L., McLeod, C. G. Jr., and McManus, W. F.: Experimental inhalation injury in the goat. J. Trauma 21:962–964, 1981.

19. Foley, F. D., Moncrief, J. A., and Mason, A. D., Jr.: Pathology of the lung in fatally burned patients. Ann. Surg. 167:251–264, 1968.

20. Eckhauser, F. E., Billote, J., Burke, J. F., and Quimby, W. C. Tracheostomy complicating massive burn injury. A plea for conservatism. Am. J. Surg. 127:418–422, 1974.

21. Austin, R. T.: Tracheostomy and prolonged intubation in the management of trauma. Injury 2:191–198, 1971.

22. Teplitz, C., Epstein, B. S., Rose, L. E., and Moncrief, J. A.: Necrotizing tracheitis induced by tracheostomy tube: Pathogenesis. Arch. Path. 77:6–19, 1964.

23. Cooper, I. D., and Grillo, H. C.: Analysis of problems related to cuffs on intratracheal tubes. Chest 62:21S–27S, 1972.

24. Teplitz, C.: Pathology of burns and fundamentals of burn wound sepsis. *In* Artz, C. P., and Moncrief, J. A. and The Treatment of Burns, 2nd. Ed. Philadelphia, W. B. Saunders Co., 1969, p. 57.

25. McGovern, F. H., Fitz-Hugh, G. S., and Edgemon, L. J.: The hazards of endotracheal intubation. Ann. Otol. Rhinol. Laryngol. 80:556–564, 1971.

26. Sasaki, C. T., Horiuchi, M., and Koss, N.: Tracheostomy-related subglottic stenosis: Bacteriologic pathogens. Laryngoscope 6:857–865, 1979.

27. Eliachar I., Moscona, R., Joachims, H. Z., Hirshowitz, B., and Shilo, R.: The management of laryngotracheal stenosis in burned patients. Plast. Reconstr. Surg. 68:11–16, 1981.

28. Slogoff, S., Allen, G. W., Warden, G. D., and McManus, W. F.: Tracheo-esophageal fistula following prolonged tracheal intubation in a thermally injured patient. Anesthesiology 39:453–455, 1973.

29. Moylan, J. A., Jr., West, J. T., Nash, G., Bowen, J. A., and Pruitt, B. A., Jr.: Tracheostomy in thermally injured patients: A review of five years' experience. Amer. Surg. 38:119–123, 1972.

30. Majeski, J. A., and MacMillan, B. G.: Tracheo-innominate artery erosion in a burned child. J. Trauma 18:137–139, 1978.

31. Pruitt, B. A., Jr., DiVincenti, F. C., Mason, A. D., Jr., Foley, F. D., and Flemma, R. J.: The occurrence and significance of pneumonia and other pulmonary complications in burned patients: Comparison of conventional and topical treatments. J. Trauma 10:519–531, 1970.

32. Lam, V. Unpublished observations.

33. Donnellan, W. L., Poticha, S. M., and Hollinger, P. H.: Management and complications of severe pulmonary burn. J. Amer. Med. Assoc. 194:1323–1325, 1965.

34. Adams, C., Moisan, T., Chandrasekhar, A. J., and Warpeha, R.: Endobronchial polyposis secondary to thermal inhalation injury. Chest 75:643–645, 1979.

35. Williams, D. O., Vanecko, R. M., and Glassroth, J.: Endobronchial polyposis following smoke inhalation. Chest 84:774–776, 1983.

36. Perez-Guerra, F., Walsh, R. E., and Sagel, S. S.: Broncheolitis obliterans and tracheal stenosis. Late complications of inhalation burn. J.A.M.A. 218:1568–1570, 1971.
37. Seggev, J. S., Mason, U. G., III, Worthen, S., Stanford, R. E., and Fernandez, E.: Broncheolitis obliterans. Report of three cases with detailed physiologic studies. Chest 83:169–174, 1983.
38. Teplitz, C.: Pathogenesis of *Pseudomonas* vasculitis and septic lesions. Arch. Path. 80:297–307, 1965.
39. Pruitt, B. A., Jr., Flemma, R. J., DiVincenti, F. C., Foley, F. D., and Mason, A. D., Jr.: Pulmonary complications in burn patients. J. Thor. Cardiovasc. Surg. 59:7–20, 1970.
40. Spotnitz, M., Rudnitzky, J., and Rambaud, J. J.: Melioidosis pneumonitis. Analysis of nine cases of a benign form of melioidosis. J. Amer. Med. Assoc. 202:130, 1967.
41. Cox, C. D., and Arbogast, J. L.: Melioidosis. Amer. J. Clin. Path. 15:567–570, 1945.
42. Brundage, W. G., Thuss, C. J., Jr., and Walden, D. C.: Four fatal cases of melioidosis in U.S. soldiers in Vietnam. Amer. J. Trop. Med. & Hygiene 17:183–191, 1968.
43. Flemma, R. J., DiVincenti, F. C., Dotin, L. N., and Pruitt, B. A., Jr.: Pulmonary melioidosis: A diagnostic dilemma and interesting threat. Ann. Thoracic Surg. 17:491–499, 1969.
44. Greenawald, K. A., Nash, G., and Foley, F. D.: Acute systemic melioidosis. Autopsy findings in four patients. Amer. J. Clin. Path. 52:188–198, 1969.
45. Patterson, M. C., Darling, C. L., and Blumenthal, J. B.: Acute melioidosis in a soldier home from South Vietnam. J. Amer. Med. Assoc. 200:117–121, 1967.
46. Piggott, J. A., and Hochholzer, L.: Human melioidosis. A histopathologic study of acute and chronic melioidosis. Arch. Path. 90:101–111, 1970.
47. Mendelson, C. L.: The aspiration of stomach contents into the lungs during obstetric anesthesia. Amer. J. Obstet. Gynec. 52:191–205, 1946.
48. Spencer, H.: Pathology of the Lung, 3rd. ed. (2 vol.), Philadelphia, W. B. Saunders Co., 1977, p 171.
49. Panke, T. W. Personal observation, burn autopsies August 1976 through July 1978.
50. Coleman, J. B., and Chang, F. C.: Pulmonary embolism. An unrecognized event in severely burned patients. Amer. J. Surg. 130:697–699, 1975.
51. Warden, G. D., Wilmore, M. W., and Pruitt, B. A., Jr.: Central venous thrombosis: A hazard of medical progress. J. Trauma 13:620, 1973.
52. Nash, G., Foley, F. D., and Langlinais, P. C.: Pulmonary interstitial edema and hyaline membranes in adult burn patients. Human Path. 5:149–160, 1974.
53. Cooney, W., Dzuira, B., Harper, R., and Nash, G.: The cytology of sputum from thermally injured patients. Acta Cytol. 16:433–437, 1972.
54. Spencer, H.: *In* Pathology of the Lung, 3rd. ed. (2 vol.) Philadelphia, W. B. Saunders Co., 1977, p 171.
55. Teplitz, C.: Pathology of burns and fundamentals of burn wound sepsis. *In* Artz, C. P., and Moncrief, J. A. The Treatment of Burns, 2nd. ed. Philadelphia, W. B. Saunders Co., 1969, p. 54.

56. Sevitt, S.: Reactions to Injury and Burns and Their Clinical Importance. Philadelphia, Lippincott, 1974, pp 53–56.
57. Teplitz, C.: The pathology of burns and the fundamentals of burn wound sepsis. *In*: Artz, C. P., Moncrief, J. A., and Pruitt, B. A. Burns. A Team Approach, 1st. ed. Philadelphia, W. B. Saunders Co., pp 45–94, 1979.
58. Langlinais, P., and Panke, T. W. Unpublished data.

Chapter 11

HEMATOPOIETIC SYSTEM

The hematopoietic system of burn patients has many morphologic and functional changes that may have diagnostic or, in some cases, prognostic importance. (See Table 11–1.) These alterations reflect the profound physiologic and pathobiologic responses that attend and follow severe trauma in general, and burns specifically. In addition, several poorly understood hematologic changes are described, hopefully to stimulate additional investigative activity in these areas. This chapter represents our interpretation of these changes as we have observed them in 75 autopsies performed over a 2-year period and our review of the literature and other selected autopsies from our files.

BONE MARROW AND PERIPHERAL BLOOD

Data on the alterations in bone marrow of burn patients have been scanty. In general, a marked, progressive, granulocytic hyperplasia or a trilineage (erythrocytic, granulocytic, megakaryocytic) hyperplasia occurs in the severely burned patient. The degree of bone marrow hyperplasia may be profound (80–100 percent). (See Fig. 11–1.) Although a marked hematopoietic response in the bone marrow may occur in the elderly, it is most commonly seen in children and young adults. Normal cellularity of the bone marrow beyond the seventh to tenth postburn day in a severely burned patient suggests suppression of the bone marrow, and such patients should be evaluated for possible drug toxicity or gram-negative bacillary infection. When the combined stresses of severe burns and infection are sustained, a striking granulocytic hyperplasia of bone marrow may occur, and cellular maturation beyond the myelocyte may be markedly diminished. The maturational arrest of bone marrow elements in the terminal phase of *P. aeruginosa* septicemia has been demonstrated experimentally.[1,2]

Table 11–1 *Summary of Hematologic Alterations in Burn Patients*

Erythrocytes—anemia due to

  Acute postburn intravascular hemolysis
  Bone marrow depression by
    drugs
    infection
  Hemorrhage

Granulocytes

  Early granulocytopenia
  Depressed bactericidal activity
  Depressed chemotaxis
  Depressed NBT reduction
  Depressed phagocytosis

Monocytes

  Early monocytopenia
  Diminished phagocytosis

Reticuloendothelial system

  Decreased clearance of colloidal fat
  Impaired degradation of immunogens
  Diminished bactericidal capacity

Depressed opsonic activity

  Impaired C1, C2, C3, C4, C5, C3 activator
  Reduced C3 conversion
  Decreased properdin concentration

Humoral (B-lymphocytic) changes

  Depression of immunoglobulin, especially IgG

Altered cell-mediated (T-lymphocytic) immunity
  Decreased number of T-lymphocytes
  Depressed skin test responsiveness
  Prolonged allograft "take"
  Altered PHA transformation
    Increased early postburn period
    Decreased 5–7 days after burn

Drug-related hematopoietic impairments

  Immunosuppression by steroids
  Altered leukocytic function by antibiotics
  Impaired neutrophil candidicidal activity by sulfa drugs

DIC—See Table 11–4

Key: PHA—phytohemagglutinin; NBT—Nitroblue tetrazolium; C1–C5—components of complement; DIC—disseminated intravascular coagulopathy.

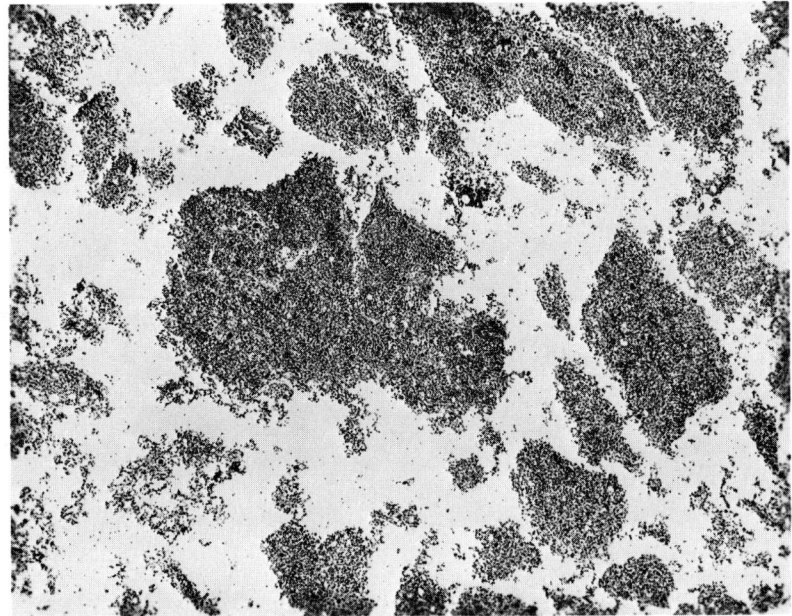

Fig. 11–1. Bone marrow aspirate particles showing marked hypercellularity. (× 18)

In severely burned patients, *megakaryocytes* are often enlarged and have hyperlobated nuclei (16 or more lobes) after the first postburn week (Fig. 11–2); this change suggests accelerated megakaryopoiesis[3,4] and thrombopoiesis and is associated with increased numbers of megakaryocytes after the second postburn week. Selective suppression of megakaryocytic proliferation has been demonstrated experimentally in *Pseudomonas*-infected animals.[2]

A mild to moderate degree of *hemosiderosis* in bone marrow, spleen, and liver is commonly seen in severely burned patients (Fig. 11–3). Hemosiderosis is usually associated with the transfusion of a large volume of blood (often more than 100 units) in a short period of time (often less than 3 weeks).

Anemia

Anemia is common in patients with burns on greater than 30 percent of the total body surface[5,6] and is seen in virtually all patients with burns on greater than 40 percent.[5] The causes of anemia are multifactorial and are

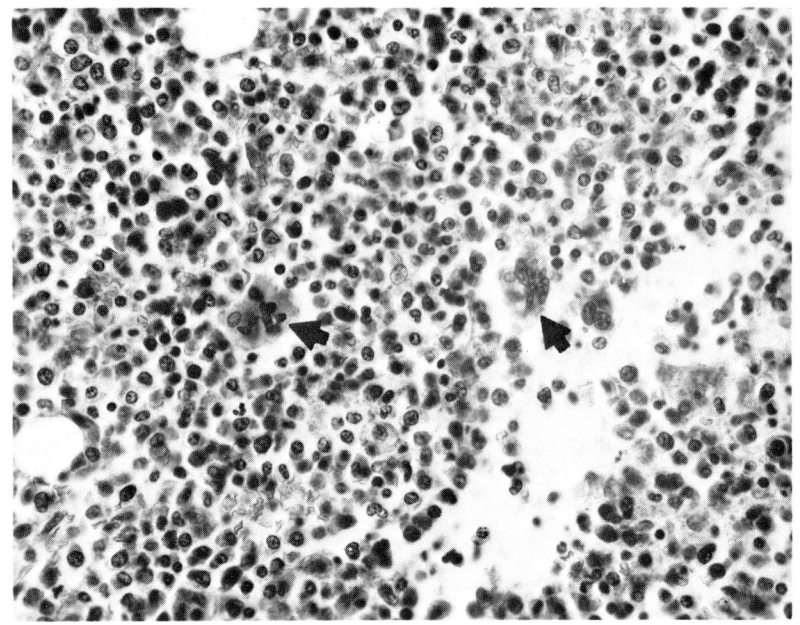

Fig. 11–2. Hypersegmented megakaryocytes (arrows) from a tissue section of bone marrow particles. (× 270)

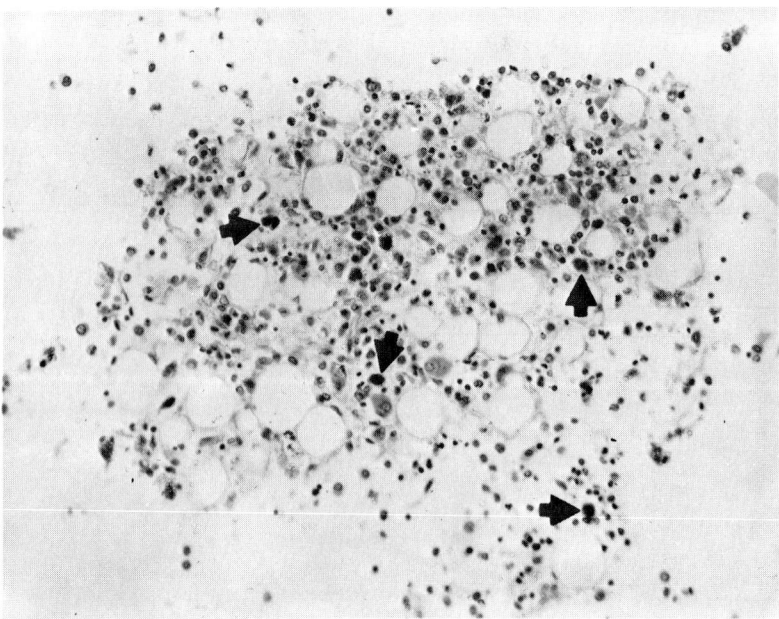

Fig. 11–3. Photomicrograph of postmortem bone marrow tissue section demonstrating an increased number of hemosiderin-laden macrophages (arrows). (× 135)

listed in Table 11–2. In children, the type of thermal injury is important. Flame burns are often larger and deeper than scald burns and more often produce anemia necessitating transfusions. A burn-initiated intravascular hemolytic process which occurs only during the immediate postburn period appears to account for loss of 10–15 percent of the preburn erythrocytic mass.[7,8] Although the exact mechanisms by which this hemolysis occurs remain speculative, it appears that some erythrocytes in the immediate postburn period are usually fragile, that aged red cells are initially affected, and that a variety of biochemical alterations may be present.[8–10] The appearance of spiculated erythrocytes has recently been associated with decreased levels of plasma lipoproteins.[11] Mildly to moderately damaged erythrocytes are predominantly sequestered in the spleen. Normally, mildly damaged red cells may be subsequently released.[12] Severely injured erythrocytes are usually phagocytized by the liver.[13] Other sites for erythrocyte sequestration may include traumatized tissue or burn sites, as well as dilated vessels of the lungs, liver, and lower extremities.[10,12] When microspherocytes initially constitute more than 2 percent of erythrocytes in the peripheral blood (Fig. 11–4), a severe erythrocytic loss may be anticipated. Large numbers of these microspherocytes may spuriously elevate the platelet count (pseudothrombocytosis).[14] One morphologic manifestation of accelerated hemolysis in severely burned patients is seen as abundant reticuloendothelial iron in liver, spleen, and bone marrow.

Later in the postburn course, several factors may be responsible for anemia (Table 11–2). In the burn patient, superimposed infection is associated with a hyporesponsive bone marrow and a depressed erythropoietin level.[7] The hematocrit may plummet 10–20 percent over a 24-hr period in patients with large burns heavily infected with *P. aeruginosa*,[15] and this abrupt erythrocyte loss is probably related to a superimposed hemolytic anemia.

Although data about erythropoietin levels in burn patients are conflicting,[7] current studies indicate that the erythropoietin level is elevated in the nonseptic burn patient in spite of marrow erythroid hyporesponsiveness.[5,16] Decreased erythroid production is indicated by the lack of medullary erythroid hyperplasia, absence of reticulocytosis and ^{59}Fe incorporation.[5,7] Several indices also suggest superimposed poor iron utilization in severely burned patients (Table 11–3), which may simulate anemia of chronic disease. Loss of such iron-binding protein through the burn wound may partially explain diminished iron-binding capacity reported in burn patients.

Evidence of marrow suppression has been reported in both burn patients[17] and experimental burn models.[18] Bone marrow erythroid stem cells gradually diminished following thermal injury; this decline parallels the appearance of serum substance(s) which inhibit normal red cell colony formation. Because the bone marrow lacks increased erythropoiesis and

Table 11–2 *Causes of Anemia in the Burn Patient[a]*

Hemorrhage related to

    Initial trauma
    Surgical procedures
      Amputations
      Debridement/escharectomy
      Intubation/bronchoscopy
      Tracheostomy
    Diagnostic phlebotomies
    Gastrointestinal
      Ulcers
      Ischemic enterocolitis

Hemolytic process due to

    Burns
    Drugs
    Infection

Aregenerative phase in bone marrow

    Burn wound toxins
    Bacterial infections

Disturbed iron metabolism

    Iron deficiency anemia
    Anemia of chronic disease

Varying degrees of malnutrition

Major impairment in renal function

Disseminated intravascular coagulopathy

Drug-induced

    Bone marrow suppressant
    Heparin, excess

---

[a] For further information, see references 5–7, 15, 24, and 105–107, from which this table was compiled.

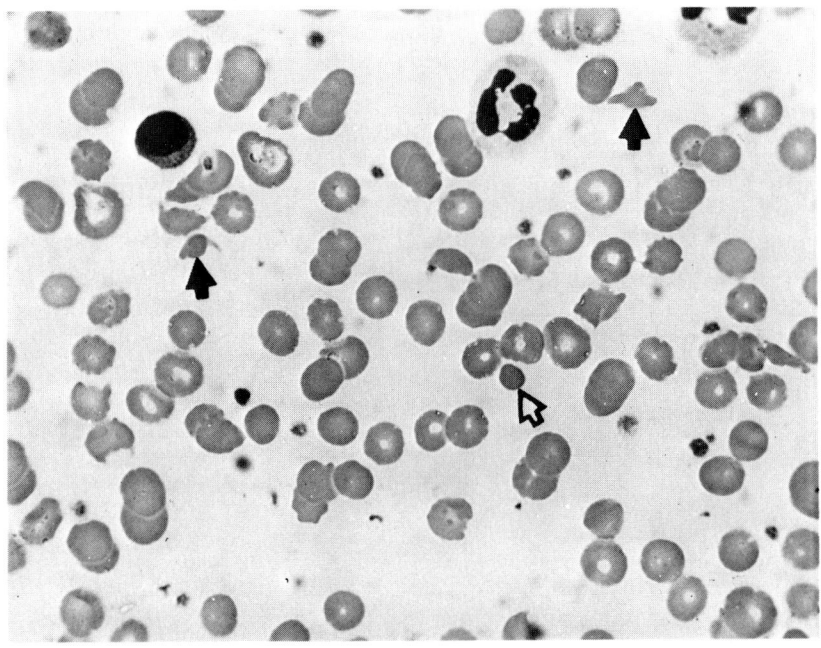

Fig. 11–4. Photomicrograph of peripheral blood with a microangiopathic hemolytic changes consisting of microspherocytes (open arrow) and schistocytes (closed arrows). (× 675)

Table 11–3 *Indices Suggesting Poor Iron Utilization in the Burn Patient*

Decreased iron 59 incorporation into erythrocytes

Decreased serum iron and serum iron-binding capacity

Decreased sideroblasts in bone marrow

Increased iron stores in reticuloendothelial cells

reticulocytosis is minimal at best, the process is probably best termed an aregenerative anemia. Undoubtedly, the above iron studies suggesting anemia of chronic disease indicate several suppressive factors acting simultaneously on erythropoiesis.

Burn-Related Changes in Granulocyte Levels

A marked granulocytosis appears shortly after thermal injury.[19-21] Granulocytosis peaks during the first postburn day and returns to "normal" during the second to third postburn day. In severely burned patients, granulocytopenia is prolonged and is related to a marked peripheral margination of granulocytes in the microcirculation of burned tissue and subsequent emigration into injured tissue.[22] A mild granulocytosis may occur again between the fifth to tenth postburn days.[21]

Gram-negative infections have been observed in burn patients, especially in *granulocytopenia*. Low levels of colony stimulating factor for granulopoiesis may partly explain the neutropenia that often accompanies septicemia in the burn patient.[23]

In septicemia, a dissemination of microorganisms may also lead to widespread visceral injury and necrosis. In severely burned patients, the bone marrow reserve is already compromised by the burn wound's demand for the granulocytic cells. At some point, the marrow is probably unable to respond adequately to the rapidly spreading infection.

Emerson et. al. prospectively evaluated the hematologic changes in 61 consecutive patients with septicemia at Boston City Hospital.[24] They described nonspecific hematologic changes associated with septicemia, including leukocytosis (31 percent) or leukopenia (16 percent). Patients with gram-negative septicemia more often failed to develop leukocytosis than patients with other types of infections. Intracytoplasmic vacuolation, toxic granules, and Dohle's bodies were often observed in granulocytes on the first day of septicemia (Fig. 11–5). Twelve percent of patients had thrombocytopenia (nadir at 2–4 days) solely related to septicemia. A modest fall in the hematocrit was noted in 85 percent of patients, and 79 percent also had a decreased number of reticulocytes. Ten percent of patients had overt hemolytic anemia. In the burn patient with gram-negative sepsis, "colony stimulating factor" for granulocyte production is significantly decreased.[25] This represents one additional mechanism for granulocytopenia in the burn patient.

Rarely, drug toxicity may cause granulocytopenia which is reflected in the bone marrow by relatively few granulocytic precursors.[21] A transient absolute neutropenia has been described in patients with topical silver sulfadiazine applied to the burn wound.[26-29] However, such reports appear to be anecdotal since controlled studies in a large group of patients (304)

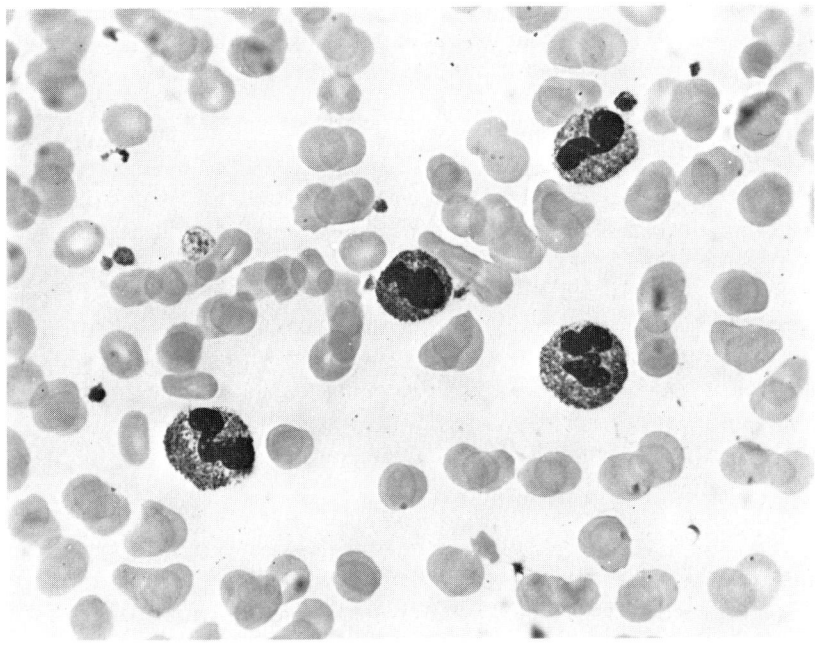

Fig. 11–5. Photomicrograph of peripheral blood neutrophils demonstrating toxic granules in the cytoplasm. (× 675)

failed to demonstrate cases of leukopenia related to silver sulfadiazine.[30] Recently, Onyiah reported leukopenia occuring in two burn patients after plasma infusion.[31]

Functional Defects in Granulocytes

A reversible chemotactic deficit has been demonstrated by in vitro studies of granulocytes. Deficiency of a heat labile factor in sera of burn patients accounts for this chemotactic deficit, the severity of which appears to be predictive of ultimate mortality.[32-34] In addition, decreased granulocytic lysosomal content,[35] decreased microbial activity,[36-39] diminished oxygen consumption following phagocytic stimulation,[40] and decreased nitroblue tetrazolium test response[41] in granulocytes have been identified. Therefore, burn patients, especially with concurrent gram-negative septicemia, are compromised not only by leukopenia but also by several well-characterized functional deficiencies in the residual granulocytes.

Defective leukocyte function (chemotaxis, phagocytosis, bacterial killing) also appears to correlate with severity of *hyperglycemia*. Insulin appears to improve leukocyte function to the extent that insulin controls hyperglycemia.[42] Although insulin has no direct effect on collagen sythesis, complete wound healing *does* appear to depend on a quantitatively "normal" inflammatory response which appears to be insulin dependent. If the expected inflammatory response occurs, wound healing is timely and complete. If significant hyperglycemia persists beyond 10 days, late administration of insulin may control hyperglycemia, but wound healing remains depressed.[42]

Experimentally, the addition of glucose to granulocytes depressed granulocytic function.[43,44] This in vitro observation raises questions about neutrophilic function in burn patients not only with diabetes mellitus, but also with pseudodiabetes mellitus (See Chapter 17).

Reticuloendothelial System

The reticuloendothelial system may be divided simplistically into fixed and mobile compartments. Those reticuloendothelial cells which are "fixed" are exemplified by Kupffer's cells of the liver sinuses and the sinusoidal lining cells of the spleen and lymph nodes. Monocytes may be viewed as "mobile" or "wandering" reticuloendothelial cells which migrate in the blood stream to a site of injury. At that point, they traverse the microvascular wall and function essentially as histiocytes or macrophages in tissue.[45] Several factors associated with tissue destruction and inflammatory reactions are responsible for the chemotactic attraction to the site and subsequent activation of macrophages.[46]

The reticuloendothelial system plays a profoundly varied and vital role in the body. Some of its functions include phagocytosis of bacteria, cell debris, and toxic materials; removal of certain activated blood clotting factors; activation of fibroplasia; stimulation of collagenase; interaction with lymphocytes; and participation in hemoglobin catabolism and iron metabolism.[47,48] If the integrity of the macrophage is altered, wound healing is remarkably retarded.[49] In the burn patient, the reticulendothelial system has a diminished capacity to phagocytize and to destroy or otherwise detoxify materials.[47,48,50–52] A similar functional suppression of monocytes was seen in an experimental mouse model.[53] Certain antibiotics, anesthetics, and analgesics have been shown experimentally to depress phagocytosis.[47] Splenic sequestration or erythrocytes may also play some role in the increased susceptibility to sepsis in the burn patient.[54]

In burned experimental animals, Alexander demonstrated a diminished bactericidal capacity of peritoneal macrophages and decreased serum and cellular lysozyme levels during the first 24 hr post burn.[50] At 24 hr, the activity of peritoneal macrophages is drastically diminished and remains abnormal for 9 days or more following thermal injury.[51]

Severely burned patients have an early and continuing depression in the level of circulating monocytes. In vitro studies have documented a severe depression of phagocytic function in monocytes of burn patients. Premature release of immature monocytes from the bone marrow partially explains this depressed monocytic function.[55] Monocytes are considered to be a circulating form of tissue macrophages.[46] Since they remove tissue debris and perform a variety of other functions mentioned above, they are probably important in wound healing.

Platelets

Thrombocytopenia is commonly seen in patients with severe burns; platelets generally reach their nadir at 3 days post burn.[56] Initially, thrombocytopenia is secondary to hemodilution by intravenous fluids. By day 3–4, intravascular platelet aggregation becomes the significant factor.[57] The bone marrow has no platelet reserve, and large numbers of platelets may be consumed to support the integrity of the microvasculature in areas of extensive thermal injury.[58] A reactive thrombocytosis usually occurs 15–30 days post burn.[56,59] Pronounced or prolonged thrombocytopenia post burn or poor or absent thrombocytosis is an ominous sign in the burn patient.[59] Significant thrombocytopenia is more frequently observed in severely burned patients and/or as a complication of superimposed infection.[60] Experimentally, infection impairs megakaryopoiesis and hence platelet production, but does not appear to affect platelet survival.[2]

Recently, Tian-Min et al.[61] have described degenerative changes and phagocytosis involving the megakaryocyte in injured dogs at both light and electron microscopic levels. Thrombocytopenia was closely correlated to changes in megakaryocytes. Degeneration and phagocytosis of megakaryocytes was noted with burn injuries but not blast injuries. These findings suggest an additional mechanism for thrombocytopenia in the burn patient.

Recently, platelet functional studies have been performed which indicate that platelets aggregate spontaneously in 15 percent of burn patients (usually, platelets require a stimulus, e.g., ADP, for platelet aggregation). Heparin also promotes platelet aggregation in 60 percent of burn patients. These platelet aggregation findings were unrelated to platelets count or sampling time.[62] Whether or not these in vitro studies have similar in vivo significance in the burn patient is not known.

IMMUNE SYSTEM CHANGES IN BURN PATIENTS

Abnormalities in the Humoral Immune System

The frequency of serious infections in burn patients suggests deficiencies in the humoral as well as the cellular components of the immune system.[63] Following severe thermal injury, there is a marked depression in serum immunoglobulin (Ig), most conspicuously IgG and IgA. IgM is mildly depressed but returns to normal by the first postburn week and becomes significantly elevated during the second postburn week. During the second postburn week, the levels of IgA, IgG, and IgE return to normal and IgD levels do likewise by the end of the first postburn month.[64] Marked elevations of IgM suggest invasive mycotic infection of the burn wound.[65]

In order to interpret changes in serum immunoglobulins in the burn patient, particularly hypogammaglobulinemia, one must consider the following:

1. Rate of synthesis and catabolism of immunoglobulin
2. Rate of extravascular loss of immunoglobulin to burned areas and/or redistribution of immunoglobulin in edematous tissue
3. Size of the immunoglobulin molecule.

Albumin has been intensively studied.[66] Despite elevated albumin synthesis, hypoalbuminemia persists for up to 60 days following severe thermal injury.[67] Large amounts of albumin have been measured in blister fluid. Albumin is a relatively small molecule and hence passes readily through minimally damaged or reactive blood vessels. In contrast, very little IgM traverses damaged blood vessels. The serum level of IgM is only mildly decreased post burn and, in most cases, returns to normal within 7 days.[64,67] In addition to IgM, other macromolecules, such as α-2-macroglobulin, tend to remain within blood vessels. Hence, serum levels of these macromolecules are not severely altered following thermal injury, and content of IgM or α-2-macroglobulin within blister fluid is negligible in comparison with the high levels of albumin and IgG.[64,67]

IgG is a molecule of intermediate size and is conspicuously diminished following thermal injury. Levels of IgG return to normal as early as 2 weeks postburn.[65,67] Transfusions of gamma globulin to the burn patient are associated with a rapid return of IgM levels to normal (IgM remains intravascular) in contrast to transfused IgG, which probably passes into the extracellular fluid in large amounts, leaving blood levels depressed for prolonged periods.[64]

As suggested by the above, depression of humoral immunity may be more apparent than real. Selected low immunoglobulin levels may reflect only excessive vascular "leakage." Current data now suggest that the humoral immune system responds with an increased antibody production in thermally injured animals.[68,69]

Cell-Mediated Immune Response

Following severe thermal injury, there is an immediate and unexplained lymphocytopenia (especially T-lymphocytes), decreased allograft rejection, prominently impaired mixed lymphocyte culture stimulation, depressed lymphocyte response to mitogens, activation of suppressor T-lymphocytes, depletion of thymus dependent areas of lymphoid tissue, and impaired lymphocyte responsiveness to inflammation as judged by skin windows.[70-77] This immune depression lasts for approximately 2 weeks.[78] Early excision of the burn wound has been reported not to reverse immunosuppression of T-cell function,[79] but preliminary studies utilizing thymosin therapy are promising.[78] One current theory[80,81] explains the diminished T-cell activity by the emergence of a suppressor T-cell population of lymphocytes. This activity of suppressor T-lymphocytes has been diminished in vitro by the administration of certain polynucleotides; successful therapeutic manipulation in vivo could lead to enhanced cell-mediated immune response.[82] A more recent study indicates that helper T-lymphocytes are decreased, and this largely explains the decreased helper-suppressor ratio.[83] Obviously, additional studies are needed for clarification of the immune response.

Complement (C) System

Decreased opsonic activity in the burn patient is related to a diminished level of C1q, C4, C2, C3, C5, and C3 activator prior to and during septic episodes.[84,85] Initial depression of complement levels is dependent on the severity and extent of the burn, and is thought to be related to a "leakage" phenomenon. Restoration of the complement levels is also retarded when burns are more extensive.[86]

MORPHOLOGIC CHANGES IN LYMPHOID ORGANS OF THE BURN PATIENT

Although the function of the immune system in the burn patient has been studied in detail, very little specific information is available about the morphologic changes in the lymphopoietic system of these patients. Our extensive morphologic evaluation of the lymphopoietic system in 72 patients has demonstrated a spectrum of changes which often can be correlated with the clinical history of the burn patient.

Spleen

The complex and time-related morphologic changes are most typically seen in the spleen; they are presented in a summary fashion below by analyzing specific components of the reaction as they sequentially appear in the postburn period. The only splenic change during the first 24 hr post burn is the formation of pale staining, spherical areas in the splenic white pulp (Fig. 11–6) in some burn patients. These appear at low magnification to be reactive lymphoid follicles ("follicular hyperplasia"), but closer scrutiny demonstrates only a few, small to intermediate sized lymphocytes, a variable number of phagocytic histiocytes, and abundant cellular (nuclear) debris in these pale areas (Fig. 11–7). Increasing numbers of histiocytes sometimes associated with homogeneous, eosinophilic, interstitial deposits are seen in the necrotic foci within the next 24–48 hr. These nonspecific *foci of lymphonecrosis* in burn patients usually do not evolve into typical reactive germinal centers. Most of the literature suggests that their formation is

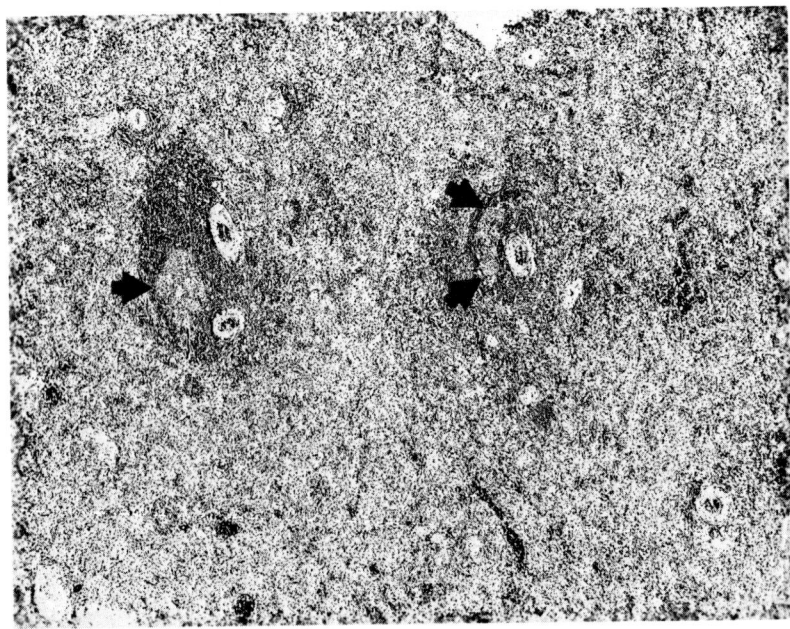

Fig. 11–6.   Photomicrograph of spleen showing pale areas of "necrosis" within the white pulp (arrows). (× 27)

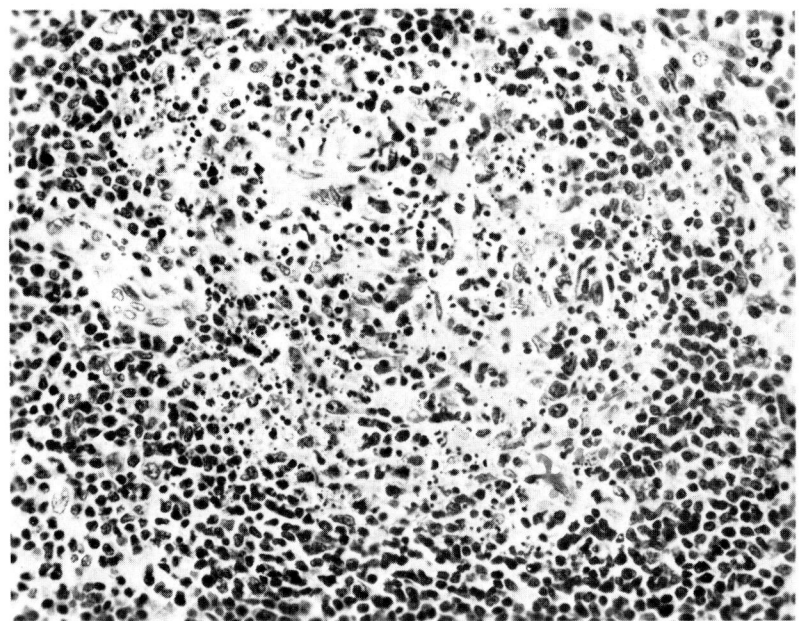

Fig. 11–7. Higher power of Figure 11–6 demonstrating polymorphonuclear leukocytes, histiocytes, and amorphous debris distributed among residual lymphocytes. (× 270)

confined to the immediate postburn period. However, in our review, we observed similar necrotic foci several days post burn in a small number of burn patients. This late development may be caused by an exacerbation of stress such as systemic infection, blood loss, or other serious complications.

Germinal centers (true follicular hyperplasia) are generally found in the splenic white pulp of young children and adolescents, and occasionally in adults sustaining burns. Undoubtedly, the development of such germinal centers precedes the thermal injury, since germinal centers are found in a fully developed state in infants who have died shortly after sustaining thermal injury. They are characterized by an outer mantle of small lymphocytes and a pale central area which is composed of many lymphocytes of variable size intermingled with a few histiocytes and scattered mitoses.

During the next stage, large lymphocytes (immunoblasts), presumably due to *lymphoid activation,* form in the splenic white pulp (Fig. 11–8). These large lymphocytes are seen as early as the third postburn day, but are more readily identified by the ninth to fourteenth day. If the stimulus is severe, cells larger than the typical immunoblasts appear; these have abundant amphophilic cytoplasm, a large nucleus, and one or more large nucleoli (Fig.

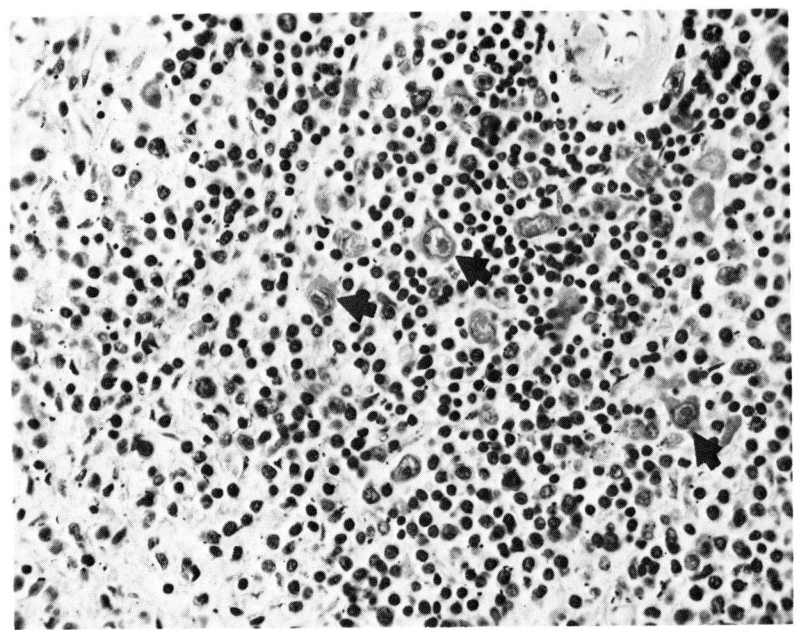

Fig. 11–8. Photomicrograph of spleen with marked lymphoid hyperplasia. Abundant immunoblasts (arrows) are evident. (× 270)

11–9). Some cells of this type simulate the mononuclear variant of a Reed-Sternberg cell.[87] These large atypical lymphocytes are usually confined to the splenic white pulp and are noted in both the spleen and lymph nodes of approximately 15 percent of autopsied burn patients who often have had severe infections.

Further morphologic evolution in this immune process is the formation of typical reactive *plasma cells* and larger atypical plasma cells initially in the white pulp. Subsequently, these cells are found in large numbers in the cords and sinuses of splenic red pulp. Teplitz[88] noted that 34 percent of burn patients die with a marked plasma cell infiltrate in the spleen. Usually, these prominent plasma cell infiltrates occur after the second postburn week. Marked lymphocytic and plasma cell infiltrates may extend into the wall of trabecular veins (Fig. 11–10). This change appears to be part of a florid lymphoid activation.

Teplitz [88]has also reported subintimal deposits of a *hyaline-like material* in splenic *arterioles* (Fig. 11–11). This material has some staining properties suggesting amyloid, lacks gamma globulin, and appears to be a mucopolysaccharide protein complex. The pathogenesis and significance of this finding are uncertain.

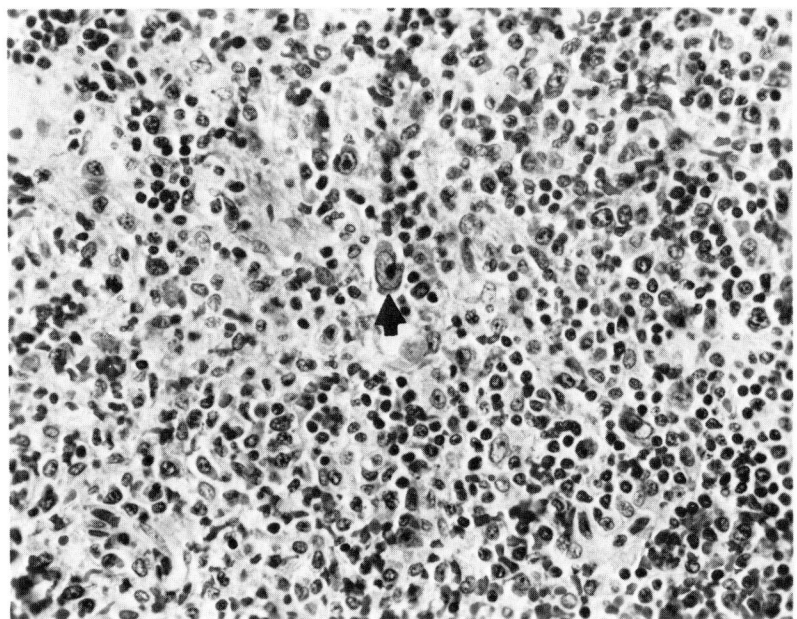

Fig. 11–9. Photomicrograph of spleen with marked lymphoid hyperplasia, including mononuclear variant of a Reed–Sternberg cell (arrow). (× 270)

Cyclic changes in splenic *eosinophils* have been described by Sevitt,[21] who felt that eosinopenia reflected adrenal cortical hyperactivity of severe stress. Such findings are clearly evident in our patients who die shortly after burns. A small number of patients have increased numbers of eosinophils which initially appear after the sixth postburn day and are associated with increased numbers of medium and large lymphocytes and plasma cells.

In burn patients with ongoing acute septicemia, many polymorphonuclear leukocytes fill the sinuses of the red pulp (Fig. 11–12). This process is termed *acute reactive splenitis*. No consistent alteration is seen in the white pulp.

So-called "lipogranulomas" are common in the spleen of burn patients and are usually seen at the interface of white pulp with red pulp. They are composed of loose aggregates of phagocytic histiocytes filled with vacuoles (Fig. 11–13). The significance of lipogranulomas is unknown.

Splenic extramedullary hematopoiesis is a common finding in most patients who survive for more than a week with severe burns and/or severe infectious complications. Components of extramedullary hematopoiesis include erythroid or granulocytic precursors or both (Fig. 11–14). A variable

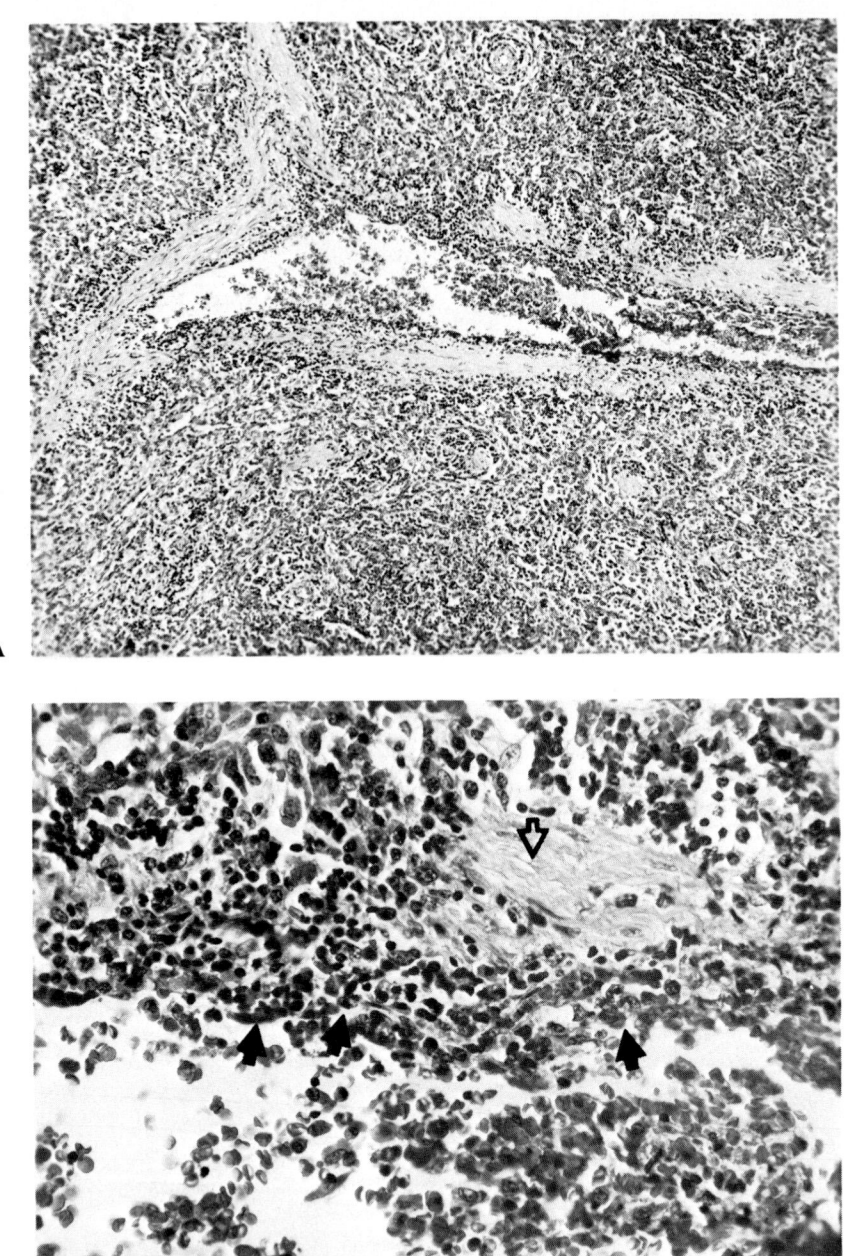

A

B

Fig. 11–10. (A): Photomicrograph of spleen showing a small trabecular vein which has a lymphocytic and plasma cell infiltrate extensively involving the wall. (× 67.5) (**B**): Higher power photomicrograph of same splenic vein demonstrating residual wall (open arrow) and marked subendothelial lymphocytic and plasma cell infiltrate (closed arrows). (× 270)

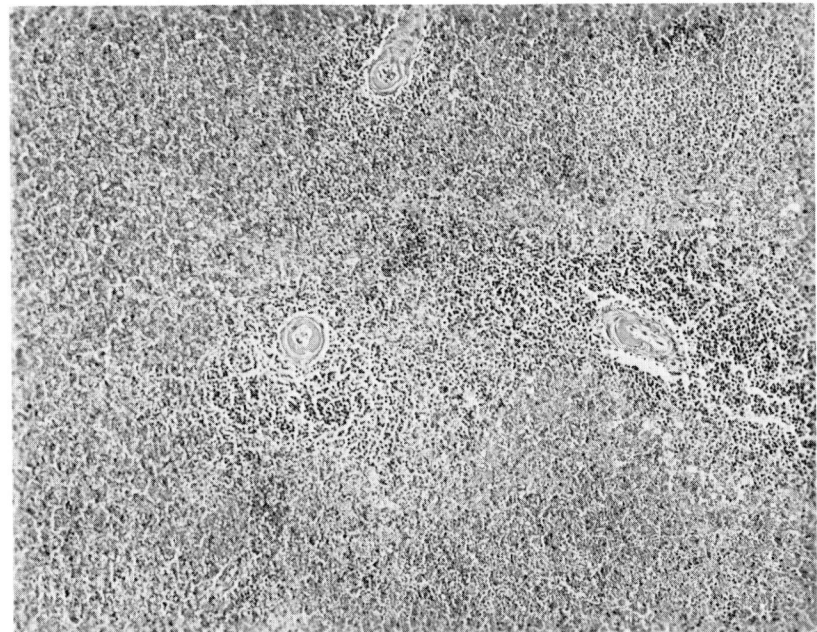

Fig. 11–11. Photomicrograph of spleen with hyaline arteriosclerosis within the white pulp. (× 67.5)

number of megakaryocytes may also be present. Acute, severe septicemia appears to be associated with extramedullary granulopoiesis.

"Metastatic" bacterial lesions, seen as small *abscesses,* are occasionally encountered in the splenic red pulp of burn patients (Fig. 11–15). These lesions usually have a perivascular distribution and probably result from hematogenous dissemination of microorganisms from the burn wound. Metastatic *P. aeruginosa* is the most common necrotizing microorganism encountered in the spleen of burn patients (Fig. 11–15).

Lymph Nodes

Morphologic alterations in the lymph nodes of burn patients reflect both local and systemic changes. In 72 consecutive autopsies on burn patients, lymph nodes from 11 anatomic sites were examined. (T. W. Panke, unpublished data) As one might anticipate, changes in the axillary, cervical, and inguinal lymph nodes reflected the extent of injury to the respective extremity and the presence of "invasive" infection of the burn wound. In some cases, the reactive lymphoid proliferation was marked. Necrosis was

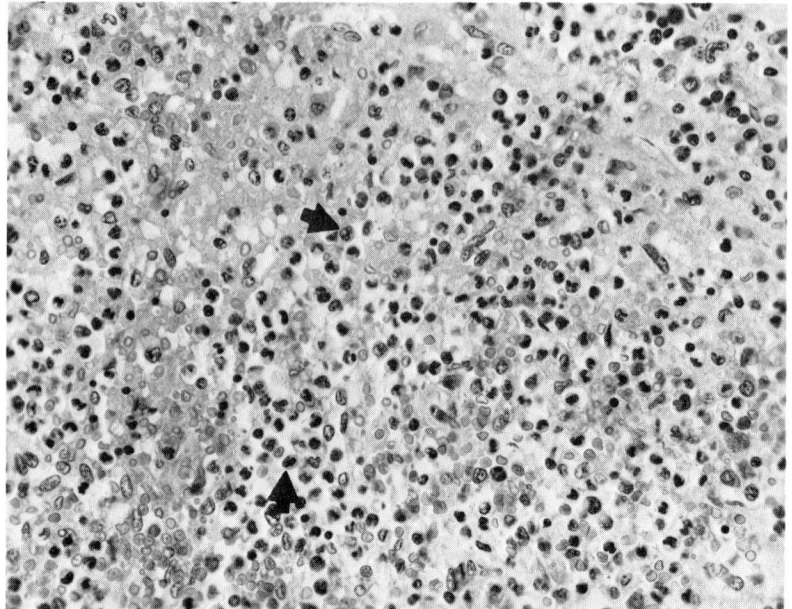

Fig. 11–12.   Acute splenitis. Polymorphonuclear leukocytes (arrows) fill red pulp sinuses and cords. (× 675)

observed in only one lymph node. This sparsity of necrotizing lesions in lymph nodes in patients with septicemia due to *P. aeruginosa* or *Staphylococcus aureus* contrasts with the findings in the lung, kidney, and heart, where necrotizing lesions are relatively common.

Recently, Linares et al.[89] have demonstrated morphologic changes in lymph nodes of children which persist up to 4 months after thermal injuries. Their studies revealed lymphoid depletion in the T-cell zones of lymph nodes in the early postburn period which corresponded with a decreased number of circulating T-lymphocytes in the peripheral circulation. They have also demonstrated lymphoid activation and the appearance of large methyl-green-pyronine-positive lymphocytes in follicular centers.[89]

Thymus

Many forms of stress cause prominent changes in the thymus. A marked lympholysis can be seen within 24 hr following severe injuries. With persistent stress over 2–3 days, the cortex may be markedly diminished due to an almost complete depletion of small lymphocytes. The medulla condenses somewhat and this reduction in size gives the impression that Hassall's corpuscles are increased in number. Many Hassall's corpuscles undergo necrosis (Fig. 11–16). The thymic medulla contains a few small and large lymphocytes, and other unrecognizable cells that have been called epithelial cells, reticulum cells, or nonphagocytic histiocytes.

Although large numbers of plasma cells and occasional germinal centers have been defined in the thymus of patients with a variety of collagen vascular and hypersensitivity diseases, such changes are rare in the burn patient.

The thymus has an excellent regenerative capacity, and dramatically enlarged thymuses have been demonstrated in a number of burned young children.[90] This observation is usually made, incidentally, with a chest x-ray and has been noted as early as 5 months after the burn. In one patient, the thymus weighed 130 g (normal for that age is less than 40 g). Microscopically, the thymic architecture was normal and the cortex was prominent. Also, thymic enlargement has been identified in some infants following the cessation of adrenal corticosteroid therapy.[91] It is speculated that when the early high levels of corticosteroids decline, a rebound thymic hyperplasia may occur.

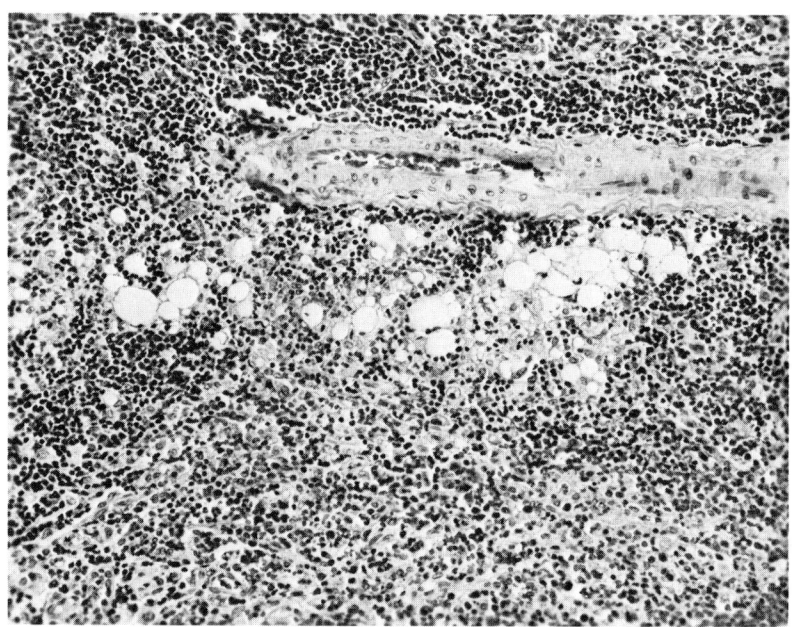

Fig. 11–13. Photomicrograph of lipogranulomas at edge of splenic white pulp which demonstrates the lipid-filled macrophages that form the lesion. ($\times$ 135)

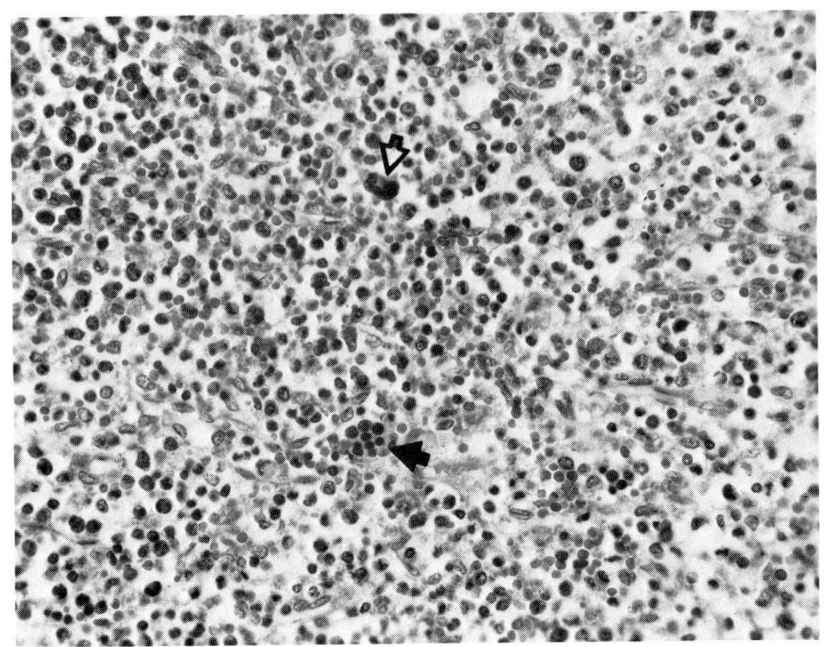

Fig. 11–14. Photomicrograph of spleen with extramedullary erythropoiesis (closed arrow) and megakaryopoiesis (open arrow). (× 270)

DISSEMINATED INTRAVASCULAR COAGULOPATHY

In the very early postburn course, there is probably a period of coagulation factor consumption which most likely occurs in the burn wound. Although a disseminated intravascular coagulopathy-like event may also be responsible, heparin does not alter these events.

Well-documented disseminated intravascular coagulopathy (DIC) is occasionally seen in the burn patient and may be associated with serious morbidity and mortality.[92,93] Common clinical manifestations of DIC in burn patients include generalized hemorrhage from the burn wound, bleeding at the site of venipunctures, and mucous membrane. (See Table 11–5.) Basically, the causes of DIC may be summarized as conditions relating to (1) gram-negative endotoxemia, (2) the release of substances having a tissue thromboplastin activity, or (3) the presence of intravascular antigen–antibody reactions (e.g., reaction from incompatible blood transfusion). Tissues with thromboplastin activity (especially injured brain tissue following head trauma) initiate intravascular coagulation by activating the extrinsic

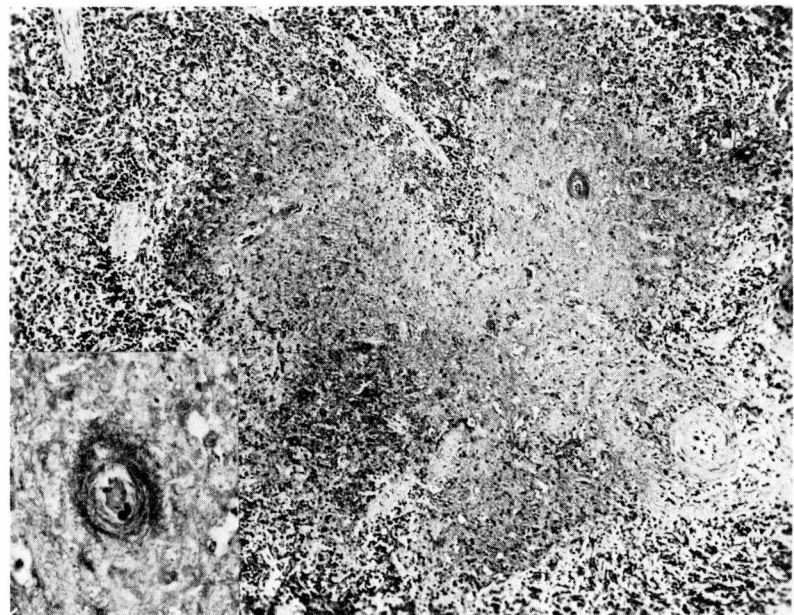

Fig. 11–15. Photomicrograph of necrotic area in spleen. (× 30) Higher power inset demonstrates vascular necrosis with minimal inflammatory reaction, typical of *Pseudomonas aeruginosa,* which was cultured from blood and other organs at postmortem examination in this patient. (× 675)

coagulation system (Fig. 11–17). Many gram-negative *endotoxins* damage platelets, granulocytes, and vascular endothelium, thereby activating Factor XII of the intrinsic coagulation system. The mechanism by which antigen–antibody complexes stimulate the production of DIC is not well known, but may also occur through activation of Factor XII and the intrinsic system.

During episodes of DIC, the reticuloendothelial (RE) function appears to be suppressed.[94] In addition to a loss of the usual phagocytic and bactericidal potential of macrophages, reticuloendothelial blockage may also attenuate immune response by inducing T-cell suppressors and by preventing T-cell recognition of, and activation by, bacterial antigens. Also, the RE system removes activated coagulation factors from the blood stream. If overwhelmed, the RE system is unable to remove activated coagulation factors from the circulation, and essentially permits intravascular coagulation to progress unopposed. The latter contributes to the evolution of DIC.

Although DIC usually develops coincident with *bacteremia* or *hypotension* in the burn patient,[92] a spectrum of clinical entities may be associated

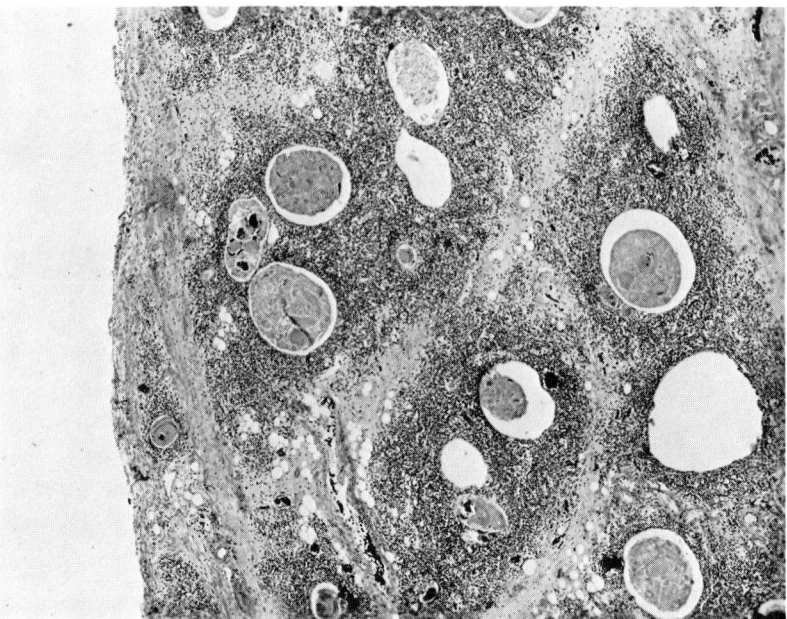

Fig. 11–16. Advanced thymic involution in child characterized by loss of thymic cortex and necrosis and cystic dilatation of almost all Hassall's corpuscles. (× 18)

with DIC (Table 11–4). It is important to note that 3–4 days after thermal injury coagulation factors are usually elevated in burn patients,[95] and normal levels of coagulation factors may be indicative of DIC. Recently, Braunstein et al.[96] reported a depression of antithrombin III or heparin cofactor (a natural, circulating anticoagulant) when *P. aeruginosa* septicemia complicates severe thermal injury. Low levels of heparin cofactor are associated with a hypercoagulable state. Heparin therapy should be adjusted to compensate for the level of heparin cofactor.[97] Low heparin cofactor levels and diminished levels of α-2-macroglobulin (another antithrombin) may explain why some patients develop DIC unresponsive to heparin therapy.[93,98]

There are a few clinical entities which may simulate DIC. One must consider *massive hepatic necrosis* whenever there is clinical suspicion of DIC. In addition to the obvious clinical and laboratory evidence of hepatic parenchymal destruction, Factor VIII, which is produced outside the liver, is usually elevated in liver disease while the other coagulation factors (all produced by the liver) are decreased. In acute DIC (Table 11–5) Factor VIII is "consumed" and is prominently decreased. A process of accelerated fibrin breakdown (*accelerated fibrinolysis*) may produce a clinical picture

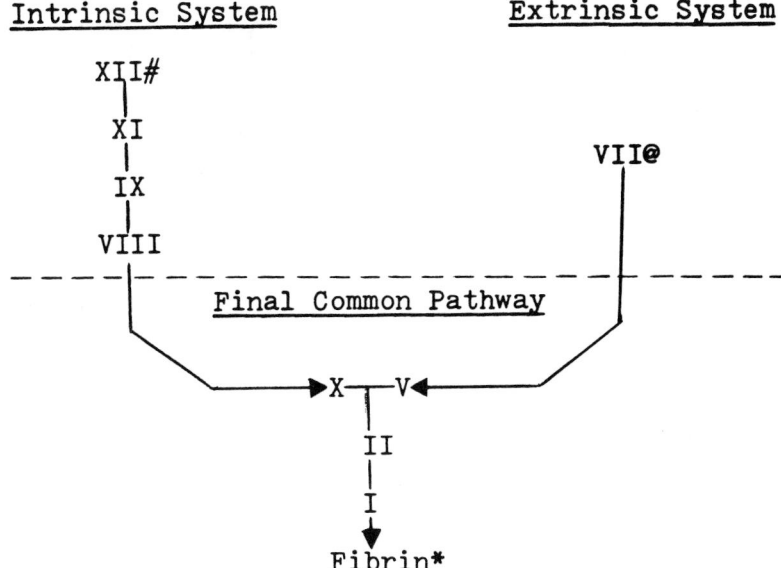

Fig. 11–17. Coagulation cascade. Hegamen factor—the intrinsic system and final common pathway is stimulated by a surface activating factor and partial thromboplastin (activated partial thromboplastin time). @ = Factor III–the extrinsic system and final common pathway are activated by tissue thromboplastin (prothrombin time); * = Fibrin is the end product of coagulation factor activation.

identical to DIC, but the two processes may be distinguished by certain laboratory tests (Table 11–6). Isolated fibrinolysis is unusual; more often, DIC will coexist with accelerated fibrinolysis.

Tissue Changes in Disseminated Intravascular Coagulopathy

Macroscopically, one should consider DIC when lack of hemostasis (Table 11–5) becomes a conspicuous clinical feature. Lesions vary from petechia to purpura; they are seen most readily on the external body surfaces, but may extensively involve deep tissues.

Thirty percent of burn patients at postmortem examination have DIC.[99] Eeles and Sevitt[99] divided pulmonary microvascular thromboemboli into two categories. *Capillary microthrombi* are composed of fibrin or platelets or both and are found in capillaries and arterioles. *Arterial microthrombi* are composed of the above plus erythrocytes, occur in small arteries, and are

Table 11–4   *Clinical Entities Associated with Disseminated Intravascular*
                 Coagulopathy in the Burn Patient

Tissue injury

  Large traumatic injury especially to the central nervous system
  Extensive burns
  Surgery
  Fat embolism secondary to bone fractures
  Acute pancreatitis

Intravascular hemolysis

  Incompatible transfusion
  Hemolytic uremia

Some viral and mycotic infections

Bacterial infections

  Pyogenic bacteria, miscellaneous
  Gram-negative bacteria

Foreign body—cannula

Severe liver disease, especially cirrhosis

Diabetic ketoacidosis

---

commonly associated with microvascular pulmonary thromboemboli originating in deep led veins. Recently, Wells et al. have utilized quantitation of
pulmonary megakaryocytes in addition to fibrin thrombi to document the
presence of a disseminated intravascular coagulopathy.[100]
    Microscopically, widespread microvascular thromboses are character-
istic in DIC. Commonly, kidney (especially renal glomeruli), lungs, liver,
adrenal, heart, etc. may be involved (Fig. 11–18).[101] Most burn patients have
occasional, small microvascular thrombi in the skin as well as in other sites.
Obviously, a few microvascular thrombi, without associated parenchymal
injury, are meaningless. We reserve the morphologic diagnosis of DIC for
cases where multiple organs contain microvascular thrombi which have
caused parenchymal necrosis.
    In addition to the thrombi in the microvasculature, similar thrombi may
occur in both large arteries and veins.[101,102] Coronary artery thrombosis has
previously been described in Chapter 9, Cardiovascular System. Similar
thrombi have been described in the renal artery, intracranial venous sinuses
and Rolandic vein, arteries and veins of all extermities, and aorta.[101-103]
Because the large vessels often lack significant atherosclerosis, it has

Table 11–5 *Clinical Manifestations Associated with Acute Disseminated Intravascular Coagulopathy*

Bleeding

  Especially around sites of intravascular cannulae

  At sites of femoral arterial puncture to obtain Specimens for blood gas determinations

Hemorrhagic necrosis of skin—related to fibrin thrombi

Waterhouse–Friderichsen syndrome—fibrin thrombi occluding adrenal medullary blood vessels, leading to hemorrhagic necrosis and marked hypotension

Shock out of proportion to blood loss

Table 11–6 *Comparative Laboratory Findings Distinguishing Between Primary Fibrinolysis and Disseminated Intravascular Coagulopathy (DIC)*

	DIC	Accelerated Fibrinolysis
Prothrombin time	Increase	Increase
Partial thromboplastin time	Increase	Increase
Fibrin degradation products	Increase	Increase
Fibrinogen	Decrease	Decrease
Platelets	Decrease	Normal
Microangiopathic hemolytic changes	Present	Absent
Euglobulin clot lysis	Slow	Rapid
Protamine sulfate test	Positive	Negative

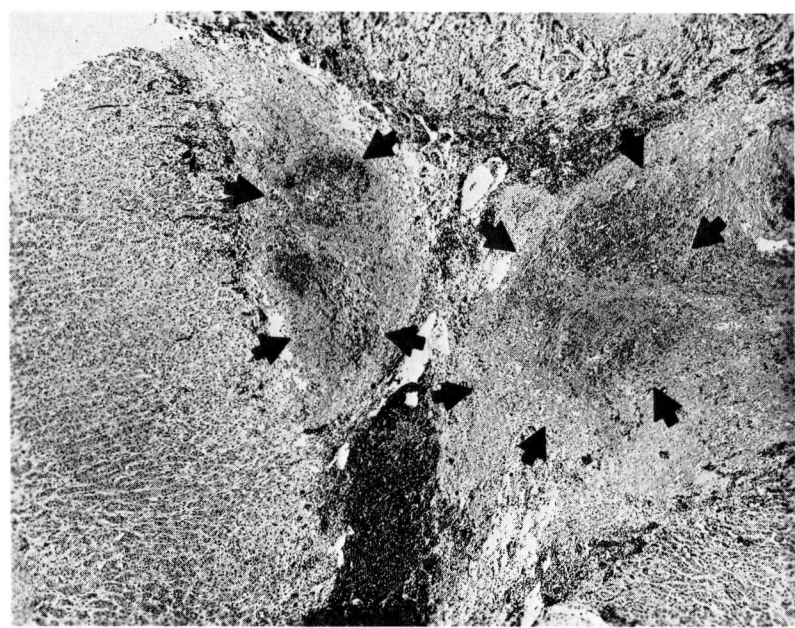

Fig. 11–18. Photomicrograph of adrenal gland demonstrating acute mixed thrombi occluding medullary veins (arrows). (× 18)

been speculated that thrombi developed in these large vessels also on the basis of a hypercoagulable state.[102]

Controversy now exists in the literature about the composition of these thrombi (fibrin versus platelets). Classically, DIC is associated with fibrin thrombi which are phosphotungstic acid–hemotoxylin (PTAH)-positive and stain negatively with periodic-acid-Schiff (PAS). However, platelet thrombi (PAS-positive and PTAH-negative) may also be seen in DIC.[104] The pathogenesis of these two types of thrombi may be quite different. The formation of platelet thrombi which may be related to spontaneous platelet aggregation,[62] has been seen in some burn patients. A postulated mechanism for platelet thrombi formation has been described in Chapter 12 under "Ischemic Enterocolitis."

References ——————————————————————————

1. Teplitz, C., Davis, D., Mason, A. D., Jr., and Moncrief, J. A.: *Pseudomonas* burn wound sepsis. I. Pathogenesis of experimental *Pseudomonas* burn wound sepsis. J. Surg. Res. 4:200–216, 1964.

2. Newsome, T. W., and Eurenius, K.: Suppression of granulocytic and platelet production by *Pseudomonas* burn wound infection. Surg. Gynecol. Obstet. 136:375–379, 1973.

3. Ebbe, S., Stohlman, F., Jr., Overcash, J., Donovan, J., and Howard, D.: Megakaryocyte size in thrombocytopenic and normal rats. Blood 32:383–392, 1968.

4. MacPherson, G. G.: Changes in megakaryocyte development following thrombocytopenia. Br. J. Haematol. 26:105–115, 1974.

5. Andes, W. A., Rogers, P. W., Beason, J. W., and Pruitt, B. A., Jr.: The erythropoietin response to the anemia of thermal injury. J. Lab. Clin. Med. 88:584–592, 1976.

6. Birdsell, D. C., and Birch, J.R.: Anemia following thermal burns: A survey of 109 children. Canad. J. Surg. 14:345–350, 1971.

7. Robinson, H., Monafo, W. W., Saver, S. M., and Gallagher, N. I.: The role of erythropoietin in the anemia of thermal injury. Ann. Surg. 178:565–572, 1973.

8. Boar, S.: Anaemia of burns. Burns 6:1–8, 1979.

9. Harris, R. L., Cottam, G. L., Johnston, J. M., and Baxter, C. R.: The pathogenesis of abnormal erythrocyte morphology in burns. J. Trauma 21:13–21, 1981.

10. Sevitt, S.: A review of the complication of burns, their origin and importance for illness and death. J. Trauma 19:358–368, 1979.

11. Harlan, W. R., Shaw, W. A., and Zelkowitz, M.: Echinocytes and acquired deficiency of plasma lipoproteins in burned patients. Arch. Int. Med. 136:71–76, 1976.

12. Sevitt, S.: Reactions To Injury and Burns and Their Clinical Importance, 1st. ed. Philadelphia, J. B. Lippincott Co., 1974, p 32.

13. Kimber, R. J., and Lander, H.: The effect of heat on human red cell morphology, fragility and subsequent survival in vivo. J. Lab. Clin. Med. 64:922–933, 1964.

14. Akwari, A. M., Ross, D. W., and Stass, S. A.: Spuriously elevated platelet counts due to microspherocytosis. Am. J. Clin. Pathol. 77:220–221, 1982.

15. Stone, H. H.: Review of *Pseudomonas* sepsis in thermal burns: Verdoglobin determination and gentamicin therapy. Ann. Surg. 163:297–305, 1966.

16. Sanders, R., Garcia, J., Sheldon, G. F., Schooley, J., Fuchs, R., and Carpenter, G.: Erythropoietin elevation in anaemia of thermal injury. Surg. Forum 27:71–72, 1976.

17. Wallner, S. F., Vautrin, R. M., Buerk, C., Robinson, W. A., and Peterson, V.: The anemia of thermal injury: Studies of erythropoiesis in vitro. J. Trauma 22:774–780, 1982.

18. Wallner, S., Vautrin, R., Murphy, J., Anderson, S., and Peterson, V.: The haematopoietic response to burning: Studies in an animal model. Burns 10:236–251, 1984.

19. Asko-Seljavaara, S.: Granulocyte kinetics in burned mice. Inhibition of granulocyte growth studied in vivo and vitro. Scand. J. Plast. Reconstr. Surg. 8:185–191, 1974.

20. Eurenius, K., and Brouse, R. O.: Granulocyte kinetics after thermal injury. Am. J. Clin. Pathol. 60:337–342, 1973.

21. Sevitt, S.: Reactions To Injury and Burns and Their Clinical Importance, 1st. ed. Philadelphia, J. B. Lippincott Co., 1974, pp 219–223.

22. Eriksson, E., Straube, R. C., and Robson, M. C.: White blood cell consumption in the microcirculation after a major burn. J. Trauma 19:94–97, 1979.

23. Peterson, V., Hansbrough, J., Buerk, C., Rundus, C., Wallner, S., Smith, H., and Robinson, W. A.: Regulation of granulopoiesis following severe thermal injury. J. Trauma 23:19–24, 1983.

24. Emerson, W. A., Zieve, P. D., and Krevans, J. R.: Hematologic changes in septicemia. Johns Hopkins Med. J. 126:69–76, 1970.

25. Peterson, V., Hansbrough, J., Buerk, C., Rundus, C., Wallner, S., Smith, H., and Robinson, W. A.: Regulation of granulopoiesis following severe thermal injury. J. Trauma 23:19–24, 1983.

26. Chan, C. K., Jarrett, F., and Moylan, J. A.: Acute leukopenia as an allergic reaction to silver sulfadiazine in burn patients. J. Trauma 16:395–396, 1976.

27. Jarrett, F., Ellerbe, S., and Demling, R.: Acute leukopenia during topical burn therapy with silver sulfadiazine. Am. J. Surg. 135:818–819, 1978.

28. Valente, P., and Axelrod, J. L.: Acute leukopenia associated with silver sulfadiazine therapy. J. Trauma 18:146–147, 1978.

29. Caffee, H. H., and Bingham, H. G.: Leukopenia and silver sulfadiazine. J. Trauma 22:586–587, 1982.

30. Kiker, R. G., Carvajal, H. F., Mlcak, R. T., and Larson, D. L.: A controlled study of the effects of silver sulfadiazine on white blood cell counts in burned children. J. Trauma 17:835–836, 1977.

31. Onyiah, I. D.: Leucopenia in two burned patients treated with plasma infusion. Burns 9:118–120, 1983.

32. McEuen, D. D., Gerber, G. C., Blair, P., and Eurenius, K.: Granulocyte function and *Pseudomonas* burn wound infection. Infect. Immun. 14:399–402, 1976.

33. Warden, G. D., Mason, A. D., Jr., and Pruitt, B. A., Jr.: Evaluation of leukocyte chemotaxis in vitro in thermally injured patients. J. Clin. Invest. 54:1001–1004, 1974.

34. Warden, G. D., Mason, A. D., Jr., and Pruitt, B. A., Jr.: Suppression of leukocyte chemotaxis in vitro by chemotherapeutic agents used in the management of thermal injuries. Ann. Surg. 181:363–369, 1975.

35. Alexander, J. W.: Serum and leukocyte lysosomal enzyme. Derangements following severe thermal injury. Arch. Surg. 95:482–490, 1967.

36. Alexander, J. W., Windhorst, D. B., and Good, R. A.: Improved tests for evaluation of neutrophil function in human disease. J. Lab Clin. Med. 72:136–148, 1968.

37. Alexander, J. W., and Wixson, D.: Neutrophil dysfunction and sepsis in burn injury: Surg. Gynecol. Obstet. 130:431–438, 1970.

38. Alexander, J. W., Dinigi, R., and Meakino, J. L.: Periodic variation in the anti-bacterial function of human neutrophils and its relationship to sepsis. Ann. Surg. 173:206–213, 1971.

39. Grogan, J. B., and Miller, R. C.: Impaired function of polymorphonuclear leukocytes in patients with burns and other trauma. Surg. Gynecol. Obstet. 137:784–788, 1973.

40. Heck, E. L., Browne, L., Curreri, P. W., and Baxter, C. R.: Evaluation of

leukocyte function in burned individuals by in vitro oxygen consumption. J. Trauma 15:486–489, 1975.

41. Curreri, P. W., Heck, E. L., Browne, L., and Baxter, C. R.: Stimulated nitro blue tetrazolium test to assess neutrophil antibacterial function. Prediction of wound sepsis in burned patients. Surgery 74:6–13, 1973.

42. Goodson, W. H., and Hunt, T. K.: Wound healing and the diabetic patient. Surg. Gynecol. Obstet. 149:600–608, 1979.

43. Mowat, A. G., and Baum, J.: Chemotaxis of polymorphonuclear leukocytes from patients with diabetes mellitus. N. Engl. J. Med. 284:621–627, 1971.

44. VanOss, C. J.: Influence of glucose levels on the in vitro phagocytosis of bacteria by human neutrophils. Infect. Immun. 4:54–59, 1971.

45. Van Furth, R., and Cohn, Z. A.: The origin and kinetics of mononuclear phagocytes. J. Exp. Med. 128:415–435, 1968.

46. Diegelmann, R. F., Cohen, K., and Kaplan, A. M.: The role of macrophages in wound repair. A review. Plast. Reconstr. Surg. 68:107–113, 1981.

47. Lemperle, G.: The role of the reticuloendothelial system in burn infection. *In* Basic Problems in Burns. R. Vrabec, Z. Konichova, J. Moserova, eds. NY, Berlin, Springer-Verlag, 1975, pp 141–144.

48. Rittenburg, M. S.: The Response of the reticuloendothelial system to the thermal injury. Surg. Clin. N. Amer. 50:1227–1234, 1970.

49. Leibovich, S. J., and Ross, R.: The role of the macrophage in wound repair. A study with hydrocortisone and antimacrophage serum. Am. J. Pathol. 78:71–100, 1975.

50. Alexander, J. W.: Effect of thermal injury upon the early resistance to infections. J. Surg. Res. 8:128–137, 1968.

51. DiMaio, A., DiMaio, D., and Jacques, L.: Phagocytosis in experimental burns. J. Surg. Res. 21:437–448, 1976.

52. Loose, L. D., and Turinsky, J.: Macrophage dysfunction after burn injury. Infect. Immun. 26:157–162, 1979.

53. Hansbrough, J. F., Peterson, V., Kortz, E., and Piacentine, J.: Postburn immunosuppression in an animal model: Monocyte dysfunction induced by burned tissue. Surgery 93:415–423, 1983.

54. Grover, G. J., and Loegering, D. J.: Effect of splenic sequestration of erythrocytes on splenic clearance function and susceptibility to septic peritonitis. Infect. Immun. 36:96–102, 1982.

55. Lloyd, R. S., and Levick, P. L.: Blood monocyte dysfunction following thermal injury. Burns 3:245–252, 1977.

56. Sevitt, S.: Reactions To Injury And Burns And Their Clinical Importance 1st. ed. Philadelphia, J. B. Lippincott Co., 1974, pp 173–175.

57. Bartlett, R. H., Fong, S. W., Marrujo, G., Hardeman, J., and Anderson, W.: Coagulation and platelet changes after thermal injury in man. Burns 7:370–377, 1981.

58. Eurenius, K., Mortensen, P. F., Meserol, P. M., and Curreri, P. W.: Platelet and megakaryocyte kinetics following thermal injury. J. Lab. Clin. Med. 79:247–257, 1972.

59. Hergt, K.: Thrombocyte counts and their relation to the clinical condition of the burned patient. *In* Research in Burns. P. Matter, T. L. Barclay, and Z.

Konickova, eds. Bern, Stuttgart, Vienna, Hans Huber Publ., 1971, pp 604–608.

60. Eurenius, K., Rossi, T. D., McEuen, D. D., Arnold, J., and McManus, W. F.: Blood coagulation in burn injury. Proc. Soc. Exp. Biol. Med. 147:878–882, 1974.

61. Tian-Min, C., Yuan, L., De-Quan, G., Chang-Kun, L., and Huai-En, Z.: Ultrastructural changes of bone marrow megakaryocytes in several types of injury. Burns 10:282–289, 1984.

62. Mims, J. A., Sarji, K. E., Kleinfelder, J. S., and Eurenius, K.: Heparin-induced platelet aggregation in burn patients. Thromb. Res. 10:291–299, 1977.

63. Deitch, E. A., Dobke, M., and Barter, C. R.: Failure of local immunity. A potential cause of burn wound sepsis. Arch. Surg. 120:78–83, 1985.

64. Arturson, G., Högman, C. F., Johansson, S. G. O., and Killander, J.: Changes in immunoglobulin levels in severely burned patients. Lancet 1:546–548, 1969.

65. Munster, A. M., Hoagland, H. C., and Pruitt, B. A., Jr.: The effect of thermal injury on serum immunoglobulins. Ann. Surg. 172:965–969, 1970.

66. Brown, W. L., Bowler, E. G., Mason, A. D., Jr., and Pruitt, B. A., Jr.: Protein metabolism in burned rats, Am. J. Physiol. 231:476–482, 1976.

67. Daniels, J. C., Larson, D. L., Abston, S., and Ritzmann, S. E.: Serum protein profiles in thermal burns. I. Serum electrophoretic patterns, immunoglobulins and transport proteins. J. Trauma 14:137–152, 1974.

68. Mortensen, R. F., and Eurenius, K.: Enhanced hemolytic antibody response following thermal injury. Int. Arch. Allergy 43:321–326, 1972.

69. Rapaport, F. T., and Bachvaroff, R. J.: Kinetics of humoral responsiveness in severe thermal injury. Ann. Surg. 184:51–59, 1976.

70. Baker, C. C., Miller, C. L., and Trunkey, D. D.: Predicting fatal sepsis in burn patients. J. Trauma 19:641–647, 1979.

71. Daniels, J. C., Sakai, H., and Ritzmann, S. E.: Lymphoid response of the burn patient. South. Med. J. 68:865–870, 1975.

72. McCabe, W. P., Rebuck, J. W., Kelly, A. P., Jr., and Ditmars, D. M., Jr.: Leukocytic response as a monitor of immunodepression in burn patients. Arch. Surg. 106:155–159, 1973.

73. Miller, C. L., and Baker, C. C.: Changes in lymphocyte activity after thermal injury. The role of suppressor cells. J. Clin. Invest. 63:202–210, 1979.

74. Munster, A. M., Eurenius, K., Katz, R. M., Canales, L., Foley, F. D., and Mortensen, R. F.: Cell mediated immunity after thermal injury. Ann. Surg. 177:139–143, 1973.

75. Ninnemann, J. L.: Immunosuppression following thermal injury through B-cell activation of suppressor T-Cells. J. Trauma 20:206–213, 1980.

76. Wood, G. W., Volenec, F. J., Mani, M. M., and Humphrey, L. J.: Dynamics of T-lymphocyte subpopulations and T-lymphocyte function following thermal injury. Clin. Exp. Immunol. 31:291–297, 1978.

77. Munster, A. M.: Immunologic response of trauma and burns. An overview. Am. J. Med. 865:142–145, 1984.

78. Ishizawa, S., Sakai, H., Sarles, H. E., Larson, D. L., and Daniels, J. C.: Effect of thymosin on T-lymphocyte functions in patients with acute thermal burns. J. Trauma 18:48–52, 1978.

79. Fried, D. A., and Munster, A. M.: Does immunosuppression by thermal injury

depend on the continued presence of the burn wound? J. Trauma 15:483–485, 1975.

80. Constantian, M. B.: Impaired reactivity of burn patients lymphocytes to phytohemagglutinin in autologous serum; failure to improve responsiveness by washing in vitro. J. Surg. Res. 27:84–92, 1979.

81. Munster, A.M.: Post traumatic immunosuppression is due to activation of suppressor T-cells. Lancet 1:1329–1330, 1976.

82. Keeling, P., Winchurch, R. A., and Munster, A. M.: Immunomodulation in major burns: An in vitro system for assaying the effect of modulators. J. Trauma 20:830–832, 1980.

83. Antonacci, A. C., Reaves, L. E., Calvano, S. E., Amand, R., De Riesthal, H. F., and Shires, G. T.: Flow cytometric analysis of lymphocyte subpopulations after thermal injury in human beings. Surg. Gynecol. Obstet. 159:1–8, 1984.

84. Bjornson, A. B., Altemeier, W. A., Bjornson, H. S., Tang, T., and Iserson, M. L.: Host defense against opportunist microorganisms following trauma. I. Studies to determine the association between humoral components of host defense and septicemia in burn patients. Ann. Surg. 188:93–101, 1978.

85. Bjornson, A. B., Altemeier, W. A., and Bjornson, H. S.: The septic burned patient: A model for studying the role of complement and immunoglobulins in opsonization of opportunist microorganisms, Ann. Surg. 189:515–527, 1979.

86. Dhennin, C., Pinon, G., and Greco, J. M.: Alterations of complement system following thermal injury: Use in estimation of vital prognosis. J. Trauma 18:129–133, 1978.

87. Panke, T. W., Langlinois, P. C., and Goyette, R. E.: An "abnormal" lymphoid proliferation simulating Hodgkins disease in a burn patient. Human Pathol. 9:716–723, 1978.

88. Teplitz, C.: The pathology of burns and the fundamentals of burn wound sepsis. *In* Burns. A Team Approach. C. P. Artz, J. A. Moncrief, and B. A. Pruitt, Jr., eds. Philadelphia, W. B. Saunders Co., 1979, pp 45–94.

89. Linares, H. A., Beathard, G. A., and Larson, D. L.: Morphological changes of lymph nodes of children following acute thermal burns. Burns 4:165–170, 1978.

90. Gelfand, D. W., Goldman, A. S., Law, E. J., MacMillan, B. G., Larson, D., Abston, S., and Schreiber, J. T.: Thymic hyperplasia in children recovering from thermal burns. J. Trauma 12:813–817, 1972.

91. Caffey, J., and Silbey, R.: Regrowth and overgrowth of the thymus after atrophy induced by the oral administration of adrenocortisteroids to human infants. Pediatrics 26:762–770, 1960.

92. McManus, W. F., Eurenius, K., and Pruitt, B. A., Jr.: Disseminated intravascular coagulation in burned patients. J. Trauma 13:416–421, 1973.

93. Alkjaersig, N., Fletcher, A. P., Peden, J. C., Jr., and Monafo, W. W.: Fibrinogen catabolism in burned patients. J. Trauma 20:154–159, 1980.

94. Effeney, D. J., Blaisdell, F. W., McIntyre, K. E., and Graziano, C. J.: The relationship between sepsis and disseminated intravascular coagulation. J. Trauma 18:689–694, 1978.

95. Bartlett, R. H., Forg, S. W., Marrujo, G., Hardeman, J., and Anderson, W.:

Coagulation and platelet changes after thermal injury in man. Burns 7:370–377, 1981.

96. Braunstein, K. M., Dodds, K. A., Stewart, G., Shull, K. C., and Eurenius, K.: Heparin cofactor activity following thermal activity. Am. J. Clin. Pathol. 70:632–636, 1978.

97. Rosenberg, R. D.: Actions and interactions of antithrombin and heparin. N. Engl. J. Med. 292:146–151, 1975.

98. Klein, H. G., Bell, W. R.: Disseminated intravascular coagulation during heparin therapy. Ann. Int. Med. 80:477–481, 1974.

99. Eeles, G. H., and Sevitt, S.: Microthrombosis in injured and burned patients. J. Pathol. Bact. 93:275–293, 1967.

100. Wells, S., Sissons, O. M., and Hasleton, P. S.: Quantitation of pulmonary megakaryocytes and fibrin thrombi in patients dying from burns. Histopathol. 8:517–527, 1984.

101. Sevitt, S.: Reactions to Injury and Burns and Their Clinical Importance, 1st. ed. J. B. Lippincott Co., Philadelphia, 1974, pp 180–189.

102. Sevitt, S.: Coronary thrombosis following injury and burns. Med. Sci. Law 13:185–191, 1973.

103. Hume, M., Sevitt, S., and Thomas, D. P.: Venous Thrombosis and Pulmonary Embolism. London, Oxford University Press, 1970.

104. Neame, P. B., Lechago, J., Ling, E. T., and Koval, A.: Thrombotic thrombocytopenic purpura: Report of a case with dissemmated intravascular platelet aggregation. Blood 42:807–814, 1973.

105. Bernat, I., Dozsan, G., Novak, J., and Elek, S.: Anaemia after thermal injury. II. Iron metabolism. Acta Med. Acad. Scient. Hung. Tomus 22:253–258, 1966.

106. Sevitt, S., Jackson, D., Stone, P., Baar, S., and Pollack, A.: Acute Heinz-body anemias in burned patients. Lancet 2:471–475, 1973.

107. Wallner, S. F., Vautrin, R. M., Buerk, C., Robinson, W. A., and Peterson, V. M.: The anemia of thermal injury: Studies of erythropoiesis in vitro. J. Trauma 22:774–780, 1982.

Chapter 12

GASTROINTESTINAL COMPLICATIONS

Gastrointestinal hemorrhage and perforation are potentially life-threatening complications in burn patients. Acute superficial erosive lesions of the stomach may be present as early as 5 hr post burn in patients with burns involving more than 30 percent of the total body surface. Gastroduodenoscopy has demonstrated an apparent progression from mild erosive mucosal disease to discrete ulcers in burn patients.[1] These lesions are generally regarded as acute stress ulcerations, and in the burn patient, they are referred to as Curling's ulcers.

General Morphologic Features of Gastroduodenal Mucosal Lesions

The superficial gastric and duodenal erosive lesions appear grossly as rounded or elongated hemorrhagic foci of variable size (Fig. 12–1). They are more commonly seen in the gastric fundus or body than in the antrum. Deeper mucosal lesions which extend to the muscularis mucosa are termed ulcers and have a distribution similar to that of the more superficial lesions. When the lesions extend into the muscularis mucosa, they have a circumscribed or "punched-out" appearance (Fig. 12–1). They are often quite numerous and range in size from 2–3 mm up to a centimeter in diameter. Duodenal ulcerations have a similar appearance but tend to be larger.

Microscopically, both the gastric and duodenal lesions are characterized by acute mucosal necrosis, usually with venular congestion and hemorrhage (Fig. 12–2A,B). Generally, both the shallow erosive lesions and deeper ulcers of the stomach and duodenum are free of significant inflammatory infiltrates. The appearance of the mucosa in the acute lesions is often that of autolysis. This feature probably represents mucosal ischemic infarction.[2,3] Small vessels may contain fibrin–platelet thrombi. Mural ero-

Fig. 12–1. Several hemorrhagic erosions (arrows) in the gastric mucosa of a burn patient. A single circumscribed ulcer (U) is also present.

sions of large arteries at the base of deep ulcers explain severe hemorrhage in some cases.

Partial occlusion of these large vessels by fibrin thrombi may also be noted. It is not uncommon to find superficial bacterial or fungal colonization of these lesions.

Etiology of Erosive and Ulcerative Gastrointestinal Disease

While it is not our purpose to evaluate all the proposed mechanisms of ulcer development, we must mention some of the broad concepts that aid in understanding this disease. There have been innumerable explanations[2,4] for the syndrome of gastroduodenal ulceration. Some of these interrelated factors are listed in Table 12–1.

Acid

The apparent success of antacid therapy in preventing serious gastrointestinal bleeding in burn patients strongly supports the concept that acid, even in normal quantities, plays an important role in production of GI

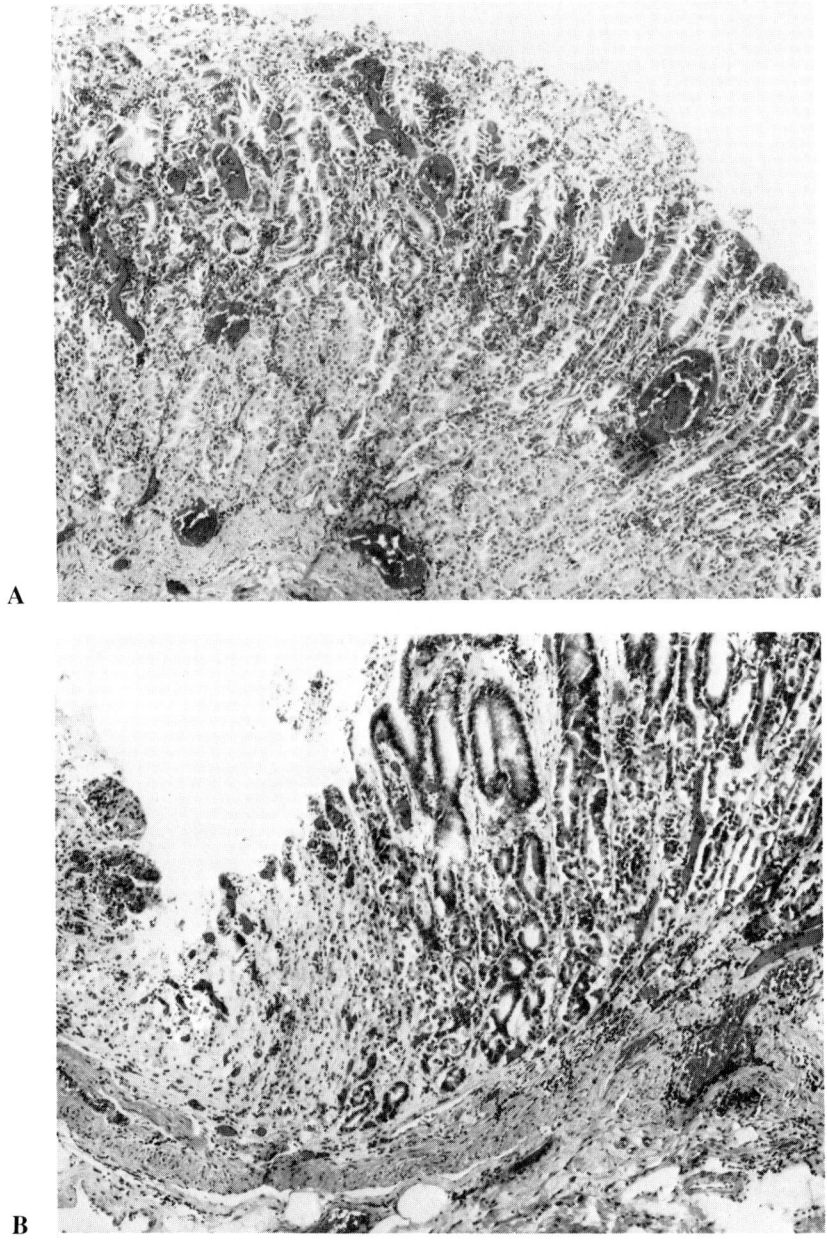

Fig. 12–2. **(A):** Gastric mucosal defects usually begin as small superficial zones of necrosis with capillary and venular congestion. Only the surface epithelium in this section is necrotic. (H & E, × 54) **(B):** More advanced mucosal defects (ulcerations) extend to the muscularis mucosa. (H & E, × 60)

Table 12–1 *Factors Reported to Contribute to Gastrointestinal Ulcerations*

Acid

   Increased acid[1,2,4,21–25]
   Normal to reduced amounts of acid[1,26–31]
   Compromised "feed back" control of acid secretion[1]
   Impaired ability of duodenum to neutralize acid[32]
   Hydrogen ion back diffusion[25]
   Acid injury to compromised mucosa[1,33–35]

Mucosal factors

   Burn-induced inhibition of repair[7,36,37]
   Alterations in composition or quantity of surface mucoproteins[5,38–44]
   Alterations of gastric mucus by adrenal cortical hormones[45–46]

Ischemia of mucosa

   Submucosal shunting, decreased mucosal blood flow, contraction of connecting
      arteries[47–50]
   Muscle contraction, vagal stimulation, vessel compression[33]
   Local action of vasoconstrictors[51]
   Undefined "vascular factor"[22]
   Hemoconcentration, resultant mucosal congestion[26]
   Hypercoagulability[52–54]
   Low blood flow contributing to glucose deficiency[55]
   Gastric mucosal energy deficit[56]
   Vasospasm induced by increased blood levels of catecholamines[57]

Sepsis

   Compromised repair rate of mucosa[7]
   Sepsis induced adrenal cortical responses[7]
   Sepsis as an additive stress effect[11,58]
   Septic embolism[59,60]

Miscellaneous factors

   Bacterial flora or bacterial products[33,61–65]
   Compromised enterogastrone release[66]
   CNS or neural stimulation[67,68]
   Vagal overactivity[33]
   Detergent action of refluxed bile on gastric mucosa[1,69–71]
   Vitamin A deficiency[9]

erosions and ulcers.[5] Moreover, the H_2 receptor blocking agents reduce the incidence of stress ulceration in experimental animals[6] and show promise as prophylactic and therapeutic agents clinically.

Mucosal Factors

It has been hypothesized that mucosal healing may be compromised in severely burned patients.[7] Also, it is known that burns inhibit connective tissue repair even in unburned tissue.[8] Surface mucopolysaccharides are reduced and changed in composition in gastric mucosa in burned experimental animals.[4] This is felt to make surface mucosal cells more susceptible to gastric acid injury. Vitamin A levels have been found depressed in stressed patients, and this deficiency may be important in the genesis of gastric lesions by inhibiting mucosal repair.[9]

Mucosal Ischemia

It is also apparent that during burn-related stress, several factors act to reduce the resistance of mucosa to the effects of acid. The "low flow state," local action of vasoconstrictors, hemoconcentration, and possibly hypercoagulability may all contribute to focal mucosal ischemia and subsequent ulceration.[4] A rat model of burn injury[10] developed gastrointestinal lesions which are considered to be comparable to acute mucosal lesions seen in burn patients. These lesions are bland, nonsuppurative, mucosal erosions or ulcers that may be found with or without acute hemorrhage in the stomach (Fig. 12–3) and occasionally in the upper small intestine. In the nonseptic rat, these lesions usually heal when stress factors are removed.

Sepsis

The relationship between sepsis and gastrointestinal lesions has been mentioned throughout the literature. Some investigators have postulated a direct association between septicemia and gastrointestinal ulcerations, while others have considered the two conditions unrelated. In patients with less than 50 percent total body surface burns, preexisting septicemia has been shown to have an addititive effect which apparently predisposed the burn patient to gastrointestinal ulceration.[11] At least part of the confusion is caused by the various definitions given to the terms "sepsis" and "ulcer." Characteristically, burn-associated or Curling's ulcers are more common during septic episodes after the first postburn week. Sepsis stimulates adrenal cortical activity, disturbs the tone of blood vessels, reduces nutritional resources, interferes with oxygen transport, and stimulates vasoactive amine and polypeptide release.[7] By any of these mechanisms, sepsis may

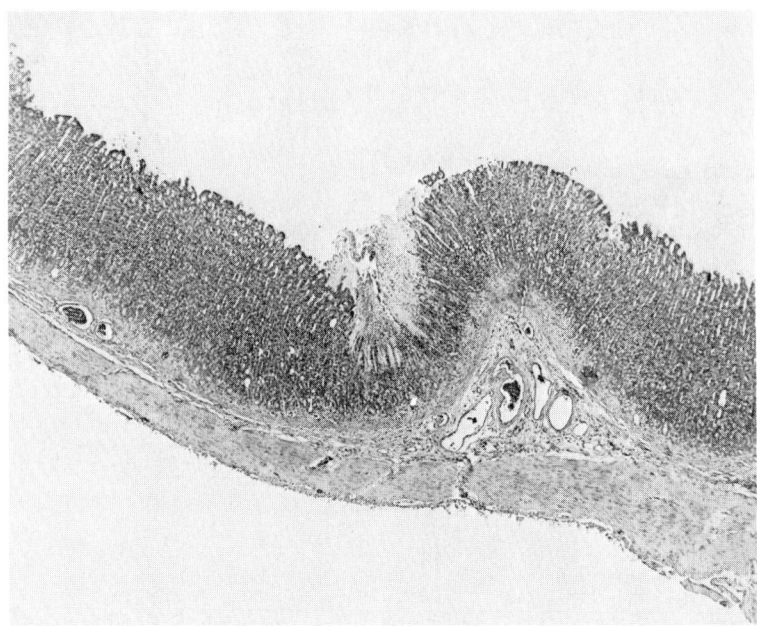

Fig. 12–3. Stress ulcerations may be found in the stomachs of experimentally burned rats. These lesions usually heal unless rats become septic. (H & E, × 54)

directly or indirectly affect the blood flow and energy potential of the gastrointestinal mucosa.

Septic burn patients may occasionally have ulcerative gastric or intestinal lesions which are thought to be caused by hematogenous *Pseudomonas aeruginosa*. These have been mentioned in several morphologic studies.[12–16] Review of autopsy data at the Institute of Surgical Research (ISR) has also demonstrated cases which have morphologic evidence of hematogenous gastrointestinal lesions. This concept has been studied in an animal model.[17] In this rat model of burn injury in which rats are inoculated topically with *P. aeruginosa*,[18] two types of ulcerative lesions occur in the gastrointestinal tract. The first type is the well-described, noninfectious mucosal erosion or "stress ulcer" (Fig. 12–3). The second lesion is apparently an infectious lesion of the gastrointestinal tract which is morphologically identical to the ulcerative *cutaneous* lesion known as "ecthyma gangrenosum" characteristic of *P. aeruginosa* infection. These lesions have been seen in the *P. aeruginosa*-infected burn rat at an early stage in which the mucosa remains intact (Fig. 12–4A). They apparently progress quickly to ulcerated circumscribed zones of necrosis and hemorrhage (Fig. 12–4B). When sectioned serially and examined microscopically, the necrotic foci appear centered

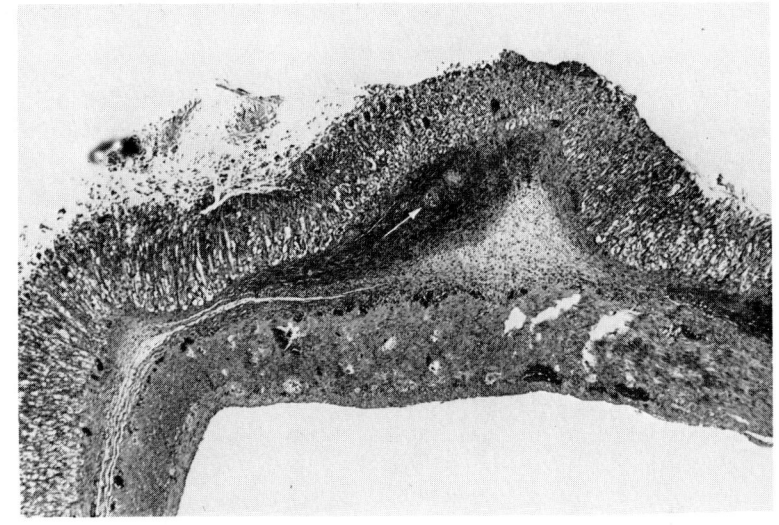

A

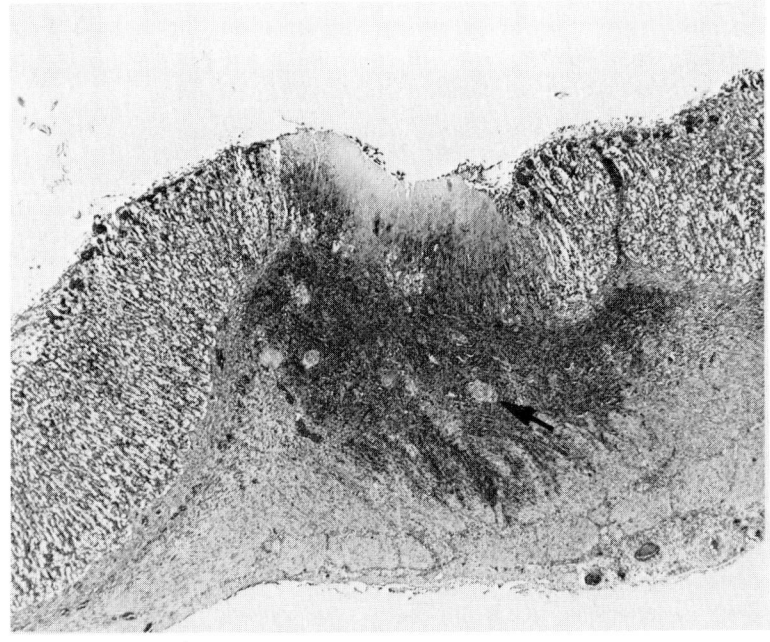

B

Fig. 12–4. **(A):** Infectious preulcerative lesion in the gastric mucosa of a *Pseudomonas*-infected burned rat. This lesion progresses rapidly to an ulcerated circumscribed zone of necrosis and hemorrhage. (H & E, × 36) **(B):** Advanced infectious ulceration in a septic burned rat. Necrosis has extended into both the mucosa and muscularis. Several submucosal blood vessels are affected by a gram-negative bacterial vasculitis characteristic of *Pseudomonas aeruginosa* infection. (H & E, × 36)

around one or more small blood vessels which are infiltrated by gram-negative bacilli in a pattern typical for classical *Pseudomonas* vasculitis of the skin or lung (Fig. 12–4A). In deep mucosal lesions of this type, there may be lateral extension of necrosis and suppuration which appears to undermine viable mucosa (Fig. 12–4B). Early studies utilizing the septic burned rat weighing less than 200 g have not emphasized gastric or enteric hematogenous lesions because they are seldom seen in young rats. In studies utilizing rats weighing more than 350 g, hematogenous gastroenteric lesions are more commonly seen.

Gastrointestinal Ulcers: Curling's versus Hematogenous Pseudomonas Lesions

It has been reported that metastatic pseudomonas lesions of the skin (ecthyma gangrenosum) may resolve with treatment.[19] Likewise, it may be anticipated that similar lesions in the gastrointestinal tract would heal spontaneously or with systemic treatment. Resolving metastatic lesions of the gastrointestinal tract may lose their characteristic histologic features of infection and resemble so-called Curling's ulcers.

As Table 12–2 and Figure 12–5 indicate, the early stages of these two types of ulcers can usually be clearly distinguished. There has been considerable confusion in the literature concerning these two entities and it is at times difficult to differentiate between the two.

Peptic Ulcers versus Stress Ulcers

Proposed pathogenic factors for benign peptic ulceration of the duodenum and the stomach overlap with those of acute stress ulceration. Peptic ulceration is a common disease; the incidence is estimated to be up to 14 percent of the population.[20] Teplitz and Skillman mention that new peptic ulcers could possibly develop during the course of a difficult illness in a susceptible patient. This view could place at least some of the lesions presently classified as stress ulcers into a category unrelated to burn-associated stress. Table 12–3 lists some of the basic differences between burn-associated stress ulcers and peptic ulcers. The potential for overlap between these two entities is obvious.

ISCHEMIC ENTEROCOLITIS

With appropriate, vigorous, antacid, and anticholinergic or H_2 antagonist therapy, the occurrence of serious gastroduodenal ulceration and hemorrhage has been essentially eliminated in burn patients.[5,6,72] This has

Table 12–2 *Comparison of Curling's Ulcers with Ulcers Caused by Hematogenous Pseudomonas Aeruginosa*

	Curling's Ulcer	Ulcer Due to Pseudomonas Species
Sites	Stomach and duodenum	Stomach thru colon
Patterns of necrosis (in ulcer)	"V"- or "U"-shaped necrotizing lesions	Deep mucosal spherical or inverted "V"-shaped ulcer
Inflammation	Minimal	Marked acute suppurative inflammatory process, later abscess formation
Blood vessels	Arteries may contain fibrin thrombi	Bacterial vasculitis possibly with acute, mainly platelet thrombi
Bacteria	Usually none. If present, are colonizing bacteria and fungi	Many—often with vasocentric pattern typical of *Pseudomonas* vasculitis

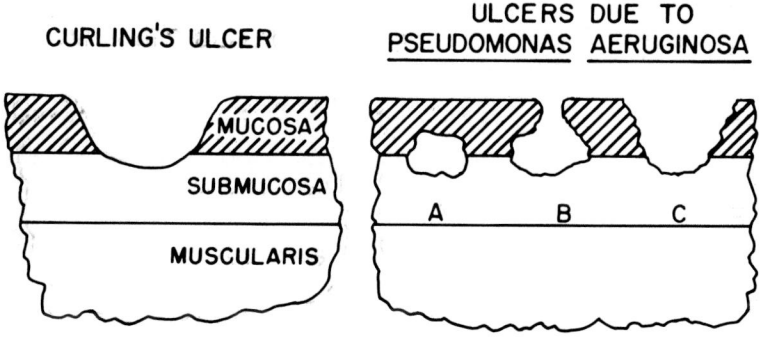

CURLING'S ULCER

ULCERS DUE TO PSEUDOMONAS AERUGINOSA

A. EARLY LESION

B. EVOLVING ULCER

C. FULLY DEVELOPED ULCER

Fig. 12–5. Comparison between stress ulcers and those due to hematogenous *Pseudomonas* infection. A = Early lesion; B = Evolving ulcer; C = Fully developed ulcer.

Table 12–3 *Comparison of Peptic and Burn-Associated Stress Ulcers*

Characteristic	Peptic Ulcer	Stress Ulcer
Numbers	Solitary	Usually multiple
Location	Pylorus, anterior duodenum	Fundus, body of stomach
Size	Variable	Usually small
Appearance	Chronic inflammation	Acute inflammation, usually minimal
Mucosa	"Moth-eaten," erosive	Autolytic appearance in acute stage
Bacteria	None	Variable number; usually superficial colonization of necrotic debris

focused attention on other diseases of the intestinal tract, such as ischemic enterocolitis.

Ischemic enterocolitis may occur as a result of occlusive disease in the main stem arterial supply, thrombosis of the main stem venous structures or in a "nonocclusive" form. By far the most commonly observed cases in burn patients are due to "nonocclusive" ischemic enterocolitis associated with states of marked hypotension or disseminated intravascular coagutopathy or both. We have defined "nonocclusive ischemic enterocolitis," for the purpose of this discussion, as coagulative necrosis of the intestinal mucosa in the absence of significant atherosclerotic disease or occlusive thrombosis of major blood vessels.

A retrospective study of ischemic enterocolitis in autopsied burn patients demonstrated an incidence of 7.5 percent (24 cases in 321 autopsies) over a 52-month period.[73] In this study, a wide variation in extent, severity, histomorphology, and suspected pathogenesis of ischemic enterocolitis was demonstrated. An additional, unrelated, prospective study carried out over an 18-month period indicated that the incidence of ischemic enterocolitis exceeded 45 percent.[73] Most of the lesions were mild and very likely insignificant in terms of morbidity or mortality or both.

The etiology and pathogenesis of these ischemic lesions in the burn patient are complex and multifactorial. With lack of disease in the major blood vessels, we propose that the condition is caused by a hypoperfusion state (many of the burn patients experienced episodes of marked hypotension) terminally or a disseminated intravascular coagulopathy or both. Clinical conditions that may predispose to the above physiologic and hematologic abnormalities are categorized in Table 12–4. To determine which factors are operative at a given point in time and the degree of each factor's influence is often impossible clinically. At autopsy, the pathologist likewise faces a dilemma in determining the significance of his findings.

Burn patients who develop ischemic enterocolitis are usually severely ill and seldom complain of colicky abdominal pain and abdominal tenderness.[74] Concurrent septicemia may impair the interpretation of fever and leukocytosis associated with the ischemic enterocolitis. Useful diagnostic

Table 12–4 *Clinical Conditions and Factors Which Predispose to Nonocclusive Ischemic Enterocolitis*

Hypotension[77,78,82,103–105]

    Cardiac failure
    Shock, neurogenic
    Endotoxemia
    Hypovolemia, hemorrhage
    Digitalis

Vasoconstriction[77,82,106]

    Epinephrine
    Norepinephrine
    Ephedrine

Infections[81,84,87,104,106]

    Bacterial
    Fungal
    Viral

Collagen vascular disease[84]

    Rheumatoid arthritis
    Necrotizing vasculitis

Intussusception and volvulus[84,87,89]

Neoplastic Disease[81,87,107]

    Obstruction
    See hypercoagulable state

Microangiopathy[77,84]

    Diabetes mellitus
    Chronic hypertension

Hypercoagulable state[76,77,82,106]

    DIC
    Oral contraceptives

Toxins[108,109]

    Clostridia
    Undefined

Drugs[110–114]

    Clindamycin
    Lincomycin
    Ampicillin
    Penicillin
    Amoxicillin
    Tetracycline
    Chloramphenicol
    Cephalothin
    Conjugated estrogens

Miscellaneous

    Recent surgery[107,115]

tools are limited. Colonoscopy may demonstrate the obliteration of haustral folds by edema and hemorrhage.[74] Proctoscopy and barium enema are useful adjuncts for diagnosis.

Spectrum of Severity of Ischemic Lesions in the Burn Patient

Ischemic lesions may be found throughout the small intestine and colon. The earliest and least severe lesion identified in the spectrum of this entity is mucous cell metaplasia of the intestinal mucosa with excessive mucin production (Fig. 12–6). This alteration is commonly seen adjacent to necrotic (ischemic) foci. Mucous cell metaplasia probably represents a sublethal, nonspecific response to ischemic injury and is suggestive of ischemic disease elsewhere in the burn patient. With more severe ischemia, the lesion may progress to hemorrhage and/or mucosal necrosis.

Lesions could be generally classified (Fig. 12–7) grossly by the extent of intestinal involvement and by the degree of transmural extension. Ischemic lesions of a mild degree (Fig. 12–8) (29 percent of cases) were considered to be focal and confined to the mucosa (Table 12–5, Fig. 12–7). Such lesions most commonly were observed on protruding mucosal surfaces, such as colonic haustral ridges, protruding surfaces of the plicae circulares of the jejunum or the ileocecal valve. Lesions of moderate severity (38 percent) were multifocal and involved both the mucosa and the submucosa. Severe lesions (33 percent) were not only multifocal, but also extensive (Fig. 12–9) and often transmural. Although no intestinal perforation was seen in such lesions, this potential hazard existed in the group with deep transmural lesions.

Microscopically, the earliest and mildest lesions were confined to the mucosa and included vascular dilatation and congestion with petechiae. Many of these lesions apparently occurred terminally (within hours of death). In older lesions "bland" coagulative necrosis of the mucosa was noted (Fig. 12–10). If evidence of necrosis was obvious (suggesting the elapse of 1–2 days), acute inflammatory cells bordered necrotic tissue. Lesions of moderate severity involved larger areas of the mucosa and extended into the submucosa.

Three major features were used to classify ischemic lesions as "severe": (1) extensive and multifocal involvement of the small intestine or colon or both, (2) transmural necrosis, and (3) marked hemorrhagic appearance with transmural involvement. The presence of abundant mural hemorrhage in the intestine was often associated with intraluminal bleeding and the formation of erythrocytic pseudomembranes (Fig. 12–11). These latter findings indicated a severe ischemic process often associated with widespread intestinal lesions capable of causing severe morbidity or mortality or

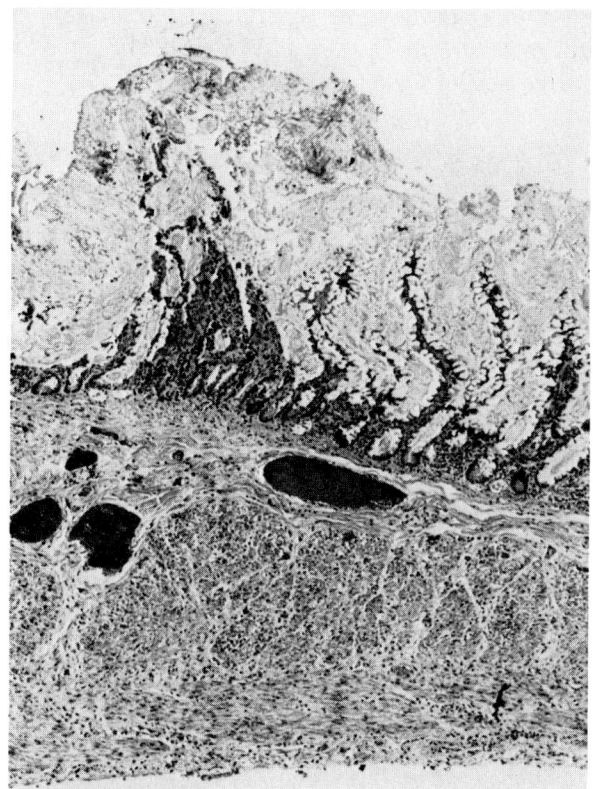

Fig. 12–6. Metaplasia of intestinal mucous cells is an early indication of ischemic injury. (H & E, × 54)

both. Severe hemorrhagic ischemic enterocolitis is probably a precipitous event. This is suggested by the lack of bacterial colonization of the lesions in some of our cases.[73] and in experimental animals[75] and the sparsity of reactive acute inflammatory cells in contrast to the extensive necrosis. Other conditions may simulate or produce ischemic intestinal lesions and are discussed in Table 12–6.

Thrombi

Despite the term "nonocclusive," 50 percent of our cases with ischemic infarcts had microvascular thrombi. These thrombi were related to an episode of disseminated intravascular coagulopathy in only a small number of cases. This raises the conjecture that microvascular thrombosis may be a

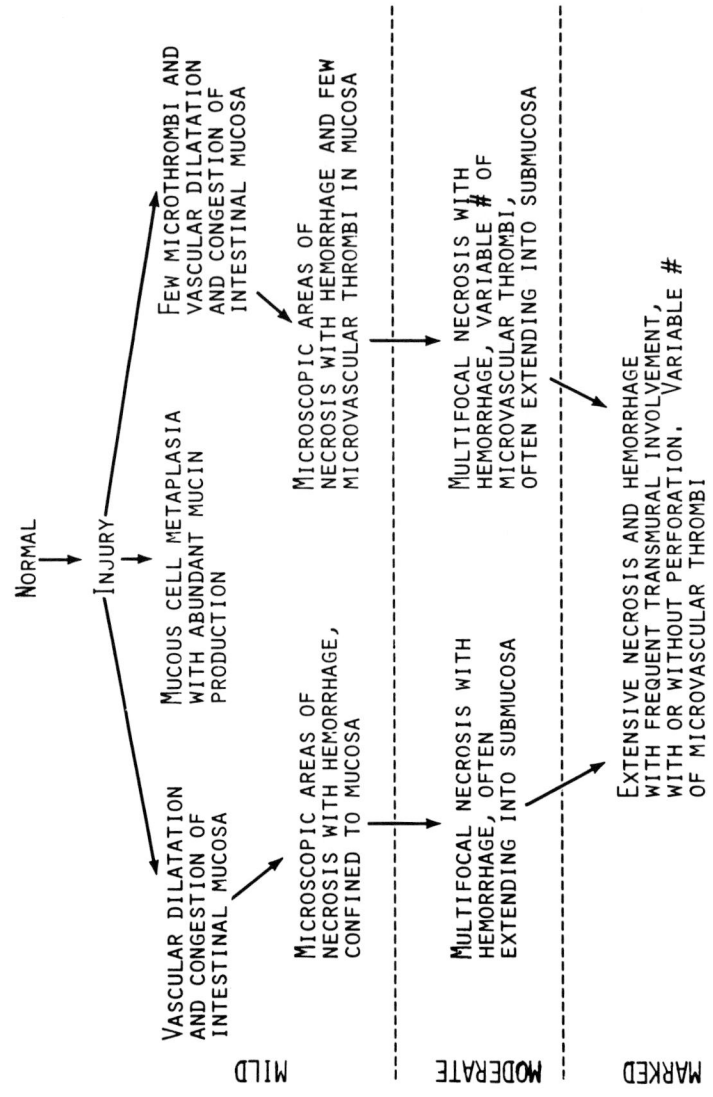

Fig. 12–7. Schematic representation of severity of ischemic enterocolitis lesions.

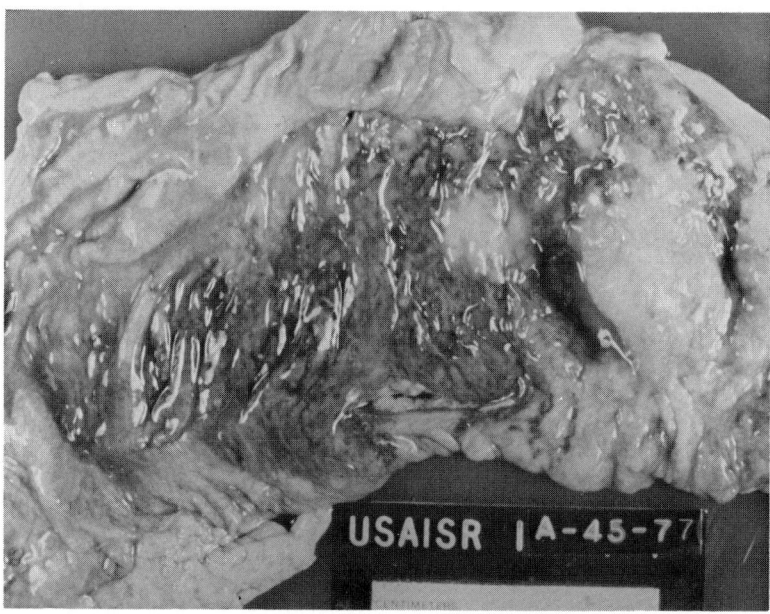

Fig. 12–8. A mild form of ischemic necrosis is often seen on protruding mucosal surfaces such as the colonic haustral ridges or on the ileocecal valve.

secondary event in most cases of ischemic enterocolitis, although this point has been highly controversial in the literature.[53–54,76–78]

Special stains demonstrated that these thrombi were composed almost entirely of platelets[79] (Fig. 12–12). Other thrombi consisted of both fibrin and platelets. This feature suggested that, at least in some patients, the initial event is probably microvascular injury with platelet aggregation–agglutination and not a hypercoagulable state with widely disseminated fibrin thrombi in the microvasculature. We were unable to relate the type of thrombus with the severity of disease, but noted a paucity and often a complete absence of thrombi in patients with extensive hemorrhagic ischemic enterocolitis.

The finding of cases without thrombi in the hemorrhagic ischemic lesions suggests the presence of a hypoperfusion state. It may be that hypoperfusion causes ischemic injury to the microvascular endothelium and exposed collagen and/or basement membrane leads secondarily to a platelet aggregation–agglutination and subsequent fibrin deposition.

A variety of hormones, drugs, and other substances may predispose to a hypercoagulable state. This includes epinephrine, norepinephrine, serotonin, histamine, free potassium, endotoxins, and miscellaneous vasoactive polypeptides.[53,77,80–81] Some of these hormones and drugs have been reported to be elevated in the portal

Table 12-5 *Degree of Severity of Ischemic Enterocolitis*

Distribution of Lesions	Depth of Necrosis	Hemorrhage	Pseudomembranes
Mild—few, small, foci	Confined to mucosa Often no necrosis	Petechia and small areas— confined to mucosa	"Classical"
Moderate—several areas	Mucosal and superficial mucosa	Small to moderate size areas— usually confined to mucosa	"Classical" and "nonclassical"
Severe—many large areas with coalescence of lesions and multifocal segmental involvement	Deep submucosa and muscularis, often transmural	Mucosal and submucosal, often transmural	"Nonclassical" and "erythrocytic"

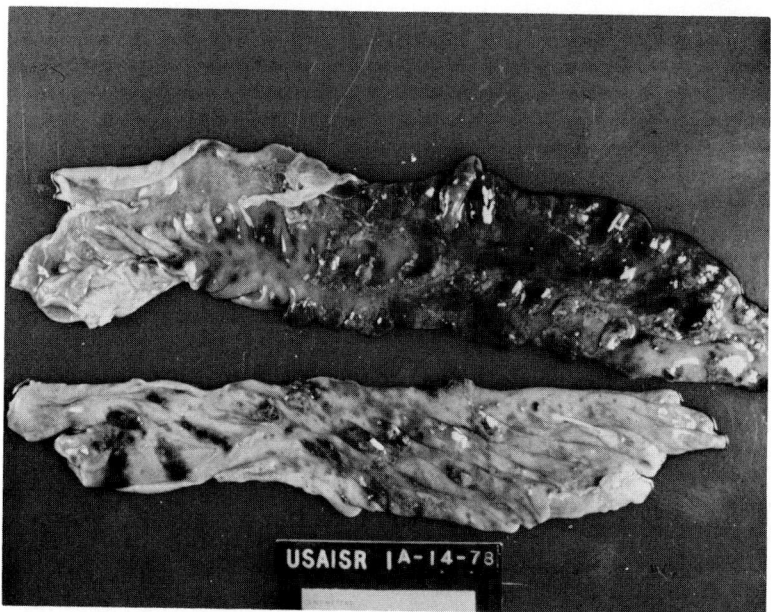

Fig. 12–9.   Severe ischemic necrosis of this type is often found histologically to be transmural.

circulation under experimental and clinical conditions of intestinal ischemia.[81] This suggests that the appropriate microvascular milieu is present in the intestine for a hypercoagulable state. Further experimental support for this concept is the protective effect of antihistamines or antiadrenergic compounds or both on the intestinal mucosa.[82]

Microbiology of Ischemic Intestinal Lesions

Bacterial colonization of ischemic intestinal lesions was common (15/24 cases) and there were other histologic features which suggested that the lesions had evolved over a period of *days*. These histologic criteria included: a prominent component of acute inflammatory cells, the presence of "classical pseudomembranes" (Fig. 12–13A), immature granulation tissue at the base of the lesions, and early mucosal regenerative changes. In almost all instances, the bacteria were gram-negative bacilli. Two cases had large colonies of gram-positive cocci (*Staphylococcus* sp.) confined to the necrotic tissue. Other sections of these lesions demonstrated that both cases were in fact lesions related to ischemia but could have been misdiagnosed as staphylococcal enterocolitis.[83] Similar problems have been presented by

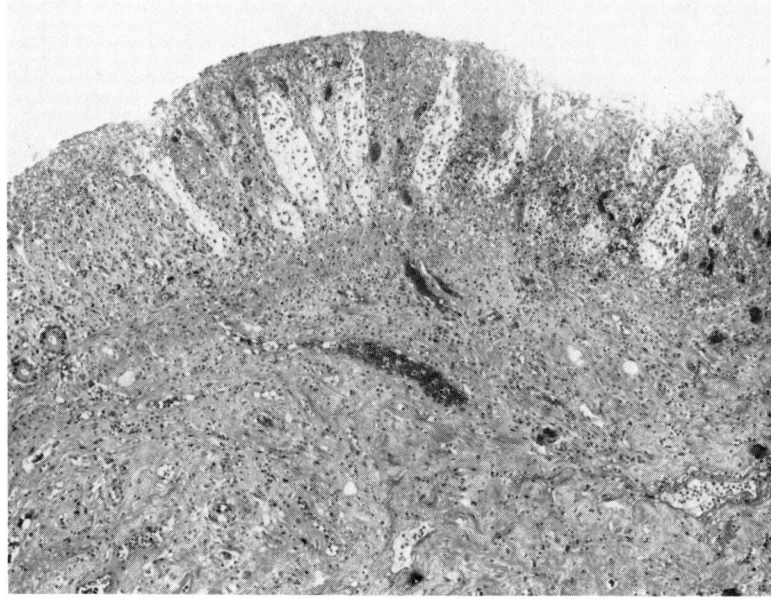

Fig. 12–10. Ischemic mucosal lesions of several days duration often have a bland form of coagulative necrosis with only mild acute inflammatory infiltrates. (H & E, × 36)

enteric clostridial lesions.[83] All lesions of this type should be examined closely with special stains for histologic evidence of bacterial extension into viable tissue (invasive infection). Twenty-one percent of our cases had definitive evidence of invasive gastrointestinal infection, and several other cases had very suggestive findings of the same (bacteria extending to, but not involving, viable tissue). These findings document a new portal from which bacterial septicemia may occur in the burn patient.

The most conspicuous bacterial growth in these 15 cases was always in the superficial necrotic debris and was not associated with underlying blood vessels. This indicated that bacterial flora in the intestinal lesions arose within the intestinal lumen and were not secondary to an episode of septicemia (although the latter does occur—see below).

Pseudomembranous Enterocolitis

Pseudomembranes were common (16 of 24 cases) in ischemic intestinal lesions. As indicated by Table 12–5, the type of pseudomembrane present in the intestine was predictive of the severity of ischemic intestinal injury.

"Classical pseudomembranes" were composed of amorphous debris and fibrin containing necrotic cells, leukocytes, and a few erythrocytes. Pseudomembranes of this type were loosely attached to the surface of minimally damaged mucosa (Fig. 12–13A). These pseudomembranes were characteristic of milder lesions[76] and probably evolved over a period of *days* as suggested by the frequent bacterial colonization. If pseudomembranes of similar composition (but containing more necrotic cells) were attached directly to the denuded surface of submucosa, they were termed "nonclassical pseudomembranes" and were more commonly associated with lesions of moderate or marked severity (Fig. 12–13B). "Erythrocytic pseudomembranes" were composed almost entirely of red cells, suggested severe hemorrhagic necrosis, and were a cardinal sign of very severe and life-threatening ischemic intestinal injury (Fig. 12–11).

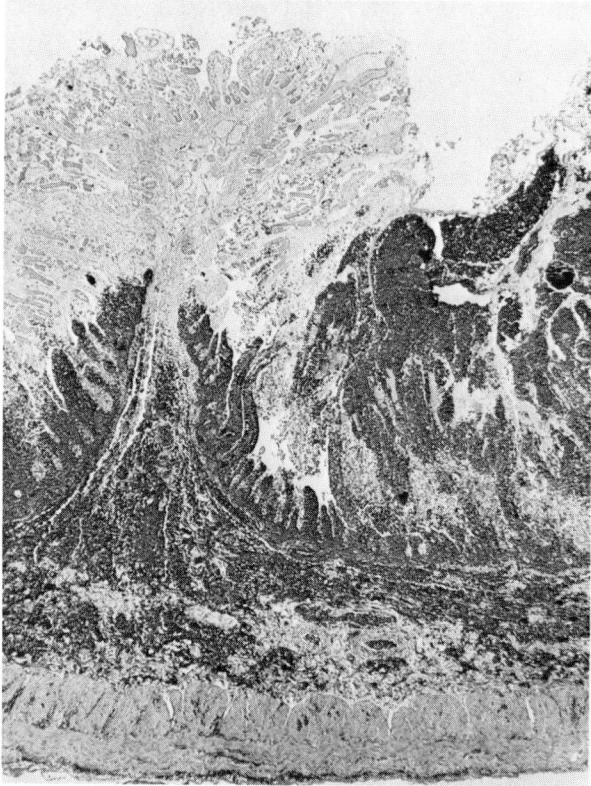

Fig. 12–11. Patients with ischemic necrosis and significant intraluminal bleeding often develop thick erythrocytic pseudomembranes. (H & E, × 40)

Table 12–6 *Differential Diagnosis of Lesions Simulating Ischemic Enterocolitis*

	Mucosa	Hemorrhage	Bacteria	Clinical Picture
Ischemic enterocolitis	Usually necrotic	Most prominent in mucosa	Variable number, colonize ischemic, necrotic mucosa; usually gram-negative bacilli	Negative to abdominal distention, diarrhea, bleeding
Anticoagulation	Usually intact	Confined to submucosa	None	Often asymptomatic
"Metastatic" *P. aeruginosa*	Usually necrotic	Focal—centered around necrotic vessels invaded by bacteria	Conspicuous around blood vessels and perivascular tissue; gram-negative bacilli	Septicemia and endotoxic shock
Infectious thromboembolus	Usually necrotic	Very little	Centered within infected thrombus; usually gram-positive cocci	Recurrent episodic fever
High dose vasopressor Rx	Usually necrotic	Moderate	Superficial colonization	Shock

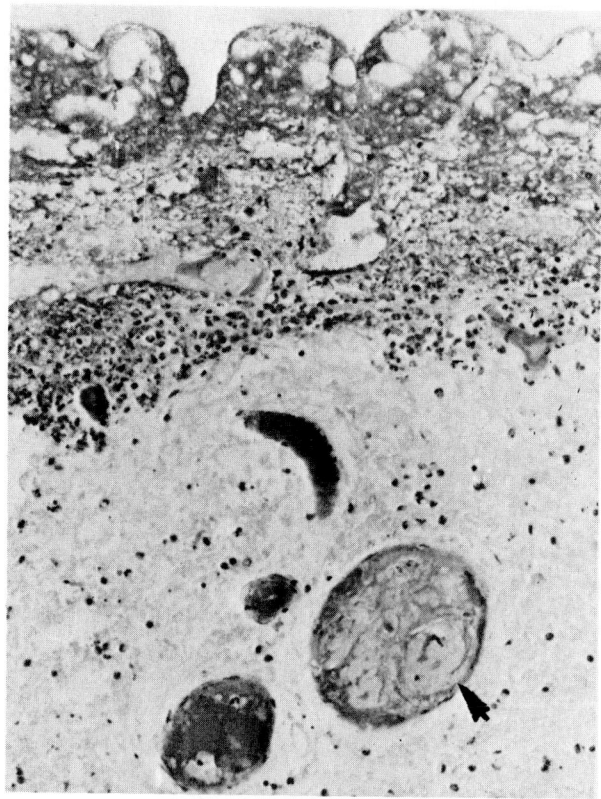

Fig. 12–12. Small submucosal vessels in segments of bowel affected by ischemic enterocolitis are occasionally found occluded by masses of platelets (arrow). (PAS stain, × 135)

Regeneration in Ischemic Lesions

Often the earliest evidence of regeneration is the presence of hyperplastic, intestinal mucosal cells forming irregular and distorted glands (Fig. 12–14). These first appear in a few days to 1 week following injury.[83] With the passage of 2–3 weeks, the cellular atypicality of the hyperplastic epithelium disappears; subsequently, the glands become more differentiated and uniform, although the underlying granulation tissue may persist for several weeks. After complete healing (months later), the only evidence of prior ischemic injury is the absence of the muscularis mucosa.[84–86] If the initial ischemic lesion extended into the muscularis propria (externa), focal fibrosis, disruption of muscle fibers, and loss of ganglion cells may be seen in the healed lesion.[86] Healing in severe ischemic enterocolitis may lead to

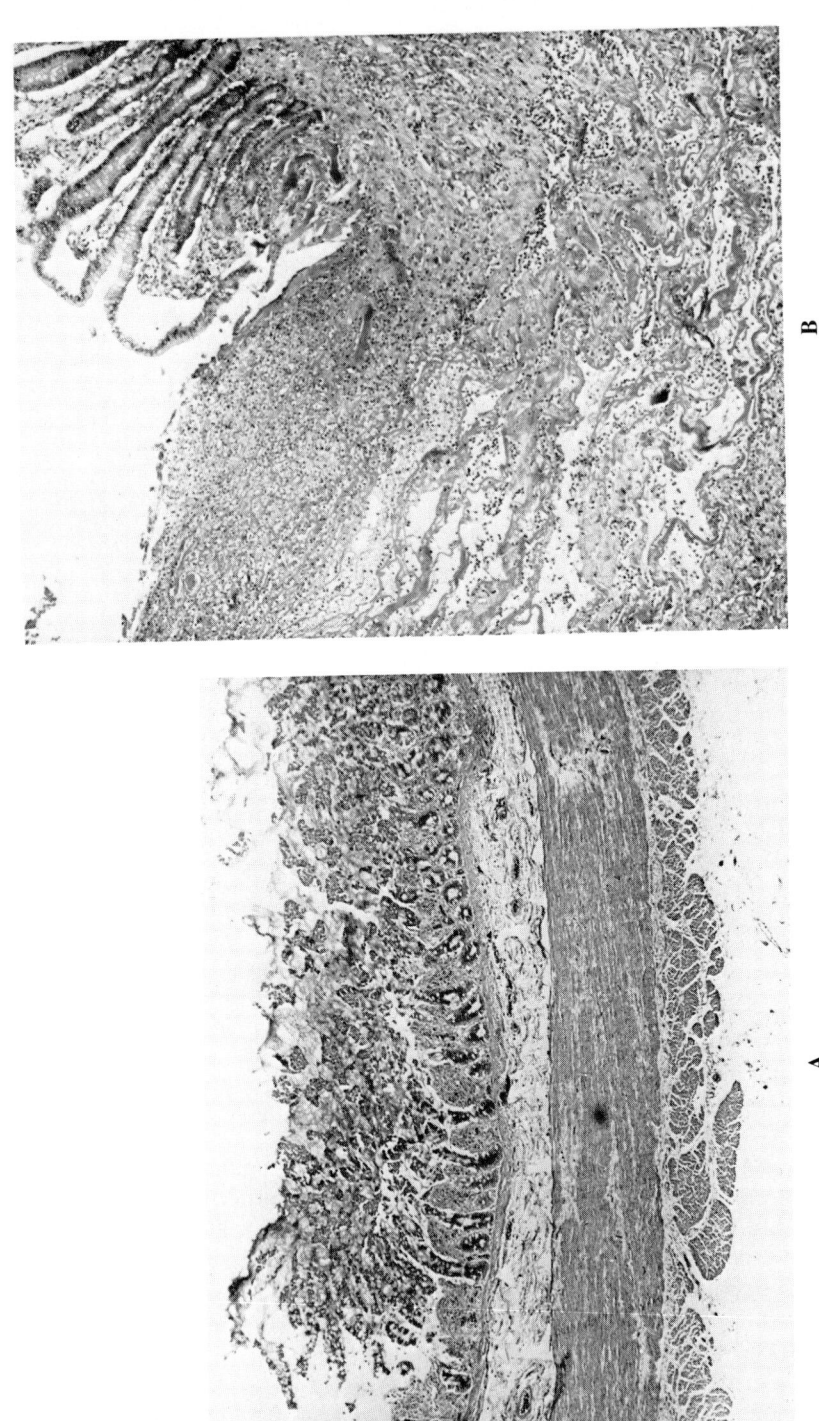

Fig. 12–13. (**A**): Pseudomembranes composed of fibrin and amorphous cellular debris are often attached to the surface of minimally damaged mucosa. (H & E, × 40) (**B**): Fibrinonecrotic pseudomembrane attached directly to denuded submucosa of colon. (H & E, × 36)

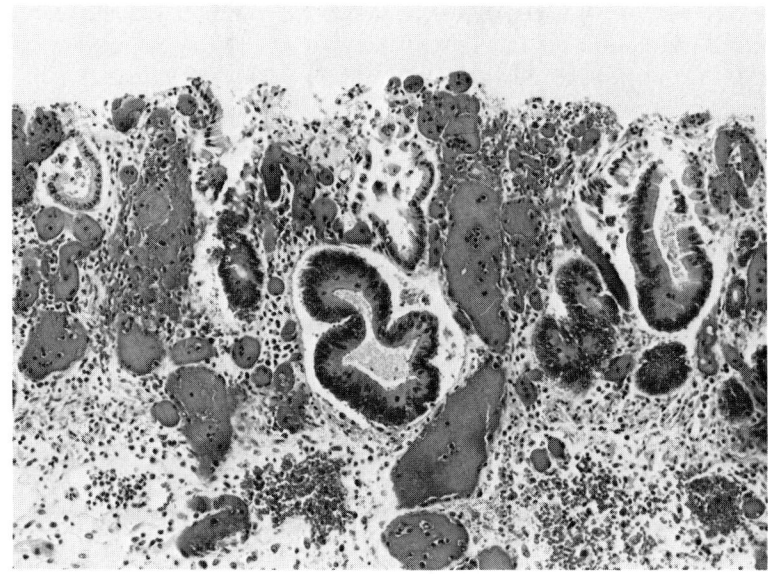

Fig. 12–14. Within a few days following ischemic mucosal injury, evidence of regeneration may be seen as mucosal epithelium forms irregular and distorted glands. (H & E, × 54)

fibrosis and stricture formation in some patients.[87] Evolution of such lesions have been documented roentgenographically in humans,[83,88–89] and this has been supported by experimental work in dogs.[90]

SUPERIOR MESENTERIC ARTERY SYNDROME

Superior mesenteric artery syndrome (SMAS) has been defined as the obstruction of the distal duodenum by surrounding structures.[91] SMAS, although uncommon (1.1 percent incidence) in the burn patient, is a life-threatening condition which may not be amenable to medical or surgical intervention or both.[92] The patients usually have severe thermal injury.[93,94] In nonburned patients, SMAS is generally chronic and more benign.[91]

Anatomically, the distal (transverse and ascending) duodenum is firmly fixed in a retroperitoneal position and is essentially "trapped" in the angle between posterior structures (vertebral column and aorta) and anterior vessels (superior mesenteric artery and middle colic artery). Postulated mechanisms which explain the acute narrowing of this angle include (1) acute weight loss* leading to loss of mesenteric and retroperitoneal fat and increased traction of the intestine on the blood vessels, (2) weak abdominal muscles, (3) prolonged bed rest in a supine position, (4) anterior displacement of the duodenum caused by increased lumbar lordosis.[92]

Clinically, the burn patients were often slender or of average build and presented with painless upper abdominal distention, recurrent emesis, excessive gastric drainage and excessive weight loss. The entity is confirmed by roentgenographic studies.[93] Upper gastrointestinal studies demonstrate abrupt, oblique obstruction of the third portion of the duodenum. Contrast material moves "to and fro" in the duodenum and refluxes back into the stomach. Placement of the patient in a knee–elbow position usually relieves the obstruction, but the latter maneuver may be impossible in the severely burned patient.

In patients who succumb to this condition, the small bowel is dilated and has a dusky, pink-blue color. Necrosis is not seen. Some cases had associated gastric or duodenal ulcers or both, but these were in areas distant from the point of compression.[92,94] Marked duodenal enlargement has been associated with the absence of ganglionic cells in Auerbach's plexus,[95] but limited studies of the duodenum in cases of SMAS have demonstrated an intact Auerbach's plexus.[94] Hence, all data suggest as an etiology physical compression of the duodenum by surrounding structures.

PSEUDO-OBSTRUCTION OF THE COLON

Approximately 1 percent of burn patients suffer from pronounced distention of the colon. There are no significant gross lesions other than marked distention and resultant thinning of the bowel wall. Microscopic findings are likewise unremarkable.

Clinically, these patients complain of crampy abdominal pain and have prominent abdominal distention and obstipation. A variety of acute or chronic conditions may be associated with pseudo-obstruction of the colon,[96] but those which may be seen in the burn patient include: pregnancy, pelvic trauma, congestive heart failure, renal failure, hepatic failure, electrolyte imbalance, sepsis, peritonitis, circulatory collapse, and alcoholism. Laboratory changes, including anemia, hypokalemia, and hypoalbuminemia, are those most commonly seen in the debilitated patient. Monitoring the leukocyte count may suggest when perforation is impending or has occurred.

Small bowel distention may occur with colonic distention, and disten-

---

* Current aggressive hyperalimentation may abolish this entity. Only one probable case was noted during a recent 2-year period.

tion in both sites is associated with a better prognosis.[96] The cecum is the most likely site for perforation, as suggested by the law of LaPlace (T = Pr).*[96,97] The mortality rates for cecal perforation have been reported to vary from 25 percent to 46 percent.[96–98]

From the variety of conditions associated with pseudoobstruction of the colon, one might suspect that the pathogenesis is uncertain. "Hinge-type" kinks at the colonic flexures have been reported,[99] but are almost certainly part of the late developmental stage of this lesion. Very likely, the pathogenesis is complex and may include ischemia, infection, and metabolic and hormonal abnormalities. Modes of therapy for pseudoobstruction of the colon vary considerably and depend on the degree and duration of pseudoobstruction of the colon and the age and general condition of the patient.

Pseudomonas aeruginosa-Induced Enterocolitis

In the past, hematogenously-disseminated infectious gastroenteric lesions were noted in certain burn patients. They were characterized by a bacterial vasculitis (bacterial invasion of the walls of small arteries and veins) with necrosis of the surrounding tissue. Similar lesions were readily reproduced in the burned rat model (Fig. 12–4A) in which the burn wound has inoculated with certain strains of *Pseudomonas aeruginosa*. These infectious lesions can be distinguished microscopically from ischemic lesions by the characteristic suppurative and necrotic nature, as well as bacterial vasculitis that is typical of hematogenous *Pseudomonas aeruginosa*.

Infected Thromboembolus

When discrete intestinal infarctions are identified in a patient with infectious thrombophlebitis or endocarditis, the possibility of metastatic infected thromboemboli in intestinal vessels is raised. The identification of an infected thrombus (abundant polymorphonuclear leukocytes and bacteria, usually *Staphylococcus* sp.) in a small intestinal artery is supportive of a hematogenous origin of the thrombus. In contrast, thrombi associated with "nonocclusive" ischemic enterocolitis are bland and consist of platelets or fibrin–platelet mixtures. These bland thrombi are confined to the microvasculature (arterioles, capillaries, and venules), and contain only occasional leukocytes and no bacteria. Evidence of origin of the septic thromboembolus from infectious endocarditis is suggested by the identification of similar lesions in the kidney, lung, or other viscera.

---

* Tension equals pressure times radius.

Hemorrhagic Lesions Related to Anticoagulation

Hemorrhagic intestinal lesions in the burn patient are not always related to ischemic enterocolitis. If the hemorrhage is confined to the submucosa and the overlying mucosa is intact, one should consider the possibility of a complication associated with anticoagulation (Fig. 12–15).

In such cases listed in the literature, all of the patients have had chronic exposure to anticoagulation with coumarin. Radiologically, the presence of mesenteric hematomata gives a "stacked coin" or "picket-fence" pattern indicative of submucosal infiltration.[100] Neither of two cases seen at the ISR were treated with coumarin, but both had been transiently treated with heparin. To our knowledge, these changes have not previously been reported with heparin therapy.

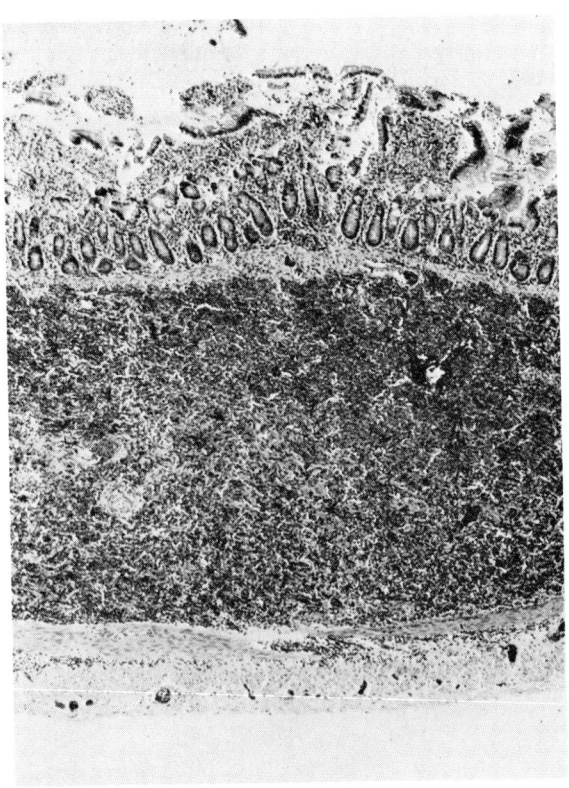

Fig. 12–15. Large submucosal hemorrhages of the intestine may be associated with anticoagulant therapy in burn patients. (H & E, × 36)

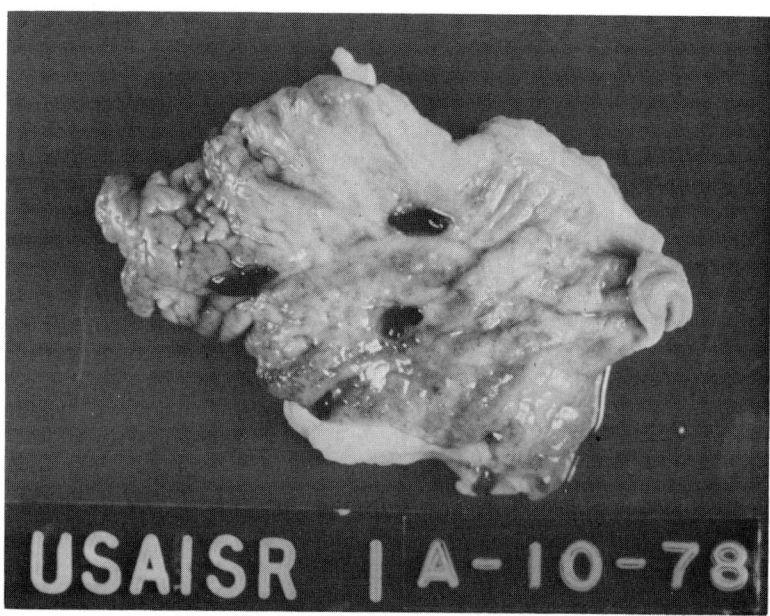

Fig. 12–16. Linear or oblong hemorrhagic erosions as shown here near the gastroesophageal junction are usually caused by nasogastric tubes.

Electrical Injury of the Intestinal Tract

Electrical injury to the intestine is quite rare. The electrical current tends to pass through solid organs (liver) or organs filled with fluid (heart), and therefore avoids porous (lung) or hollow (bowel) organs. We are aware of two previously reported cases,[101,102] and have studied one case at the ISR. In intestinal electric injury, the involved portion of bowel is usually adjacent to a severely injured part of the body wall. The bowel lesions are related to the dissipation of an extraordinary amount of thermal energy at the site adjacent to a viscus, and the bowel is only secondarily involved.

IATROGENIC ESOPHAGEAL LESIONS

Nasogastric tubes regularly cause multiple linear erosions (retropharyngeal and gastroesophageal junction) (Fig. 12–16) or circular areas of hemorrhage in the gastric mucosa measuring 4–6 mm in diameter. Significant bleeding from these lesions is uncommon; however, mural edema and even abscess formation may develop and interfere with deglutition and passage of esophageal contents.

References ───

1. Czaja, A. J., McAlhany, J. C., Jr., Andes, W. A., and Pruitt, B. A., Jr.: Acute gastric disease after cutaneous thermal injury. Arch. Surg. 110:600–605, 1975.
2. Pruitt, B. A., Jr., and Goodwin, C. W., Jr.: Stress ulcer in the burned patients. World J. Surg. 5:209–222, 1981.
3. Teplitz, C.: The pathology of burns and the fundamentals of burn wound sepsis. *In* Burns, A Team Approach. C. P., Artz, J. A., Moncrief, and B. A. Pruitt, Jr., eds. Philadelphia, W. B. Saunders Co., 1979, p 88.
4. Szabo, S.: Biology of disease. Pathogenesis of duodenal ulcer disease. Lab. Invest. 51:121–147, 1984.
5. McAlhany, J. C., Jr., Czaja, A. J., and Pruitt, B. A., Jr.: Antacid control of complications from acute gastrointestinal disease after burns. J. Trauma 16:645–649, 1976.
6. Levine, B. A., Teegarden, D. K., McLeod, C. G., Jr, Sirinek, K. R., and Pruitt, B. A., Jr.: Cimetidine prevents stress-induced gastric erosions. Surg. Forum 28:359–361, 1977.
7. Hunt, T.: Injury and repair in acute gastroduodenal ulceration. Amer. J. Surg. 125:12–18, 1973.
8. Levenson, S. M., Pirani, C. L., Braash, J. W., and Waterman, D. F.: The effect of thermal burns on wound healing. Surg. Gynec. Obstet. 99:74–82, 1954.
9. Chernov, M. S., Cook, F. B., Wood, M., and Hale, H. W., Jr.: Stress ulcer: A preventable disease. J. Trauma 12:831–833, 1972.
10. Walker H. L., and Mason, A. D., Jr.: A standard animal burn. J. Trauma 8:1049–1051, 1968.
11. Pruitt, B. A., Foley, F. D., and Moncrief, J. A.: Curling's ulcer: A clinical–pathology study of 323 cases. Ann. Surg. 172:523–539, 1970.
12. Fraenkel, E. Uber allgemeininfektionendurch den *Bacillus pyocyaneus.* Virchows Arch. Path. Anat. 183:405–440, 1906.
13. Markley, K., Gurmendi, G., Chavez, P. M., and Bazan, A.: Fatal *Pseudomonas* septicemias in burned patients. Ann. Surg. 145:175–181, 1957.
14. Sevitt, S.: Duodenal and gastric ulcerations after burning. Br. J. Surg. 54:32–41, 1967.
15. Rabin, E. R., Graber, C. D., Vogel, E. H., Finkelstein, R. A., and Tumbusch, W. A.: Fatal *Pseudomonas* infections in burned patients. New Engl. J. Med. 265:1225–1231, 1961.
16. Teplitz, C. Pathology of burns. *In* The Treatment of Burns. C. P. Artz and J. A., Moncrief, eds. Philadelphia, W. B. Saunders Co., 1969, p 78.
17. McLeod, C. G., Jr., McManus, A. T., Panke, T. W., and Mason, A. D., Jr.: Gastrointestinal alterations and complications in burned troops—Ulcerative gastric disease in the septic burned rat: A model for the study of non-healing ulcers in the military population. Annual Research Report, October 1, 1977–September 30, 1978, Fort Sam Houston, Tex., pp 120–124.
18. Walker, H. L., Mason, A. D., Jr., and Raulston, G. L.: Surface infection with *Pseudomonas aeruginosa.* Ann. Surg. 160:297–305, 1964.
19. Loebl, E. C., Marvin, J. A., Curreri, P. W., and Baxter, C. R.: Survival with

ecthyma gangrenosum, a previously fatal complication of burns. J. Trauma 14:370–377, 1974.

20. Kirsner, J. B.: Peptic ulcer: A review of the recent literature on various clinical aspects. II. Gastroenterology 54:945–975, 1968.

21. Necheles, H., and Olson, W. H.: Experimental investigation of gastrointestinal secretions and motility following burns and their relation to ulcer. Surgery 11:751–765, 1942.

22. Bonfils, S., LieFooghe, G., Rossi, G., and Lambling, A.: Lulcere experimental de contrainte du rat blanc. Analyse des principaux facteurs determinants. Arch. Mal. Appar. Dig. 48:449–459, 1959.

23. Brodie, D. A., Marshall, R. W., and Moreno, O. M.: Effect of restraint on gastric acidity in the rat. Am. J. Physiol. 202:812–814, 1962.

24. Czaja, A. J., McAlhany, J. C., and Pruitt, B. A., Jr.: Gastric acid secretion and acute gastroduodenal disease after burns. Arch. Surg. 111:243–245, 1976.

25. Rosenthal, A., Czaja, A. J., and Pruitt, B. A., Jr.: Gastric levels and gastric acidity in the pathogenesis of acute gastroduodenal disease after burns. Surg. Gynec. Obstet. 144:232–234, 1977.

26. Friesen, S. R.: The genesis of gastroduodenal ulcer following burns: An experimental study. Surgery 28:123–158, 1950.

27. Menguy, R.: Effect of restraint stress on gastric secretion in the rat. Amer. J. Dig. Dis. 5:911–916, 1960.

28. O'Neill, J. A., Pruitt, B. A., Jr., Moncrief, J. A., and Switzer, W. E.: Studies related to the pathogenesis of Curling's ulcer. J. Trauma 7:275–287, 1967.

29. O'Neill, J. A., Jr.: The influence of thermal burns on gastric acid secretion. Surgery 67:267–271, 1970.

30. Robbins, R., Idjadi, F., and Stahl, W., et al.: Studies on gastric secretion in stressed patients. Ann. Surg. 175:555–562, 1972.

31. Harrison, A. M., Gaisford, J. C., and Wechsler, R. L.: Gastric secretion in the burned patient. J. Trauma 12:1041–1043, 1973.

32. Greenberg, L., and Himal, H. S.: Hydrochloric acid clearance and endotoxin induced duodenal erosions in dogs. Surg. Gynec. Obstet. 139:561–565, 1974.

33. Goldman, H., and Rosoff, C. B.: Pathogenesis of acute gastric stress ulcers. Am. J. Pathol. 52:227–243, 1968.

34. Hase, T., Anderson, P. R., and Mehlman, B.: Significance of gastric secretory changes in the pathogenesis of stress ulcers. Am. J. Dig. Dis. 20:443–448, 1975.

35. Zinner, M. J., Turtinen, L., and Gurll, N. J.: The role of acid and ischemia in production of stress ulcers during canine hemorrhagic shock. Surgery 77:807–816, 1975.

36. Max, M., and Menguy, R.: Influence of adrenocorticotropin, cortisone, aspirin, and phenylbutazone on the rate of exfoliation and the rate of renewal of gastric mucosal cells. Gastroenterology 58:329–336, 1970.

37. Rasanen, T.: The role of the mucosal mast cells in gastric ulceration. *In* Peptic Ulcer. C., Pfieffer, ed. Philadelphia, Lippincott Co., 1971, pp 237–243.

38. Wolf, S., and Wolff, H. G.: Human Gastric Function. NY, Oxford University Press, 1943, p 167.

39. Hollander, F.: The two-component mucous barrier; its activity in protecting

the gastroduodenal mucosa against peptic ulceration. Arch. Intern. Med. 93:107–120, 1954.

40. O'Neill, J. A., Jr., Ritchey, C. R., and Mason, A. D., Jr.: Influence of thermal burns on gastric mucous production. Surg. Forum 17:293–295, 1966.

41. Menguy, R., and Desbailetts, L.: The gastric mucous barrier: Influence of protein-bound carbohydrate in mucus on the rate of proteolysis of gastric mucus. Ann. Surg. 168:475–482, 1968.

42. Chung, R. S. K., Field, N., and Silen, W.: Gastric permeability to hydrogen and lithium: A new method for quantitation of the gastric mucosal barrier. Surg. Forum 21:297–299, 1970.

43. Chiu, C. J., McArdle, A. H., and Brown, R. A.: Gastric mucosal changes following burns in rats. A morphological and metabolic approach to the stress ulcer problem. Arch. Surg. 103:142–152, 1971.

44. McAlhany, J. C., Jr., Czaja, A. J., and Villarreal, Y., et al.: The gastric mucosal barrier in thermally injured patients: Correlation with gastroduodenal endoscopy. Surg. Forum 25:414–416, 1974.

45. Menguy, R., and Masters, Y. F.: Effect of cortisone on mucoprotein secretion by gastric antrum of dogs: Pathogenesis of steroid ulcer. Surgery 54:19–28, 1963.

46. Desbaillets, L., and Menguy, R.: Inhibition of gastric mucous secretion by ACTH: An experimental study. Amer. J. Dig. Dis. 12:582–588, 1967.

47. Barclay, A. E., and Bentley, F. H.: The vascularization of the human stomach: A preliminary note on the shunting effect of trauma. Br. J. Radiol. 22:62–67, 1949.

48. Barlow, T. E., Bentley, F. H., and Walder, D. N.: Arteries, veins and arteriovenous anastomoses in the human stomach. Surg. Gynec. Obstet. 93:657–671, 1951.

49. Womack, N. A.: Blood flow through the stomach and duodenum. Amer. J. Surg. 117:771–780, 1969.

50. Hase, T., and Moss, B. J.: Microvascular changes of gastric mucosa in the development of stress ulcer in rats. Gastroenterology 65:224–234, 1973.

51. Myren, J.: On secretion and cell structure of gastric mucosa. Scand. J. Clin. Lab. Invest. 15:30–39, 1963.

52. Sethbhakdi, S., Roth, J. L., and Pfeiffer, C. J.: Gastric mucosal ulceration after epinephrine. A study of etiologic mechanisms. Amer. J. Dig. Dis. 15:1055–1065, 1970.

53. Linder, M. M., and McKay, D. G.: An experimental study of thrombotic ulceration of the gastrointestinal mucosa. Surg. Gynec. Obstet. 133:21–29, 1971.

54. Margaretten, W., and McKay, D. G.: Thrombotic ulcerations of the gastrointestinal tract. Arch. Intern. Med. 127:250–253, 1971.

55. Chiu, C. J., Scott, H. J., and Gurd, F. N.: Intestinal mucosal lesion in low flow states II. The protective effect of intraluminal glucose as energy substrate. Arch. Surg. 101:484–488, 1970.

56. Menguy, R., Desbaillets, L., and Masters, Y. F.: Mechanisms of stress ulcer: Influence of hypovolemic shock on energy metabolism in the gastric mucosa. Gastroenterology 66:46–55, 1974.

57. Birke, G., Duner, H., Liljedahl, S. O., Pernow, B., Plantin, L. O., and Troell,

L.: Histamine, catecholamines and adrenocortical steroids in burns. Acta. Chir. Scand. 114:87–98, 1958.

58. Bruck, H. M., and Pruitt, B. A., Jr.: Curling's ulcer in children: A 12-year review of 63 cases. J. Trauma 12:490–496, 1972.

59. Billroth, T.: Ueber duodenalgeschwure bei septicamie. Wien. Med. Wchnschr. 17:705–709, 1867.

60. Moynihan, B. G. A.: Duodenal Ulcer, Chapter 2. W. B. Saunders Co., Phila 1912, pp 24–43.

61. Hartman, F. W.: Curling's ulcer in experimental burns. II. The effect of penicillin therapy. Gastroenterology 6:130–139, 1946.

62. Rosoff, C. B., and Goldman, H.: Effect of intestinal bacterial flora on acute gastric stress ulceration. Gastroenterology 55:212–222, 1968.

63. Nance, F. C., Kaufman, H. J., and Batson, R. C. The role of microbial flora in acute gastric stress ulceration. Surgery 72:68–73, 1972.

64. Batson, R. C. Effect of serotonin and endotoxin on acute gastric stress ulceration in conventional germ free rats. Surg. Forum 24:426–428, 1973.

65. Nichols, R. L., and Smith, J. W.: Intragastric microbial colonization in common disease states of stomach and duodenum. Ann. Surg. 182:557–561, 1975.

66. Beck, I. T., Kahn, D. S., and Lacerte, M.: Chronic duodenitis: A clinical pathological entity. Gut 6:376–383, 1965.

67. Bonfils, S., and Lambing, A.: Psychological factors and psychopharmacological actions in the restraint-induced gastric ulcer. *In* Pathophysiology of Peptic Ulcer. S. C., Skoryna, ed. Montreal, McGill Univ. Press, 1963, pp 153–171.

68. Davenport, H. W.: Physiology of the Digestive Tract, Chapters 8, 9. Chicago, Year Book Medical Publishers, 1966.

69. Guilbert, J., Bounous, G., and Gurd, F.: Role of intestinal chyme in the pathogenesis of gastric ulceration following experimental hemorrhagic shock. J. Trauma 9:723–743, 1969.

70. Hamza, K. N., and Den Beston, L.: Bile salts producing stress ulcers during experimental shock. Surgery 71:161–167, 1972.

71. Braun, S. A., Sampson, R. H., Norton, L., and Eisman, B.: Bile reflux in experimental stress ulcer. Surgery 73:521–524, 1973.

72. Watson, L. C., and Alston, S.: Prevention of upper gastrointestinal hemorrhage in 582 burned children. Amer. J. Surg. 132:790–793, 1976.

73. Panke, T. W. Unpublished data pertaining to gastrointestinal problems in burn patients from Institute of Surgical Research, Fort Sam Houston, Tex.

74. Hagihara, P. F., Parker, J. C., and Griffen, W. O., Jr.: Spontaneous ischemic colitis. Dis. Col. Rect. 20:236–251, 1977.

75. McKay, D. G., Hardaway, R. M., III, Wahle, G. H., Jr., and Hall, R. M.: Experimental pseudomembraneous enterocolitis. Production by means of thrombosis of intestinal mucosae capillaries. Arch. Intern. Med. 95:779–787, 1955.

76. Whitehead, R.: Ischaemic enterocolitis: An expression of the intravascular coagulation syndrome. Gut 12:912–917, 1971.

77. Hardie, R. R., and Nicoll, P.: Localized ulceration of the caecum due to microcirculatory thrombosis. A new concept of non-specific ulceration of the caecum. Aust. N.Z. J. Surg. 43:149–157, 1973.

78. Bonakdarpour, A., Ming, S., Lynch, P. R., Essa, N., and Reichle, F.: Superior mesenteric artery occlusion in dogs: A model to produce the spectrum of intestinal ischemia. J. Surg. Res. 19:251–257, 1975.

79. Neame, P. B., Lechago, J., Ling, E. T., and Koval, A.: Thrombotic thrombocytopenic purpura: Report of a case with disseminated intravascular platelet aggregation. Blood 42:807–814, 1973.

80. Zweifach, B. W., Nagler, A. L., and Thomas, L.: The role of epinephrine in the reactions produced by endotoxins of gram-negative bacteria. II. The changes produced by endotoxin in the vascular reactivity to epinephrine in the rat meso-appendix, and the isolated, perfused rabbit ear. J. Exp. Med. 104:881–896, 1956.

81. Renton, C. J. C.: Non-occlusive intestinal infarction. Clin. Gastroenterol. 1:655–673, 1972.

82. Bialostozky, D., Contreras, R., Tinajeros, C. A., and Franco-Browder, S.: Gastrointestinal hemorrhagic necrosis. Report of ten cases. Amer. J. Med. 46:90–95, 1969.

83. Whitehead, R.: Reversible ischaemic colitis. The Practitioner 213:54–58, 1974.

84. Whitehead, R.: The pathology of intestinal ischaemica. Clin. Gastroenterol. 1:613–637, 1972.

85. Robinson, J. W. L., Rausis, C., Basset, P., and Mirkovitch, V.: Functional and morphological response of the dog colon to ischaemia. Gut 13:775–783, 1972.

86. Ming, S. C., and Bonakdarpour, A.: Evolution of lesions in intestinal ischemia. Arch. Pathol. Lab. Med. 101:40–43, 1977.

87. Marcuson, R. W.: Ischaemic colitis. Clin. Gastroenterol. 1:745–763, 1972.

88. Dewbury, K. C.: Ischaemic and evanescent colitis. Proc. Roy. Soc. Med. 69:617, 1976.

89. Meyers, M. A., Ghahremani, G. G., and Govoni, A. F.: Ischemic colitis associated with sigmoid volvulus: New Observations. Amer. J. Roentgenol. 128:591–595, 1977.

90. Marston, A., Marcuson, R. W., Champan, M., and Arthur, J. F.: Experimental study of devascularization of the colon. Gut 10:121–130, 1969.

91. Barner, H. B., and Sherman, C. D., Jr.: Vascular compression of the duodenum. Int. Abstr. Surg. 117:103–118, 1963.

92. Reckler, J. M., Bruck, H. M., Munster, A. M., and Curreri, P. W.: Superior mesenteric artery syndrome as a consequence of burn injury. J. Trauma 12:979–985, 1972.

93. Wallace, R. G., and Howard, W. B.: Acute superior mesenteric artery syndrome in the severely burned patient. Radiology 94:307–310, 1970.

94. Ogbuokiri, C. G., Law, E. J., and MacMillian, B. G.: Superior mesenteric artery syndrome in burned children. Amer. J. Surg. 124:75–79, 1972.

95. Barnett, W. O., and Wall, L.: Megaduodenum resulting from absence of parasympathetic ganglion cells in Auerbach's plexus. Ann. Surg. 141:527–535, 1955.

96. Lescher, T. J., Teegarden, D. K., and Pruitt, B. A., Jr.: Acute pseudoobstruction of the colon in thermally-injured patients. Dis. Col. Rect. 21:618–622, 1978.

97. Norton, L., Young, D., and Scribner, R.: Management of pseudoobstruction of the colon. Surg. Gynec. Obstet. 138:595–598, 1974.
98. Wojtalik, R. S., Lindenauer, S. M., and Kahn, S. S.: Perforation of the colon associated with adynamic ileus. Amer. J. Surg. 125:601–606, 1973.
99. Kukora, J. S., and Dent, T. L.: Colonscopic decompression of massive nonobstructive cecal dilatation. Arch. Surg. 112:512–517, 1977.
100. Messinger, N. H., and Beneventano, T. C.: Submucosal hemorrhage of the small intestine: A complication of anticoagulant therapy. Amer. J. Gastroenterol. 57:74–78, 1972.
101. DiVincenti, F. C., Moncrief, J. A., and Pruitt, B. A., Jr.: Electrical injuries. A review of 65 cases. J. Trauma 9:497–507, 1969.
102. Sinha, J. K., and Roy, S. K.: Perforation of the caecum caused by an electrical burn. J. Plast. Surg. 29:179–181, 1976.
103. Ming, S. Levitan, R.: Acute hemorrhagic necrosis of the gastrointestinal tract. New Eng. J. Med. 263:59–65, 1960.
104. MacPherson, B., and Pfeiffer, C. J.: Experimental colitis. Digestion 14:424–452, 1976.
105. Rifkin, G. D., Fehety, F. R., Silva, J., Jr., and Sack, R. B.: Antibiotic-induced colitis: Implication of a toxic neutralized by *Clostridium sordellii* antitoxin. Lancet 2:1103–1106, 1977.
106. McKay, D. G., Linder, M. M., and Cruse, V. K.: Mechanisms of thrombosis of the microcirculation. Amer. J. Pathol. 63:231–241, 1971.
107. Pettet, J. D., Baggenstoss, A. H., Dearing, W. H., and Judd, E. S., Jr.: Postoperative pseudomembranous enterocolitis. Surg. Gynec. Obstet. 98:546–552, 1954.
108. Larson, H. E., and Price, A. B.: Pseudomembranous colitis: Presence of clostridial toxin. Lancet 2:1312–1314, 1977.
109. Bartlett, J. G., Chang, T. W., Gurwith, M., Gorbach, S. L., and Onderdonk, A. B.: Antibiotic-associated pseudomembranous colitis due to toxin-producing *Clostridia*. New Engl. J. Med. 298:531–534, 1978.
110. McClennon, B. L.: Ischemic colitis secondary to Premarin: Report of a case. Dis. Col. Rect. 19:618–620, 1976.
111. Buts, J., Weber, A. M., Roy, C. C., and Morin, C. L.: Pseudomembranous enterocolitis in childhood. Gastroenterology 73:823–827, 1977.
112. Miller, R. R., and Jick, H.: Antibiotic associated colitis. Clin. Pharm. Ther. 22:1–6, 1977.
113. Price, A. B., and Davies, D. R. Pseudomembranous colitis. J. Clin. Pathol. 30:1–12, 1977.
114. Read, L., and Cove-Smith, J. R.: Pseudomembraneous enterocolitis complicating ampicillin therapy. Postgrad. Med. J. 53:324–326, 1977.
115. Kay, A. W., Richards, R. L., and Watson, A. J.: Acute necrotizing (pseudomembranous) enterocolitis. Brit. J. Surg. 46:45–57, 1958.

Chapter 13

HEPATOBILIARY
SYSTEM

LIVER

The liver is unquestionably an extremely active organ in a patient who suffers a serious burn. This is true whether the burn victim is simply experiencing the usual postburn hypermetabolic state or whether the patient is withstanding the additional stresses of fluid and electrolyte imbalance, sepsis, or drug toxicity. The pathologist must be prepared to interpret and evaluate these multiple adaptive, inflammatory, and toxic hepatic changes that may occur following thermal injury.

Hepatomegaly

Dramatic enlargement of the liver is a frequent observation on postmortem examination in the burn patient. Apparently normal livers commonly weigh more than 2500 g and often exceed 3000 g in adult burn victims who die during the third postburn week. This degree of hepatomegaly contrasts strikingly with the paucity of histologic findings in the liver. Animal data[1] and preliminary postmortem evaluation in burn patients[2] have demonstrated a conspicuous increase in the dry weight and DNA content of these livers without significant histopathological change. Burn patients with hepatic disease (cholestasis, fatty metamorphosis, etc.) were excluded from this study. These findings indicate that hyperplasia of the liver rather than an increase in fluid content[3] is responsible for the postburn hepatomegaly. These changes have been interpreted by Sevitt as evidence of hepatic regeneration.[4] Histologic findings consistent with hyperplasia or regeneration include the appearance of large nuclei, with prominent nuclear membranes, large nucleoli, binucleation of hepatic cells, and scattered mitosis.

Sublethal Hepatic Changes

Hepatocellular fatty change is seen in about one-third of the fatal burn cases[3] and is usually mild. Sevitt[5] noted that the distribution of this fatty change was centrilobular in patients who died soon after sustaining thermal injury (Fig. 13–1). Later in the burn course, a periportal fatty change is more commonly observed.

Biochemical and histological evidence of hepatocellular injury is apparent by 24 hr in the severely burned patient.[6] Experimental data indicate alteration in the transmembrane potential in the early postburn course that is not related to energy depletion. This alteration results in increased membrane permeability which leads to increased sodium and decreased potassium within hepatocytes with subsequent cellular swelling.[7] In the burn patient, early postburn evidence of liver injury includes cytoplasmic vacuolation and focal crenation or "ballooning" of nuclei. Glycogen depletion and fine lipid droplets or even larger vacuoles are suggestive of other sublethal hepatocellular damage.[8] In some patients, these cellular changes apparently progress to centrilobular necrosis of hepatic cells, but the latter may also result from abnormal hemodynamics, poor tissue oxygenation, or other deficiencies related to shock or cardiac failure.

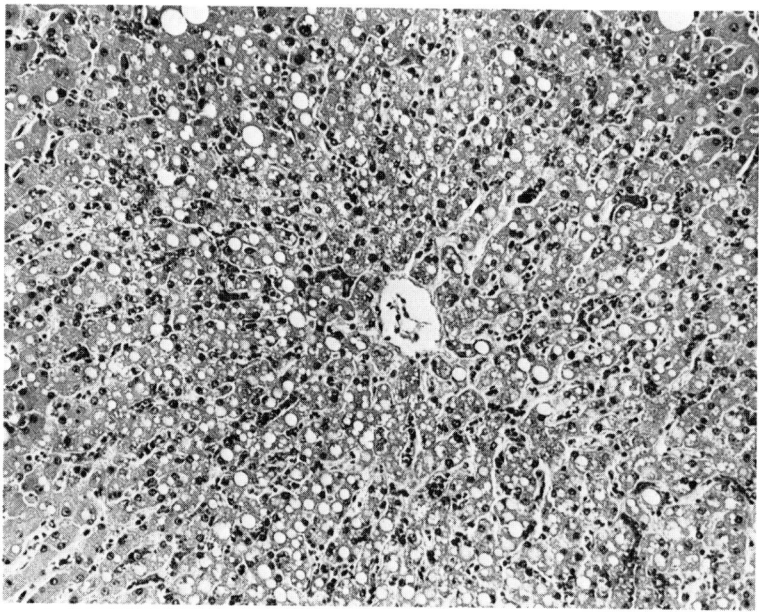

Fig. 13–1. Hepatic centrilobular fatty change is often seen in burn patients who die within the first postburn week. (H & E, × 144)

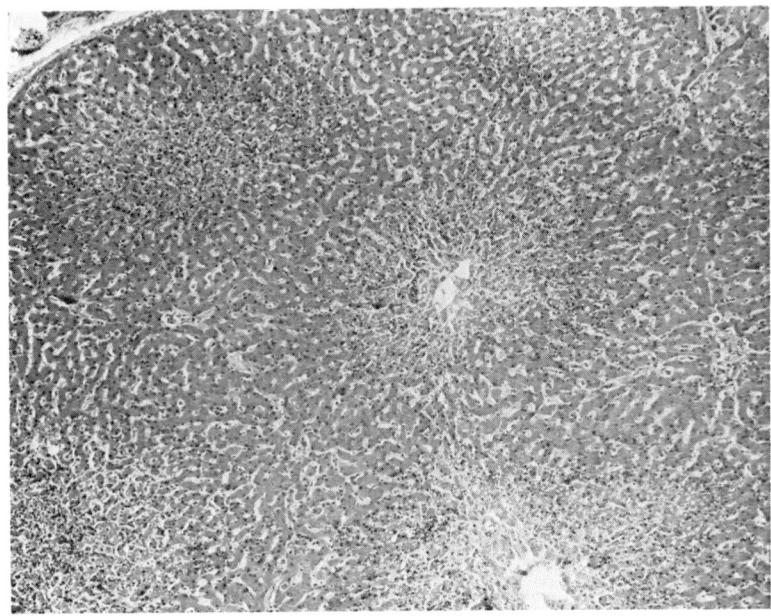

Fig. 13–2. Significant centrilobular necrosis may be found in burn patients who suffer congestive heart failure. (H & E, × 54)

Centrilobular Hepatic Congestion and Necrosis

In a series of burn patients studied by Teplitz,[3] hepatic congestion was common (50 percent of patients), but hepatic necrosis occurred in only about 12 percent of autopsies. This necrosis, usually focal and nonspecific, was most often seen in patients with sepsis and shock. Significant centrilobular necrosis may be seen in less than 4 percent of autopsies and has usually been related to significant congestive heart failure[3] (Fig. 13–2).

Jaundice

Unexplained jaundice is a common clinical and postmortem observation in the severely burned patient. It usually appears at the end of the first week or during the second week post burn.[9] Although the number of conditions which might predispose to jaundice is relatively small in the burn patient, no exhaustive study has been carried out in a large number of patients to define the responsible factors. The clinical conditions and therapeutic agents that may cause or be associated with jaundice in the burn patient are listed in

Table 13–1. A complete clinical history and chart review is paramount in the diagnostic consideration of any case of jaundice.

An immediate postburn *hemolytic anemia,* combined with a requirement for *multiple transfusions* (some patients require in excess of 100 units of blood in less than a 3-week period), leads to accelerated erythrocytic breakdown. Mild jaundice in the first 5 postburn days is related to the former.[10] Extreme erythrocytic destruction may overwhelm the capacity of the body to eliminate bilirubin. In such cases, prominent icterus may occur despite the absence of hepatic dysfunction. However, in severely burned patients, hepatic dysfunction is common,[11] especially in the immediate postburn period, and this dysfunction aggravates coexistent causes for jaundice.[6,10]

In a group of 36 severely traumatized patients with jaundice,[9] mortality was associated with marked hyperbilirubinemia and prolonged hypotension with low cardiac output. A variety of other factors (Table 13–1) known to elevate serum bilirubin were present but could not be positively correlated

Table 13–1 *Differential Diagnosis of Jaundice in the Burn Patient*

Calculous cholecystitis

Acalculous cholecystitis (causes and differential diagnosis in Table 13–4)

Drug hypersensitivity (Table 13–4)

Drug-related toxic hepatitis (Table 13–2)

Narcotics

Physical Agents

Hyperthermia
Hypothermia

Severe, sustained hypotension in early postburn course

Reduced cardiac output

Septicemia, especially gram-negative bacteria

Multiple transfusions (often greater than 100 units)

Intravascular hemolysis (immediate postburn period)

Hematoma and extravasation of blood in injured tissues

Anesthesia

Renal failure

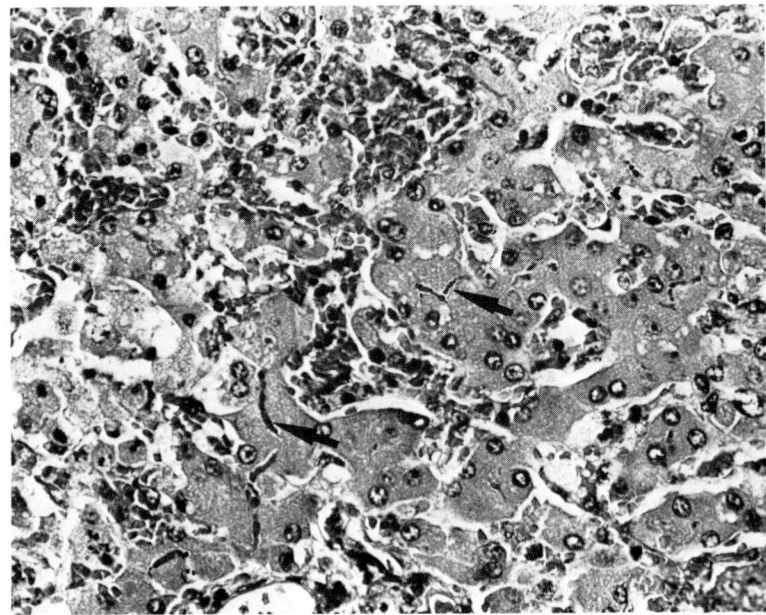

Fig. 13–3. Centrilobular intrahepatic cholestasis is often a finding in burn patients with jaundice. In this case, bile canaliculi are distended and distinctly outlined by bile contents (arrows). Adjacent centrilobular hepatocytes are necrotic. (H & E, × 342)

with patient mortality. Fatal cases had centrilobular hepatic congestion or necrosis with bile stasis (Fig. 13–3).

Heightened clinical awareness of the broad spectrum of potentially injurious drugs[12,13] which cause intrahepatic cholestasis (morbidity) or hepatic necrosis (mortality) will permit removal of any offending agent from the therapeutic regimen (Table 13–2). Direct cellular injury leading to hepatocellular necrosis is caused by yet other drugs. Tetracyclines and sulfonamides are typical of this group (Table 13–2). In drug-related *toxic hepatitis*, the latent period before symptoms appear is usually shorter (a few days instead of 2–5 weeks), and spontaneous resolution of symptoms following cessation of the drug use is less commonly noted in comparison with hepatic hypersensitivity drug reactions. Histologically, increased lipofuscin pigment (especially centrilobular) (Fig. 13–4) indicates mild damage, and more severe injury may result in multiple small random foci of hepatic necrosis. Cholestasis is usually present but is less conspicuous in toxic hepatitis. A mild, mixed inflammatory infiltrate composed of polymorphonuclear leukocytes and lymphocytes is seen in the areas of

Table 13–2 *Drugs and Clinical Conditions Associated with (Hypersensitivity—Related) Cholestasis or Toxic Hepatitis[a]*

Cholestatic (hypersensitivity reaction) agents

Chlorpromazine
17-C-substituted testosterones
Halothane
Methoxyflurane
Para-aminosalicylate
Isonicotinic acid hydrazide (isoniazid)
Methyldopa
Erythromycin
Nitrofurantoin
Diphenylhydantoin (Dilantin, Parke-Davis)
Methimazole
Pheylbutazone
Hydralazine
Oxacillin
Penicillin
Chlorpropamide

Direct hepatotoxins (toxic hepatitis)

Sulfonamides
Tetracyclines
Cytotoxic drugs
Acetaminophen
Salicylates
Ferrous sulfate
Phenobarbital

Physical agents

Hyperthermia

Hypothermia

---

[a] This partial list of drugs and conditions considers only those which might be used in the treatment of the burn patient.

necrosis and occasionally in portal triads.[12,13] Hepatotoxins all affect protein synthesis, as indicated by an early fall in prothrombin levels, and the fatty change may be related to decreased synthesis of carrier proteins.[13] Although the classical pattern of toxic hepatitis and hypersensitivity reactions to drugs is easily distinguished, overlap of histologic patterns is frequent, and good clinicopathologic correlation is necessary to insure accurate diagnosis. Table 13–3 presents the classical histologic features that differentiate drug-related

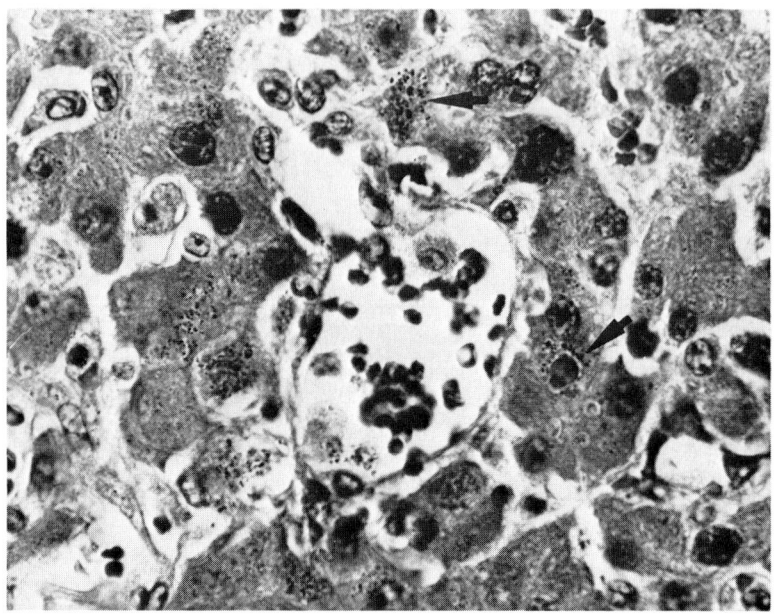

Fig. 13–4. Intracellular "wear and tear" pigment (arrows), lipofuscin, is often found in hepatocytes of burn patients. (H & E, × 540)

cholestasis from cholestasis caused by bile duct obstruction and viral hepatitis.

A variety of factors may aggravate hepatotoxic effects of drugs or other injurious agents, including ethyl alcohol, anoxia, and even ordinarily innocuous infections. In a study of 38 patients with severe nonburn trauma, Champion et al.[10] noted that jaundice was associated with systemic gram-negative bacterial infections and massive transfusions. Liver function tests (SGOT, LDH) were often abnormal. Obviously, previous liver disease (cirrhosis) impairs hepatic reserve and predisposes the severely burned patient to liver failure.

Jaundice has been associated with a very poor prognosis and a fatality rate of up to 90 percent in traumatized patients.[6,9] Extreme hyperbilirubinemia in some patients (30 mg/dl) reflects the excessive pigmentary load presented to a compromised liver when the patient suffers immediate postburn intravascular hemolysis or is treated with a large volume of infused blood.

In summary, causes of jaundice in the early postburn period most commonly are confined initially to intravascular hemolysis, and shortly thereafter, resolution of large hematomas and hemorrhage into tissue.

During the second week post burn, the causes of jaundice include a

Table 13–3 *Common Causes of Cholestatic Jaundice: Differential Histologic Features*

	Large Bile Duct Obstruction	Drug-Related Cholestasis	Viral Hepatitis
Bile Stasis			
Degree	+++ to ++++; Bile lakes are diagnostic	++ to ++++	+ to +++
Site	Ducts, ductules, centrilobular hepatic areas	Centrilobular	Panlobular
Parenchymal damage/necrosis	Minimal	Mild, focal to no change; cells may be enlarged	Distorted hepatic cell plate; panlobular irregular cell swelling/necrosis; severe condensation or centrilobular reticulin
Inflammatory cells			
Type	Acute/chronic	Mixed eosinophils, lymphocytes, polymorphonuclear	Mostly chronic, some acute
Site	Portal triad	Parenchymal/portal triad	Parenchymal
Degree	0 to +	0 to +++	++ to +++
Bile duct proliferation	Prominent with edema of portal triad and few acute inflammatory cells	Unusual	Unusual

variety of interrelated factors: severe shock (prolonged hypotension), sepsis, anesthesia or drugs, multiple transfusions, renal failure, etc. Most often these factors are difficult to separate.

Effects of Infection on the Liver

Infections in the burn patient are most commonly due to gram-negative bacteria. Death in these patients is very likely related to the endotoxemia which causes multisystem failure. Since the liver inactivates endotoxin, several investigators have studied alterations in the liver during sepsis.[6,9,14] In patients with sepsis, serum transaminases and bilirubin are elevated. Histologically, hepatic sinusoidal enlargement and congestion, mononuclear infiltrates of the portal areas, fatty metamorphosis, and fatty necrosis are seen. Severe liver necrosis with liver failure is uncommon; however, impaired liver function may contribute to morbidity and mortality in the patient with sepsis.[14]

Intrahepatic Inflammatory Reactions

Frequently the portal triads contain a prominent inflammatory infiltrate composed primarily of small lymphocytes with occasional plasma cells (Fig. 13–5). Sevitt noted lymphocytes within the portal triads in patients dying less than 24 hr post burn. However, periportal lymphocytic infiltrates were inconspicuous in patients who died between 1 and 4 days post burn. Lymphocytes were again observed after the fourth day and were sometimes accompanied by a moderate number of eosinophils.[4] Excessive lymphocytes in the portal areas may persist for years.[8] The significance of these lymphocytes is uncertain. They probably reflect activation of the immune system, but the specific stimulus remains unidentified.

A wide variety of cells are conspicuous within the hepatic sinuses. These include Kupffer's cells as well as exogenous, activated phagocytic macrophages, polymorphonuclear leukocytes, and lymphocytes (Fig. 13–6). The finding of acute inflammatory cells, most commonly correlated with septicemia in the burn patient, indicates why the liver is such a good source of tissue for culturing microorganisms at postmortem examination. Despite the common presence of microorganisms by cultural techniques, the liver is remarkably free of gross or microscopic bacterial lesions, even in cases in which there are numerous hematogenous lesions in other organs.

Intrasinusoidal Bodies

Recently, Langlinais et al. have described in fatal burn cases hepatic intrasinusoidal bodies which approximate the size of macrophages[15] (Fig. 13–7A). These bodies stain faintly with hematoxylin and eosin, and special

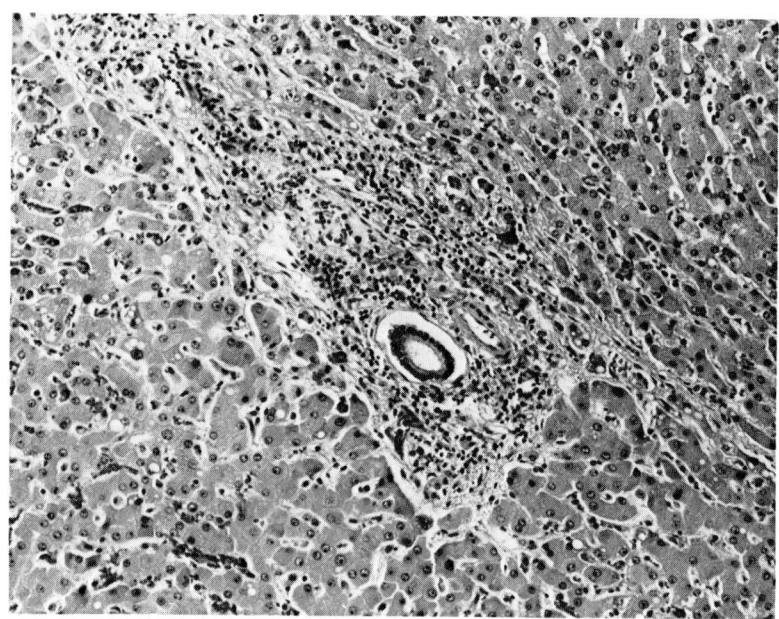

Fig. 13–5. Periportal infiltrates consisting mainly of lymphocytes and plasma cells are relatively common in burn patients. (H & E, × 126)

stains indicate a lack of mucin, fat, RNA, DNA, glycogen, or mucopolysaccharide in these structures. Ultrastructurally, these "bodies" are unit membrane-bound and composed completely of homogenously finely particulate (probably proteinaceous) material. No nuclear material or cytoplasmic organelles are noted in these bodies (Fig. 13–7B). A small number of similar structures appear within the cytoplasm of liver cells and Kupffer's cells. It has been postulated that these bodies represent an antigen–antibody response to an as yet unknown stimulant. Available data suggest that the use of silver sulfadiazine may be a common factor in these patients.[15]

Viral Hepatitis

Although viral hepatitis has been rare in our patient population, it has been demonstrated 2 or more months following thermal injury by others.[5,8] Considering the large volume of blood used in some patients (in excess of 100 units), viral hepatitis is not an unexpected sequel.

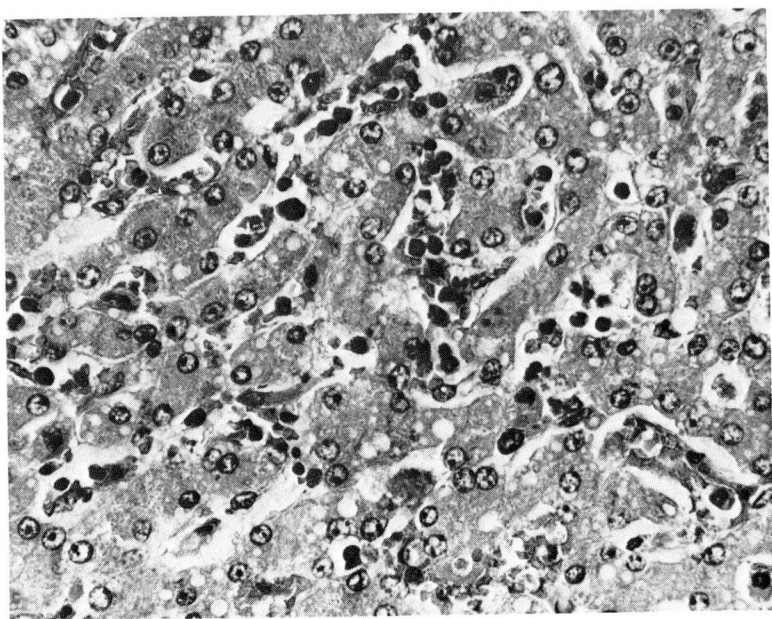

Fig. 13–6. Hepatic sinusoidal spaces often appear hypercellular. In this patient the Kupffer's cells are prominent and there is a sinusoidal infiltrate composed of macrophages and polymorphonuclear leukocytes. (H & E, × 360)

GALLBLADDER

Acute Acalculous Cholecystitis

Acute cholecystitis generally is believed to be secondary to obstruction of the gallbladder outlet or the biliary tract by gallstones in most nonburn patients. Acute calculous cholecystitis may, of course, occur in patients with burns or trauma,[16] but burn patients and other critically ill patients may also develop acute acalculous cholecystitis.[17] The incidence of acalculous cholecystitis is between 2 and 10 percent in the United States.[18–21] A more recent study by Glenn and Becker suggests that the incidence is increasing.[22] In a 10-year study of autopsied burn patients, Munster et al.[18] demonstrated 10 cases with acalculous cholecystitis. All patients were males with a mean age of 29 years. The mean burn size was 56.2 percent of the total body surface with a 33 percent third degree component. Jaundice occurred in six patients, and the serum bilirubin ranged from 2.5 to 30 mg/dl. The patients tended to fall into two distinct groups. In the *first group*, patients had

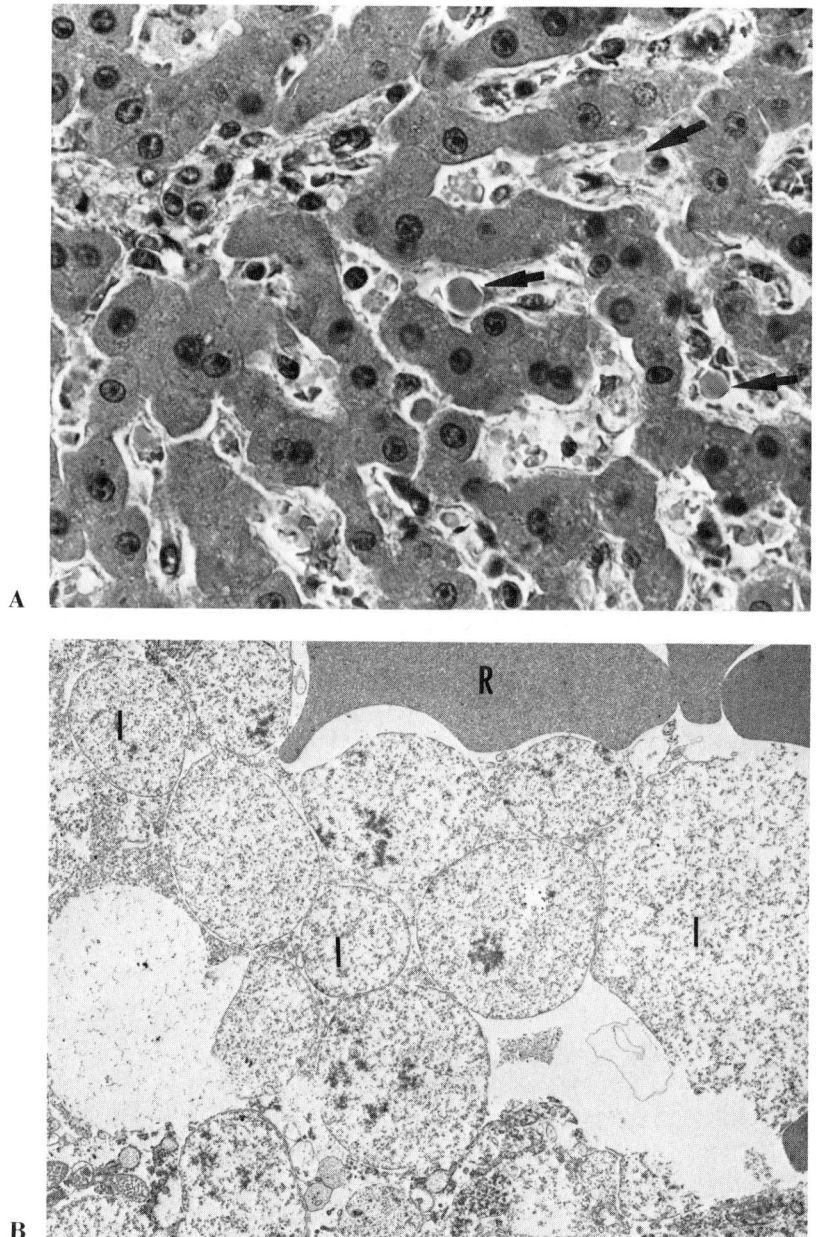

Fig. 13–7. (**A**): Hepatic intrasinusoidal bodies (arrows) have been seen in fatal burn cases. (H & E, × 360) (**B**): Transmission electron micrograph of several intrasinusoidal bodies (1). Portions of red blood cells (R) are also present within the sinus. (Uranyl acetate and lead citrate, × 7180)

overwhelming and invariably fatal sepsis. Antemortem blood cultures yielded gram-negative organisms and *Staphylococcus* species. At postmortem examination, the gallbladder was only one of many sites of metastatic bacterial infection. Hematogenous lesions were also indentified within the lung, liver, adrenal, and prostate.

In the *second group,* massive disseminated bacterial infection was *not* present, and what might be termed primary acalculous cholecystitis was felt to play a significant role in morbidity and mortality. Clinical conditions considered to predispose patients to acalculous cholecystitis have been summarized in Table 13–4. One common etiologic factor appears to be bile stasis aggravated by prolonged fasting, anesthesia, narcotics, dehydration, and fever.[23,24]

In the burn patient, the most conspicuous predisposing features of acalculous cholecystitis include overwhelming bacterial septicemia or postburn complications leading to dehydration or impaired gastrointestinal motility. Some patients receive multiple blood transfusions, and the increased blood breakdown products may play an important role in the development of acute acalculous cholecystitis.[23] In general, severely traumatized patients who subsequently develop acalculous cholecystitis usually become symptomatic between the eight and sixteenth postinjury day.[23]

The histologic features of the affected gallbladders ranged from acute necrotizing cholecystitis with a marked acute polymorphonuclear inflammatory-infiltrate and many bacterial colonies (Fig. 13–8) to a marked plasma cell and lymphocytic infiltrate with histiocytes, scant eosinophils, and blunted villi. Some cases demonstrated epithelial regeneration with underlying submucosal edema.[18] Others have shown focal inflammation and necrosis affecting arteries and veins, as well as occasional thrombi.[22] The lesions appear to be centered in the muscularis and serosa; the gallbladder mucosa is only secondarily involved. Initially, only marked edema of the muscularis and serosa is noted; later, foci of acute inflammation and necrosis appear. They hypothesize that the intrinsic portion of the coagulation cascade plays an important role in the development of these lesions.

Bilirubin Stones

At autopsy examination, small, dark, friable stones are occasionally seen in patients who die from severe thermal injury (Fig. 13–9). These stones probably develop as a result of the aforementioned excessive red cell destruction, which leads to prominent accumlation of bilirubin in the bile. No cases with ductal obstruction by such stones have been seen.

Table 13–4 *Clinical Conditions Which Predispose to Development of Acalculous Cholecystitis in Burn Patients*

Anatomic

  Distortion of cystic duct due to fibrosis
  Obstruction of cystic duct due to
    Lymphadenopathy, edema
    Hematoma

Physical Conditions

  Spasm or fibrosis of sphincter of Oddi
    Pancreatitis
    Drugs—opiates and atropine
  Ileus

Systemic diseases

  Vascular compromise due to atherosclerosis/emboli
  Hematogenous infection
  Primary infection of gallbladder

Miscellaneous

  Dehydration
  Fasting
  Generalized debilitation
  Multiple transfusions
  Hemolytic anemia
  Trauma

Severe Edema of the Gallbladder

Sevitt has described extreme distention of the gallbladder in children with thermal injury. These gallbladders had edematous walls and contained a clear, watery, "bile-free" fluid. Communication among the cystic, hepatic, and common bile ducts appeared normal in these cases. Pericholedochal lymphadenopathy was proposed to cause compression and obstruction of the cystic duct.[25]

A similar entity, so-called hydrops of the gallbladder, has been defined in nonburned children and is likewise not well understood. It has been associated with upper respiratory infections, scarlet fever, gastroenteritis, diarrhea, and cervical lymphadenopathy. The gallbladders histologically lack inflammatory cells; thus, this condition does not appear to be an inflammatory process. Clinically, the patients complain of right upper

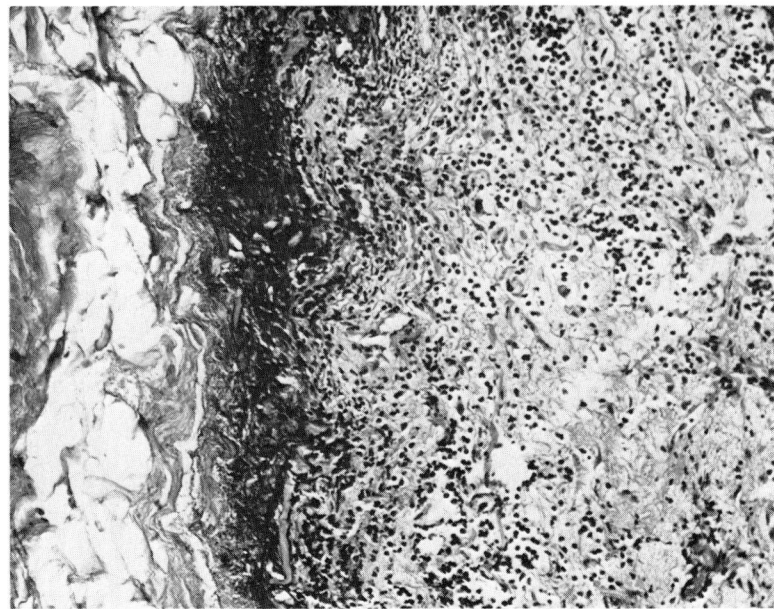

Fig. 13–8. Massive bacterial growth beneath eroded and autolyzed gallbladder mucosa. A mild suppurative inflammatory response is also present. (H & E, × 144)

Fig. 13–9. Multiple soft bilirubin stones are occasionally seen in patients who die from severe thermal injury.

quadrant or epigastric abdominal pain, fever, nausea and vomiting, and abdominal tenderness. At surgery, mesenteric adenitis is noted. Ultrasonography has been useful as a diagnostic tool.[26]

References

1. Arturson, G.: Pathophysiological aspects of the burn syndrome. Acta Chir. Scand. 274:1–135, 1961.
2. Herndon, D. N., and Panke, T. W. Unpublished data.
3. Teplitz, C.: The pathology of burns and the fundamentals of burn wound sepsis. *In* Burns. A Team Approach. C. P. Artz, J. A. Moncrief, and B. A. Pruitt, Jr., eds. Philadelphia, W. B. Saunders Co., 1979, pp 82–83.
4. Sevitt, S.: Reactions to Injury and Burns and Their Clinical Importance, 1st. ed. Philadelphia, J. B. Lippincott Co., 1974, pp 112–113.
5. Sevitt, S.: Burns. Pathology and Therapeutic Applications. London, Butterworth and Co., Ltd., 1957, p 303.
6. Czaja, A. J., Rizzo, T. A., Smith, W. R., and Pruitt, B. A., Jr.: Acute liver disease after cutaneous thermal injury. J. Trauma 15:887–894, 1975.
7. Shires, G. T., III, Albert, S. A., Illner, H., and Shires, G. T.: Hepatocyte dysfunction in thermal injury. J. Trauma 23:899–901, 1983.
8. Chlumska, A., Dobias, J., and Chlumsky, J.: Morphological findings in hepatic tissues of burned patients. *In* Basic Problems in Burns. R. Vrabec, Z. Konickova, and J. Moserova, eds. NY, Berlin, Springer Verlag, 1975, pp 207–208.
9. Sarfeh, I. J., and Balint, J. A.: The clinical significance of hyperbilirubinemia following trauma. J. Trauma 18:58–62, 1978.
10. Champion, H. R., Jones, R. T., Trump, B. F., Decker, R., Wilson, S., Stega, M., Nolan, J., Crowley, R. A., and Gill, W.: Post-traumatic hepatic dysfunction as a major etiology in post-traumatic jaundice. J. Trauma 16:650–657, 1976,
11. Aprille, J. R., Horn, J. A., and Rulfs, J.: Liver and skeletal muscle mitochondrial function following burn injury. J. Trauma 17:279–288, 1977.
12. Schiff, L.: Diseases of the Liver, 3rd. ed. Philadelphia, J. B. Lippincott Co., 1969.
13. Sherlock, S.: Diseases of the Liver and Biliary System, 5th. ed. London, Blackwell Scientific Publ., pp 340–382, 1975.
14. Gursel, S., Kandemir, B., Karacadag, S., and Telatar, H.: Liver in septic shock. Am. J. Gastroenterol. 59:250–254, 1973.
15. Langlinais, P. C., and Panke, T. W.: Intrasinusoidal bodies in the livers of thermally injured patients. Arch. Pathol. Lab. Med. 103:499–504, 1979.
16. DuPriest, R. W., Jr., Khaneja, S. C., and Cowley, R. A.: Acute cholecystitis complicating trauma. Ann. Surg. 189:84–89, 1979.
17. Alaweneh, I.: Acute noncalculous cholecystitis in burns. Br. J. Surg. 65:243–245, 1978.
18. Munster, A. M., Goodwin, N. M., and Pruitt, B. A., Jr.: Acalculous cholecystitis in burned patients. Am. J. Surg. 122:591–593, 1971.

19. Chen, P. S., and Aliapoulios, M. A.: Acute acalculous cholecystitis. Ultrasonic appearance. Arch. Surg. 113:1461–1462, 1978.

20. Fox, M. S., Wilk, P. J., Weissmann, H. S., Freeman, L. M., and Gliedman, M. L.: Acute acalculous cholecystitis. Surg. Gynecol. Obstet. 159:13–16, 1984.

21. Lee, A. W., Proudfoot, W. H., and Griffin, W. O., Jr.: Acalculous cholecystitis. Surg. Gynecol. Obstet. 159:33–35, 1984.

22. Glenn, F., and Becker, C. G.: Acute acalculous cholecystitis. An increasing entity. Ann. Surg. 195:131–136, 1982.

23. Weeder, R. S., Bashant, G. H., and Muir, R. W.: Acute noncalculous cholecystitis associated with severe injury. Am. J. Surg. 119:729–732, 1970.

24. Lindberg, E. F., Grinnan, G. L. B., and Smith, L.: Acalculous cholecystitis in Viet Nam casualties. Ann. Surg. 171:152–157, 1970

25. Sevitt, S.: Burns. Pathology and Therapeutic Applications. London, Butterworth and Co. (Publishers), Ltd., 1957, pp 307–308.

26. Kumari, S., Lee, W. J., and Baron, M. G.: Hydrops of the gallbladder in a child: Diagnosis by ultrasonography. Pediatrics 63:295–297, 1979.

Chapter 14

GENITOURINARY SYSTEM

Acute renal failure is a common terminal event in the severely burned patient despite optimal fluid therapy, but it has not been well characterized morphologically. The pathologist will often be unable to find a degree of renal disease which explains the serious loss of function that occurs prior to death. This problem is caused by the partial masking of early evidence of cellular injury by the accelerated postmortem degeneration and autolysis which may occur following septic death. Our observations have led us to believe that many of the renal diseases seen in the burn patient fit into the well-characterized forms of acute tubular necrosis. For this reason, we have presented an outline of this process based partly on the time of onset post burn (Table 14–1). The areas that apply to renal disease in burn patients are emphasized. The significance of preexisting renal disease in the burn patient is obvious and will not be further discussed.

Urinary Protein Profiles

Glomerular or tubular dysfunction or both may be demonstrated in burn patients; the severity appears comparable to the extent of injury. Although specific histologic changes are often not apparent in the kidney, analysis of urinary protein may document the exact site of renal injury. Shakespeare et al.[1] demonstrated a mild transient glomerular lesion in which large molecular weight proteins are lost. This was followed by excretion of low molecular weight proteins (<60,000 daltons) indicating a tubular lesion. Tubular proteinuria appears to be related not to a filtration abnormality but to a lack of tubular reabsorption and processing of the low molecular weight proteins which are normally present within the glomerular filtrate. Yu et al.[2] noted that patients with mild burns had little or no evidence of proteinuria. A second group had a mild, transient tubular (low molecular weight) proteinuria as measured by levels of alpha-1-microglobulin, beta-2-

Table 14–1 *Classification of Renal Disease in the Burn Patient*

Directly related to trauma

• Hyperacute (hours to a few days)

    Electrical injury—myoglobinuric nephrosis
    Severe hypotension (hypoperfusion)
      Acute renal tubular necrosis
      Acute renal cortical necrosis
    Crush injury

• Acute-to-subacute (few days to few weeks)

    Hypotension
      Acute renal tubular necrosis
      Acute renal cortical necrosis

    Infectious—hematogenous or ascending pyelonephritis (bacterial and mycotic)

Preexisting disease

• Diabetes mellitus—microvascular sclerosis

• Hypertensive cardiovascular disease—arterial, arteriolar nephrosclerosis

• Chronic pyelonephritis

---

microglobulin, retinol binding protein and *N*-acetyl-beta-D-glucosaminidase. In the third group, patients with more severe burns developed a moderate to severe proteinuria that started as a mixed glomerular (increased total protein, IgG, albumin) and tubular proteinuria. After a few days, the pattern in this third group changed to a typical tubular proteinuria (alpha-1-microglobulin, beta-2-microglobulin and *N*-acetyl-beta-D-glucosaminidase) and the latter lasted up to 200 days. A secondary tubular proteinuria was sometimes induced by aminoglycoside therapy. The severity and longevity of these dysfunctions closely correlated with severity of burns and subsequent clinical progress.

Myoglobinuric Nephrosis

Electric injuries resulting in severe muscle necrosis can lead to precipitous renal failure within hours after the injury. Massive myoglobinemia and fatal myoglobinuric nephrosis may occur despite aggressive therapeutic management. Grossly, the medulla may be darker than the cortex (Fig.

14–1A). This is related physiologically to corticomedullary shunting of blood and microscopically to the presence of numerous myoglobin casts, most of which are within medullary renal tubules and collecting ducts (Fig. 14–1B). Acute tubular necrosis may also be present. Morphologically and histochemically, renal tubular casts of myoglobin are indistinguishable from hemoglobin casts. However, historical and biochemical data easily separate the two entities (Table 14–2). Similar myonecrosis may be seen in patients with carbon monoxide poisoning.[3] Therefore, in the absence of electrical injury or deep muscle trauma, myonecrosis may suggest carbon monoxide poisoning.

Physiologic Renal Abnormalities

In burns with a calculated mortality of less than 50 percent, renal function is fairly well maintained.[4] In patients with more extensive burns, renal impairment is common and renal failure is generally fatal.[4–7] More often, tubular reabsorption of sodium in the ascending loop of Henle is impaired. Total tubular water reabsorption is also diminished despite high levels of ADH and is probably related to decreased sodium reabsorption, increased nonreasorbable solutes in tubular fluid, and decreased capillary osmotic gradient in the renal medulla. This results in an apparent maintenance of "normal" urinary excretion as measured by volume and sodium content despite an advanced state of renal failure (nonoliguric renal failure). The most reliable measurements of renal function in the severely burned patients are creatinine clearance and urine osmolality.[4] Early signs of renal failure include decreased urine/plasma creatinine ratio, decreased creatinine clearance, increased ratio between osmolality and creatinine clearance, and increase in plasma osmolality and creatinine concentration. Daily body weight measurements are a useful guide to the patient's state of hydration but are an insensitive guide to renal homeostasis after acute resuscitation is accomplished.[7]

In early onset renal failure, hypovolemia is the most common underlying cause. Later, hemodynamic alterations may be secondary to septicemia, pulmonary embolism, congestive heart failure, paralytic ileus, etc.[8]

Acute Tubular Necrosis

Depending on the severity of the clinical disease state (Table 14–3), acute tubular necrosis may present as a precipitous event, as does myoglobinuric nephrosis, or as a slowly evolving process with several apparent remissions and excerbations. Sometimes, significant morphologic changes in the kidney are absent at autopsy despite the presence of clinical and biochemical evidence of acute renal failure. An explanation for the

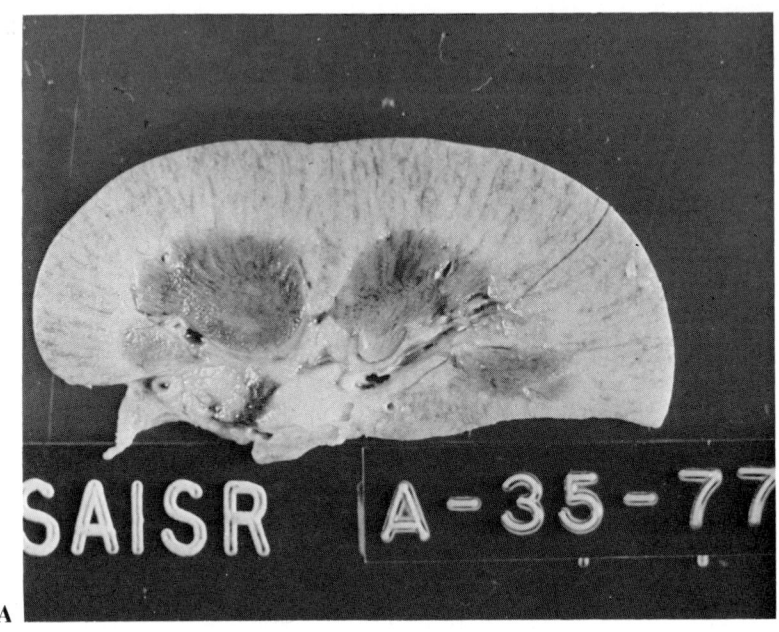

A

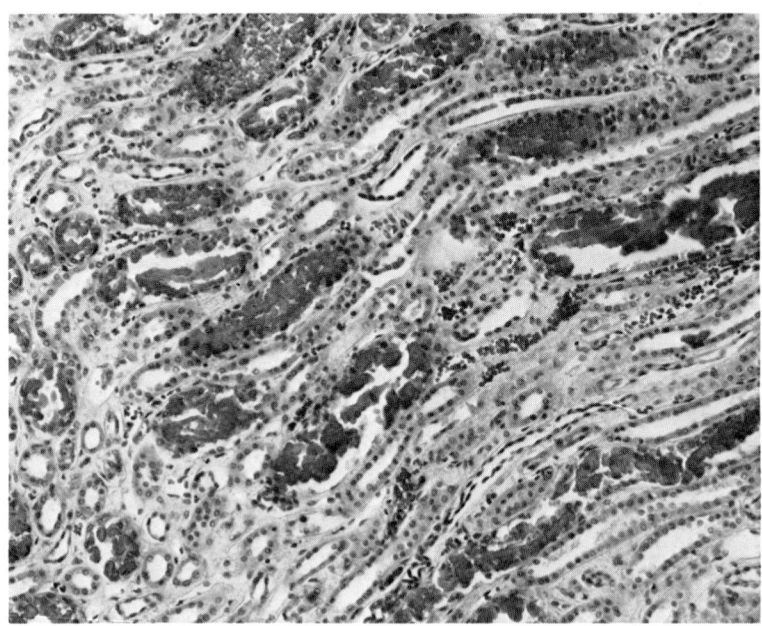

B

Fig. 14–1. (A): Gross specimen of kidney from a patient who died with massive myoglobinuric nephrosis after sustaining a severe electric injury. (B): Photomicrograph demonstrating myoglobin casts within renal medullary tubules of this patient. (× 180)

236

Table 14–2 *Features Differentiating Between Myoglobin and Hemoglobin*

	Myoglobin	Hemoglobin
Historical	Trauma, especially electrical	Incompatible blood transfusion or severe burns
Appearance of urine	Red	Red
Appearance of serum	Myoglobin rarely present, pale yellow serum	With massive hemolysis hemoglobin always present (pink to red)
Concentrated (80%) ammonium sulphate	Soluble[a]	Precipitate
Hemoglobin electrophoresis of urine[b]	Band migrating with hemoglobin C (or E)	Band with same mobility as patient's hemoglobin
Immunofluorescence (myoglobin specific antibody)	Positive	Negative

[a] Denatured myoglobin will precipitate. Test requres freshly voided urine sample.[21]
[b] In blood, hemoglobin is attached to haptoglobin, and this complex has the same electrophoretic pattern as myoglobin. Hemoglobin is "free" in the urine and may easily be distinguished from myoglobin by electrophoresis.[21]

discrepancy between clinical and pathological findings is currently not available. Histologic changes suggesting acute tubular necrosis in the burn patient are varied. As is readily apparent, these changes are certainly not specific for the burn patient; they may be seen in a variety of other conditions which are complicated by abnormal hemodynamic or toxic events. Because acute tubular necrosis is seen with such frequency in the burn patient, we have closely examined the morphologic changes and reported these below.

Acute Tubular Necrosis, Tubulorrhexic Type

One popular morphologic classification subdivides acute tubular necrosis into tubulorrhexic and nephrotoxic lesions.[9] *Tubulorrhexic* lesions usually involve the distal tubules at the "boundary zone" (corticomedullary

Table 14–3 *Factors Contributing to Renal Dysfunction*

Hypotension due to

    Hypovolemia
    Cardiac failure
    Endotoxemia/septicemia

Dehydration

Trauma

    Myoglobinuria
        Electrical injury
        Thermal injury
        Crush injury

Toxins

    Endotoxemia—gram-negative bacterial infections
    Burn wound toxin—speculative

Preexisting renal disease

    Glomerulonephritis
    Diabetic glomerulosclerosis
    Arteriolar nephrosclerosis (hypertensive cardiovascular disease)

Infections

    Hematogenous pyelonephritis
    Ascending pyelonephritis
    Periotonitis
    Miscellaneous infections

Miscellaneous

    Hepatorenal syndrome

Drugs

    Amphotericin B
    Gentamicin
    Carbenicillin
    Polymyxin
    Vancomycin
    Kanamycin
    Cephalothin
    Cephaloridine

Anesthetics

    Methoxyflurane

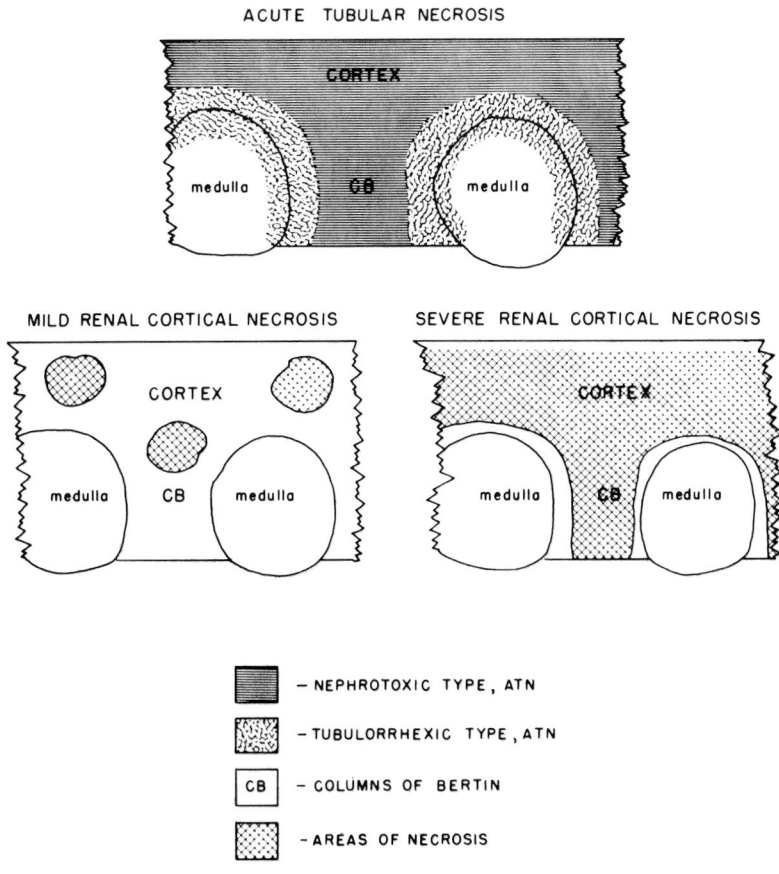

Fig. 14–2. Distributional comparison of "tubulorrhexic" and "nephrotic" acute renal tubular necrosis with renal cortical necrosis.

junction); they are multifocal and necrotizing lesions commonly associated with pigmented casts (Fig. 14–2, Tables 14–4, 14–5). It has been clearly demonstrated by nephron dissection that necrotic foci most commonly occur in the "lower nephron" (distal tubule), but may be seen occasionally in the proximal tubules as well. These lesions involve both the epithelial cells and the underlying basement membrane (Fig. 14–3). Cellular necrosis appears to be related to ischemia.

In tubulorrhexic-type of acute tubular necrosis (TATN), no significant changes appear in the glomeruli. The cells lining Bowman's capsule appear prominent in both number and size. Sometimes, epithelial cells are noted in Bowman's space and appear similar to cells lining the proximal convoluted tubules.[9] This change has caused some investigators[10] to postulate a

Table 14—4 *Appearance of Kidney in Acute Tubular Necrosis, "Tubulorrhexic" Type*

Early

Protein and heme casts in collecting ducts and distal renal tubules

Early tubular necrosis (focal) predominantly distal renal tubules

Glomerular hypermia, coarse eosinophilic granules and protein debris   in Bowman's space

Late

Kidneys enlarged and swollen; widened pale cortex; red, striated medulla

Focal renal tubular necrosis, especially at boundary zone

Granular and hyaline protein casts, heme casts

Tubular epithelial regeneration (double and triple layers of cells)

Interstitial edema and inflammatory cells (lymphocytes; some plasma cells and eosinophils)

Fatty change in many renal tubules

Tubulovenous anastomosis—rare and probably insignificant[12]

Scattered venous thrombi

Parietal cell layer of Bowman's capsule is cuboidal

"reflux" of renal tubular epithelium into Bowman's space. Frequently, an eosinophilic amorphous substance is found within the capsular space.[9]

When renal casts in patients with TATN are found to contain heme and to be confined to the renal medulla, signs of acute renal failure are usually present antemortem (Fig. 14–4). Fragmented sloughed tubular epithelial cells are commonly seen within the tubular lumens (Fig. 14–5). These cells appear to be more commonly associated with "tubular regeneration," which may be evident as early as 4 days after injury. These "regenerating" cells consist of flat basophilic tubular epithelial cells with dark nuclei and a small number of mitoses (Fig. 14–6). These regenerative changes are readily apparent, while the above necrotizing epithelial lesions are often subtle and may be overlooked.

Interstitial edema may be very prominent in both types of acute tubular necrosis and may widely separate renal tubules (Fig. 14–5, 14–6). This change is usually generalized, but may be more prominent in the boundary zone. If the predisposing abnormal (injurious) condition persists for an

Table 14—5 *Comparison of "Tubulorrhexic" and "Nephrotoxic" Types of Acute Tubular Necrosis with Renal Cortical Necrosis*

	"Tubulorrhexic" ATN	"Nephrotoxic" ATN	Renal Cortical Necrosis
Necrosis, distribution and pattern	Corticomedullary junction, diffuse	Cortex, diffuse	Discrete focus (foci) to extensive
Portion of tubule involved	DT, occasional PT	PT	PT initially, then DT&G with more severe lesions
Probable etiology	Ischemia	Toxic substances	Marked ischemia associated with shock

ATN = Acute tubular necrosis; PT = Proximal renal tubule; DT = Distal renal tubule; G = Glomerulus

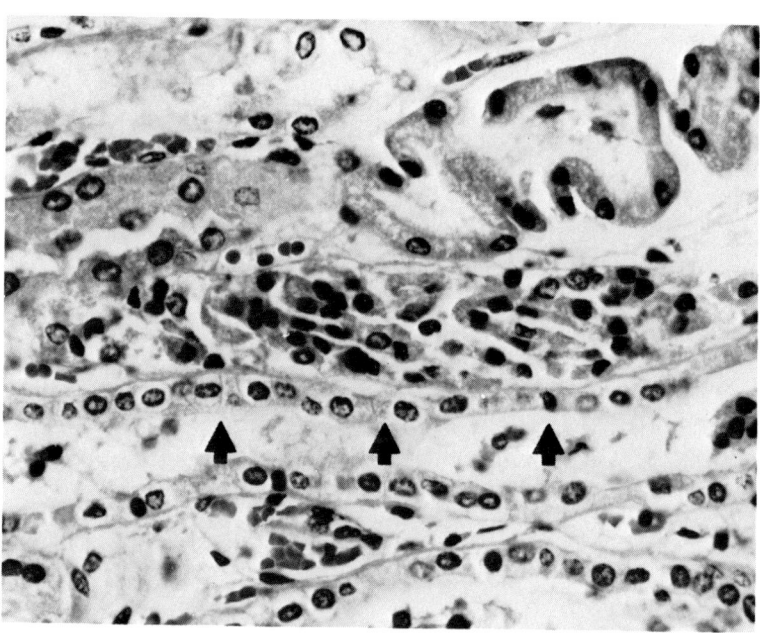

Fig. 14—3. Acute renal tubular necrosis (arrows). Necrotic, sloughed renal tubular epithelial cells fill tubular lumen. (× 351)

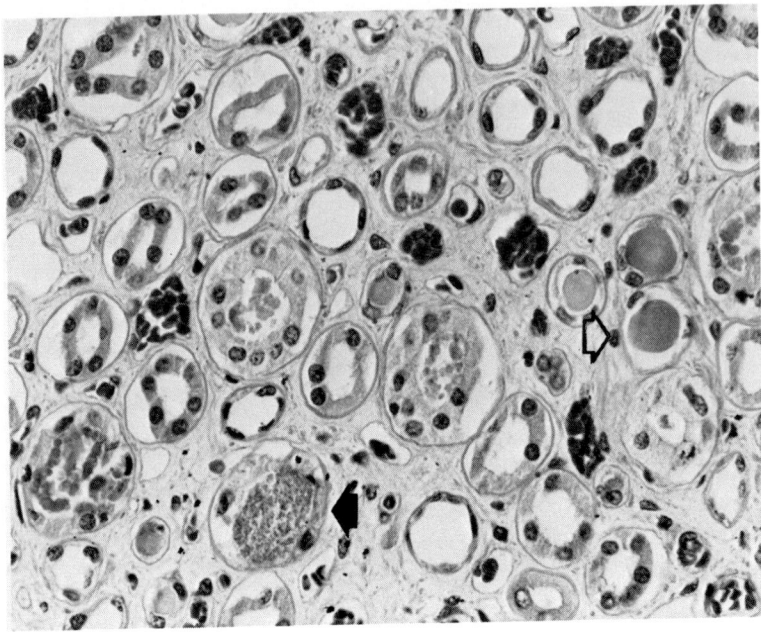

Fig. 14—4. Kidney granular (closed arrow) and heme (open arrow) casts and ? erythrocytes within renal tubules. (× 350)

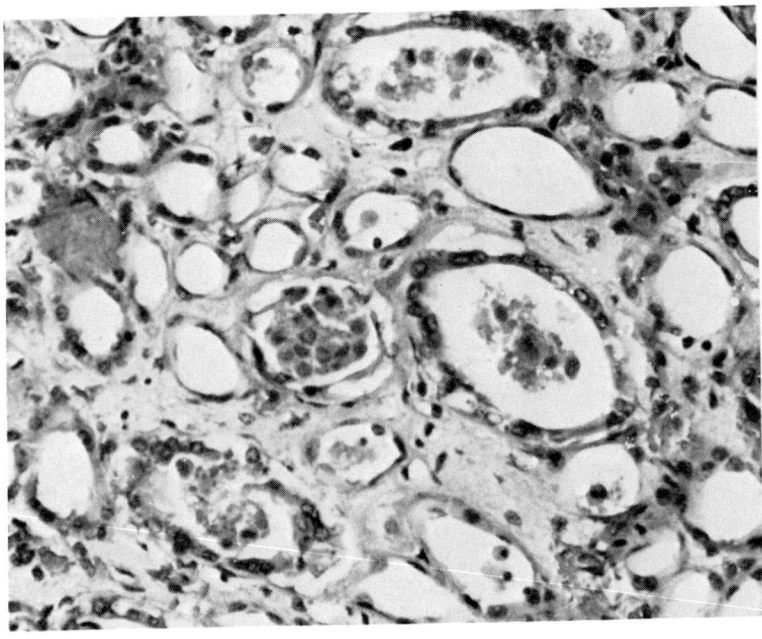

Fig. 14—5. Renal tubular necrosis. Sloughed tubular epithelial cells within renal tubules in which early regenerative changes are evident. (× 351)

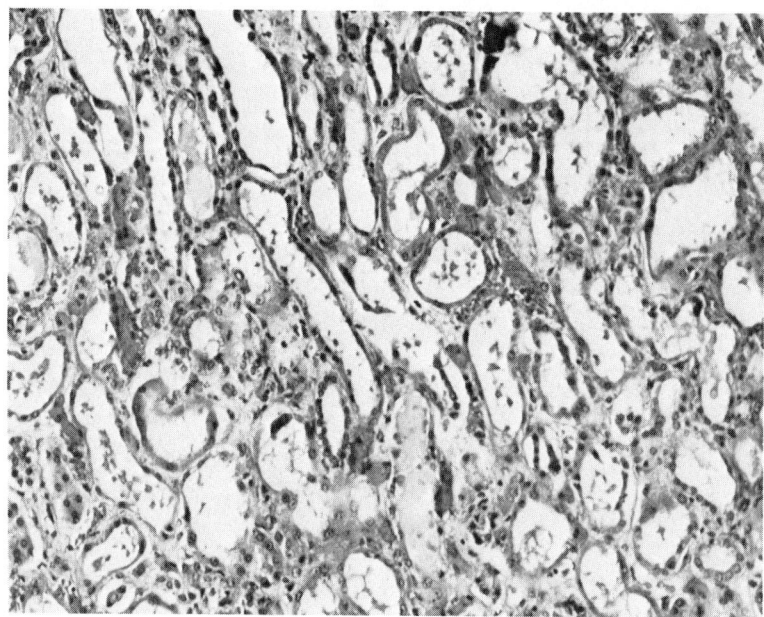

Fig. 14–6. Photomicrograph demonstrating the regenerative stage of acute renal tubular necrosis and interstitial edema. (× 180)

extended period of time, fibroblasts and reticulin fibers within the interstitium are increased in number.

Acute Tubular Necrosis, Nephrotoxic Type

In contrast to the tubulorrhexic variety of acute tubular necrosis, nephrotoxic lesions more commonly involve the proximal convoluted tubules, are confined to the renal cortex, and may be clearly secondary to toxic chemicals such as mercuric chloride in nonburn patients (Fig. 14–7A,B, Table 14–5). Since similar renal lesions occur in burn patients, it is assumed that such lesions evolve secondarily to a toxic stimulus. In the nephrotoxic type of acute tubular necrosis (NATN), changes within the proximal convoluted tubular epithelium are usually generalized rather than focal. Although areas of necrotic proximal renal tubular epithelium are easily identified, the basement membrane remains intact (in contrast to the tubulorrhexic lesion). Pathogenetically, if the action of noxious agents is localized due to impaired transit through the renal tubule, it is reasonable to surmise that the proximal convoluted tubule (PCT) should bear the most

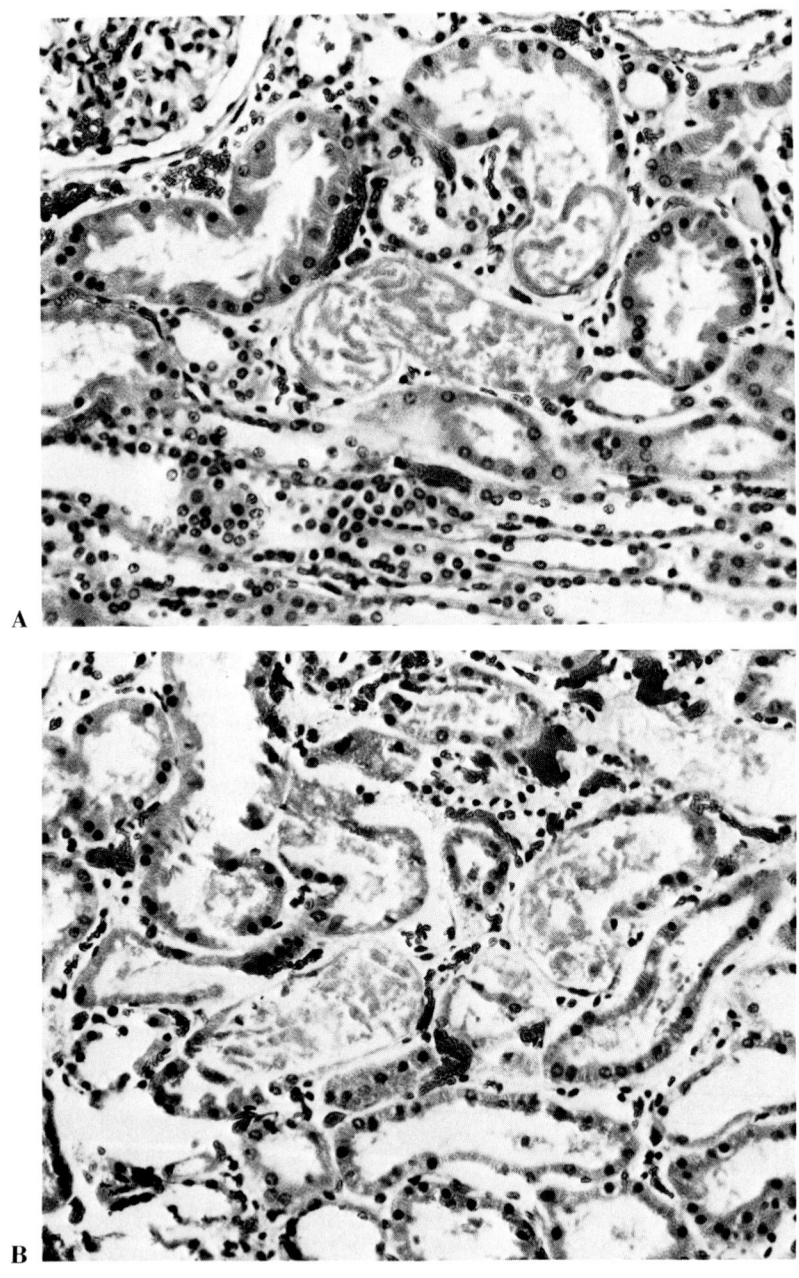

Fig. 14—7A,B. Photomicrographs of renal tubules undergoing nephrotoxic type of acute renal cortical necrosis. (× 360)

severe injury. The PCT has initial exposure to any chemicals or toxin and may be more susceptible to noxious agents.

A review of possible physiological and biochemical mechanisms which might explain the evolution of acute renal tubular necrosis is beyond the scope of this book. A partial outline of some of the conditions which may apply to the burn patient is presented in Table 14–3. Interested readers are referred to other sources which provide more complete information.[9]

Heme Casts

The significance of heme casts in the renal tubules is controversial. Heme casts are frequently seen in burn patients who die within a few days following thermal injury. Sevitt suggests that hemoglobin and myoglobin cause tubular dilatation by mechanical blockade.[11] Certainly, tubular dilatation and necrosis may occur within casts.[12] Experimental animals injected intravenously with large amounts of hemoglobin developed hemoglobin casts and subsequent medullary tubular dilation.[13] If the animals survived, tubular regeneration was prominent at 48–96 hr post hemoglobin injection and was complete by 10 days. Jaenike and Schneeberger[14] have also demonstrated casts in the renal vasa recta shortly after intravenous hemoglobin injections. Whatever the pathogenic mechanism(s) may be for renal tubular injury, the above studies present a compelling argument for the injurious effects of erythrocytic components and myoglobin.

Protein Casts

Renal tubular protein casts (Fig. 14–8) occur in both types of acute tubular necrosis and consist of hyaline and granular forms. The presence of proteinaceous casts appears to be the single morphologic feature most closely correlated with renal failure in the burn patient.[15] These protein casts are most often seen in the distal convoluted tubules and collecting ducts. A mild polymorphonuclear leukocytic infiltrate may be seen in areas that have extensive cast formation. There are frequently no significant changes in the renal tubular epithelial cells or glomeruli.

Interstitial Nephritis

An interstitial inflammatory process occurs to a variable degree in most thermally injured patients, usually involving the boundary zone (corticomedullary junction). Most commonly, the inflammatory infiltrate is composed of lymphocytes, plasma cells, and occasionally polymorphonuclear leukocytes including eosinophils (Fig. 14–9). These inflammatory cells are most con-

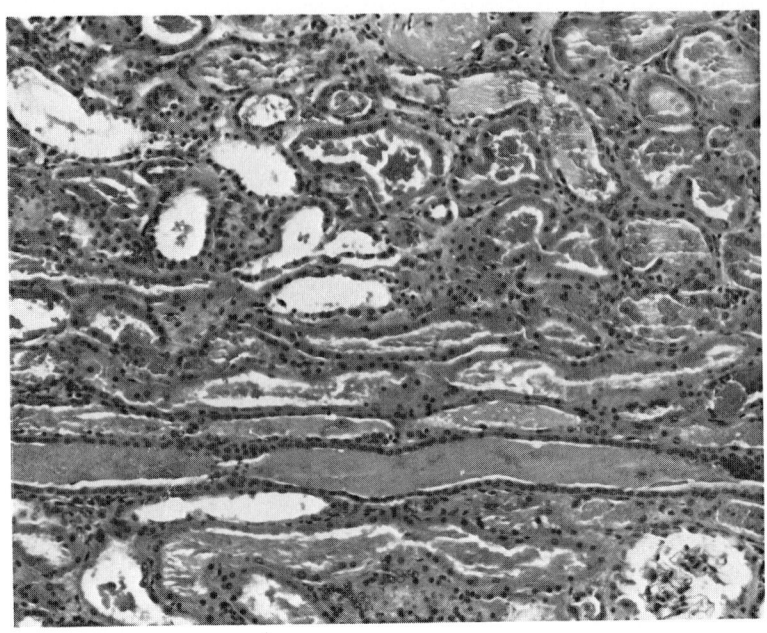

Fig. 14–8. Photomicrograph of hyalin and granular casts within renal tubules. (× 189)

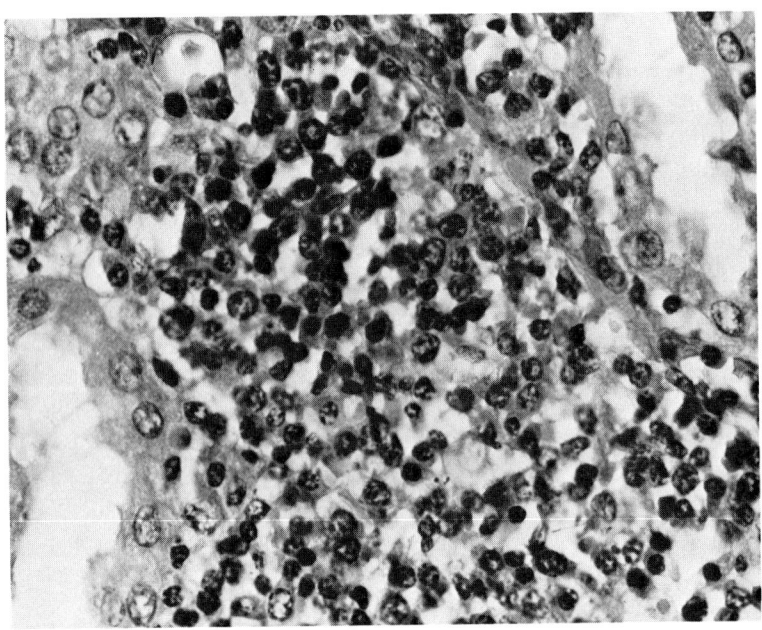

Fig. 14–9. Photomicrograph showing chronic "interstitial nephritis" with small to medium sized lymphocytes and plasma cells. (× 495)

246

spicuous near damaged renal tubules or scars. Certain therapeutic agents (cephalothin and methicillin) have been associated with prominent interstitial infiltrates.[9,16] It is uncertain whether such drugs can be implicated etiologically.

Many burn patients have interstitial nephritis without evidence of renal failure.[15] These prominent lymphoid infiltrates, when seen in the kidney, are also usually present in the adrenal and the portal triads of the liver. These infiltrates in the burn patient probably represent a reactive systemic immune response.

Intravascular Hematocytosis

In patients who die with acute renal failure, small renal vessels, particularly the vasa recta, often contain numerous mononuclear cells (Fig. 14–10). These are especially common in patients who have had renal failure for more than 1 day.[17] Lymphocytes appear to be most common in the first 24–48 hr of renal failure, then immature granulocytic cells appear, and finally, erythrocytic precursors predominate.[18] Speculation about the signif-

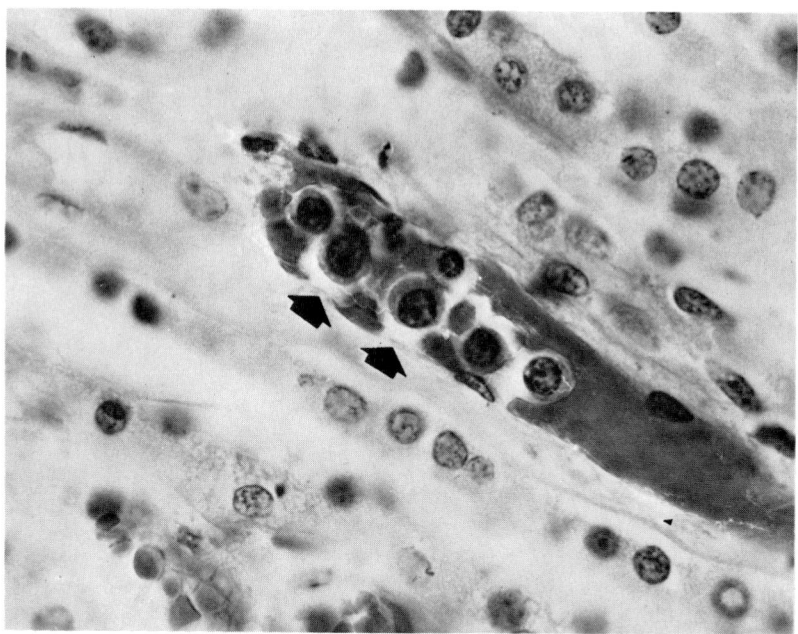

Fig. 14–10. Intravascular hematocytosis (arrows) microvasculature within the corticomedullary junction of the kidney. Some of the mononuclear cells may be granulocytic precursors which are intermingled with immature erythrocytes.

icance and cause of these cells is abundant and varied.[17,18] Heptinstall[9] suggests that this intravascular cellular accumulation may be related to a chemotactic substance in the renal medulla or to damaged endothelium of the vasa recta in acute renal failure.

Morphologic Evidence of Sublethal Renal Injury

The earliest reversible changes in the renal tubular epithelium following injury consist of hydropic and vacuolar degeneration. Hydropic degeneration consists of protein aggregation into intracytoplasmic eosinophilic droplets. This process may become so severe that renal tubular epithelial cells appear to burst and release their contents into the renal tubular lumen. It is likely that a variety of injurious substances and conditions may precipitate this change.

Vacuolar degeneration is due to an abnormal accumulation of intracytoplasmic water (Fig. 14–11). This change appears to be related to fluid and electrolyte abnormalities and is noted when hypertonic solutions (glucose or dextran) are utilized. Not uncommonly, this change is superimposed terminally on acute tubular necrosis.

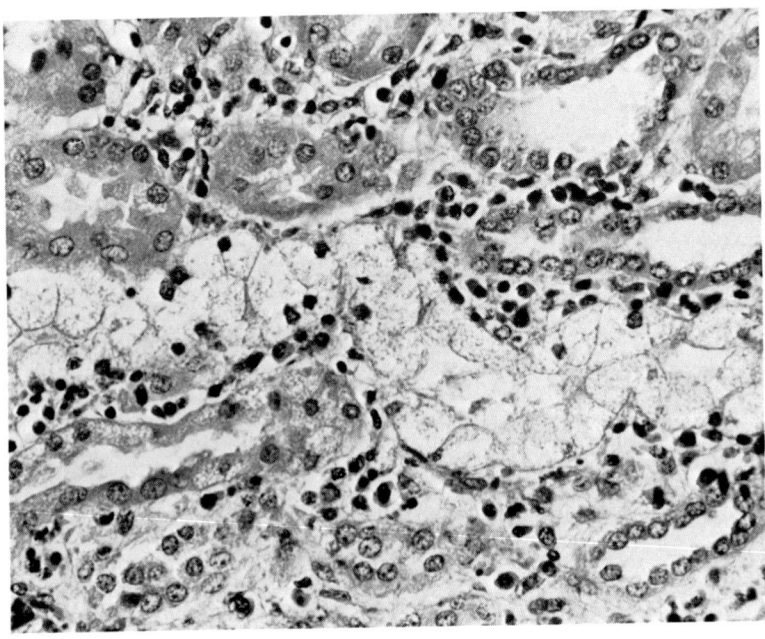

Fig. 14–11. Photomicrograph of a vacuolar degeneration of renal tubular epithelium. (× 351)

Postburn Renomegaly

Burn patients have an increased kidney weight and size that cannot be accounted for in terms of edema. Adult patients who die after the first postburn week often have kidney weights in excess of 200–250 g each. Such renomegaly is commonly seen in the total absence of any histologic renal changes. Although this phenomenon is not understood, we suspect that it represents a true hypertrophy or hyperplasia or both. Renomegaly in burn patients has been attributed to edema by another observer.[15]

Renal Cortical Necrosis

In renal cortical necrosis, the outer two-thirds of the cortex is pale and swollen; it contrasts sharply with the dark brown (markedly congested) deep cortex and medulla. The pathogenic mechanism appears to be related to corticomedullary "shunting" of blood. Focal-to-widespread necrosis is initially confined to the proximal convoluted tubules. Fragmented renal tubular epithelial cells separate from the basement membrane and lose their nuclei. The distal convoluted tubules are often not affected but may be collapsed and show mild swelling. Protein casts are easily found; hemoglobin casts are less common. A zone of viable renal tubules is often seen just below the renal capsule (Fig. 14–12A,B). These cells are thought to be spared because of the microvascular blood supply from capsular blood vessels.

The severity of the lesions varies from small discrete areas of necrosis affecting only proximal tubules to destruction of the entire renal cortex, including the columns of Bertin. When renal cortical necrosis is severe, distal convoluted tubules as well as the glomeruli are often involved.[9]

Pyelonephritis

Although both hematogenous and ascending bacterial pyelonephritis are found in the burn patient, hematogenous lesions are more prevalent. In ascending pyelonephritis, classically there is concurrent cystitis with retrograde extension of the infection into the renal pelvis. Development of large lesions irregularly distributed in both the renal medulla (predominantly) and the renal cortex suggests an ascending pyelonephritis (Fig. 14–13).

Hematogenous bacterial pyelonephritis usually consists of smaller necrotic lesions widely scattered throughout the cortex (predominantly) and the medulla (Fig. 14–14). Perivascular distribution of the smaller lesions and necrotizing glomerulitis are additional histologic findings which support a vascular origin. In a rat model of *Pseudomonas* pyelonephritis, Teplitz[19] demonstrated that only certain strains of *Pseudomonas aeruginosa* have a

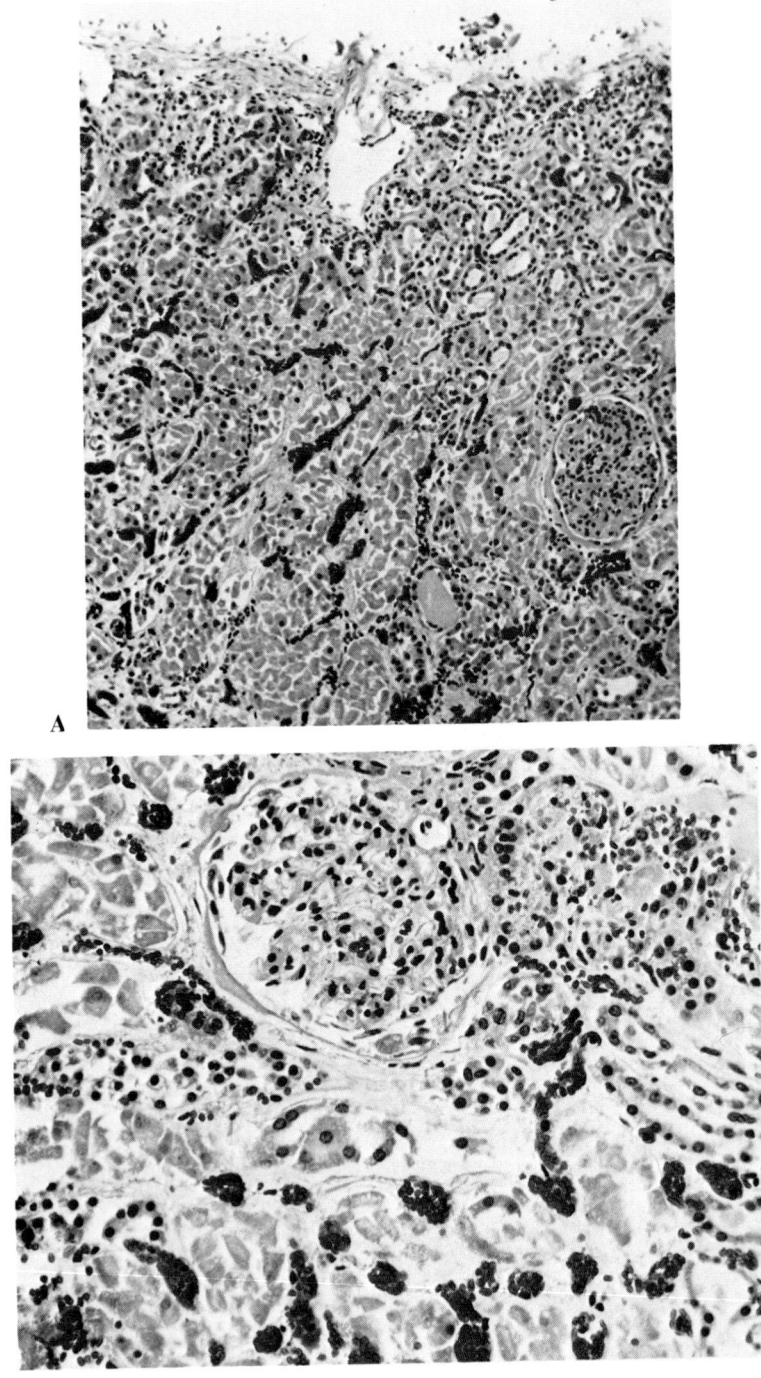

Fig. 14–12. (A): Photomicrograph of acute renal cortical necrosis with sparing of renal tubules just below the capsule. (× 72) (B): Complete necrosis of proximal renal tubules with relative sparing of distal tubules and glomeruli. (× 342)

Fig. 14–13. Ascending pyelonephritis. The patient has a necrotizing cystitis and ureteritis associated with this pyelonephritis.

predilection for causing hematogenous renal lesions. He described necrotizing lesions indentical to those seen in the skin and lung of burn patients with hematogenous *Pseudomonas* infections. Specifically, a bacterial "vasculitis" was often noted. Bacteria appeared first in glomerular tufts and intertubular capillaries, where they invaded through the capillary and venular wall. Fibrin thrombi were noted in the vessels of the interstitium at sites of bacterial invasion. Inflammation was often absent in many of these hematogenous lesions. In contrast to other gram-negative renal infections, *Pseudomonas* necrotizing infections were usually not associated with white cell tubular casts. Later, bacterial invasion of renal tubules led to tubular necrosis and large numbers of bacteria were seen throughout the tissue. Glomerular and tubular basement membranes appeared to restrict the invasion of *P. aeruginosa*. The external elastic lamina of interlobular arteries also appears to be a temporary restrictive barrier to *Pseudomonas*.

Renovascular Thrombosis

Microvascular thrombi in the renal glomeruli are seen in 5–10 percent of burn patients. In such cases, usually 10–20 percent of glomeruli contain such thrombi, which consist of fibrin, as indicated by positive phosphotungstic

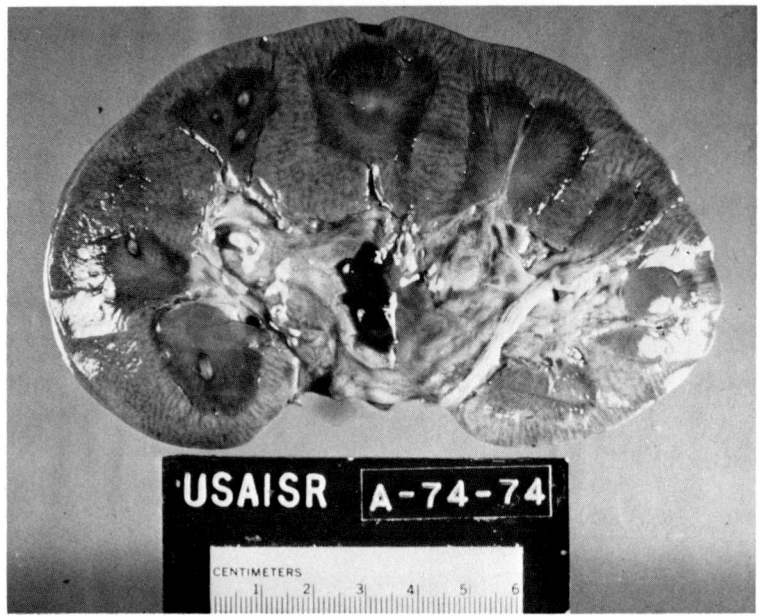

Fig. 14–14. Gross specimen of a kidney with hematogenous pyelonephritis. Note several small spherical areas of necrosis in the renal medulla.

acid–hematoxylin (PTAH) staining. The renal artery may also contain thrombi which are associated with renal infarcts. Sevitt[11,20] speculates that such thrombi develop as a result of a hypercoagulable state in the burned or injured patient.

URINARY BLADDER

Despite the widespread usage of Foley catheters and the common finding of significant concentrations of pathogenic bacteria in urine specimens, gram-negative necrotizing cystitis is remarkably uncommon. Approximately 1–4 percent of the autopsy cases a year have significant infectious cystitis. Although gram-negative bacilli predominate in these infections, gram-positive cocci and mycotic organisms are occasionally present.

Scattered areas of submucosal hemorrhages are frequently seen on the posterior wall of the bladder and are felt to be caused by mechanical trauma from the Foley catheter. This bland process is sometimes mistakingly diagnosed as "hemorrhagic cystitis," but that term should be reserved for those cases in which there is an infectious component.

PROSTATE

Very little information is available about prostatic disease in burn patients. Prostatic lesions are identified in 5–8 percent of autopsies; most can be directly related to the presence of a Foley catheter. Foci of hemorrhagic necrosis may occur in the lateral lobes of the prostate and are probably due to pressure necrosis. Focal, necrotizing infectious prostatitis may also occur (Fig. 14–15A–D). This process sometimes involves the periurethral glands and may secondarily infect foci of pressure necrosis caused by an indwelling Foley catheter. Since such infectious lesions represent potential nidi for bacteremia/septicemia, one should search carefully for prostatitis in patients with unexplained septicemia.

Other infectious lesions in the prostate can be multifocal and located deep within the gland. These lesions may represent hematogenous infectious dissemination to the prostate. A pattern of hemorrhagic necrosis with perivascular accumulations of gram-negative bacteria would be typical of *P. aeruginosa*. Metastatic lesions due to *P. aeruginosa* occasionally seen in the prostate of the septic burned rat have a similar morphologic appearance.

TESTES

Inguinal and genital burns are sometimes quite extensive and severe and may cause marked scrotal edema. Typically, gross findings in the testes are unremarkable. Histologically, severe interstitial edema, sometimes with mild inflammation and formation of early granulation tissue, is apparent. Seminiferous tubules often have prominently thickened basement membranes. Spermatogenesis is remarkably diminished, but Sertoli's cells are present in normal numbers (Fig. 14–16). Hemorrhage is uncommon. The degree of histologic change is clearly related to the severity of genital thermal injury. The significance of the above awaits clarification.

Infectious lesions, primary or hematogenous, are very uncommon in the testis. Only occasional instances of primary infections have been seen in patients with severe thermal injury or physical trauma to the external genitalia.

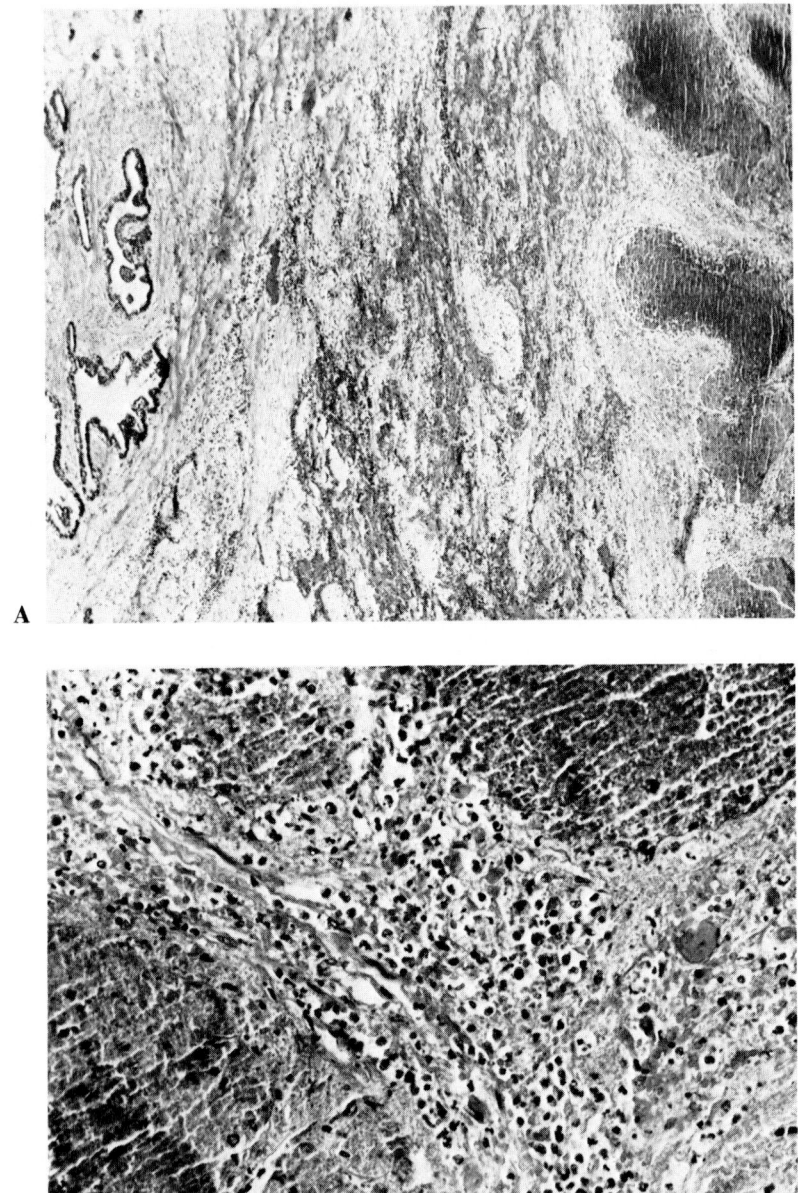

Fig. 14–15 A, B.

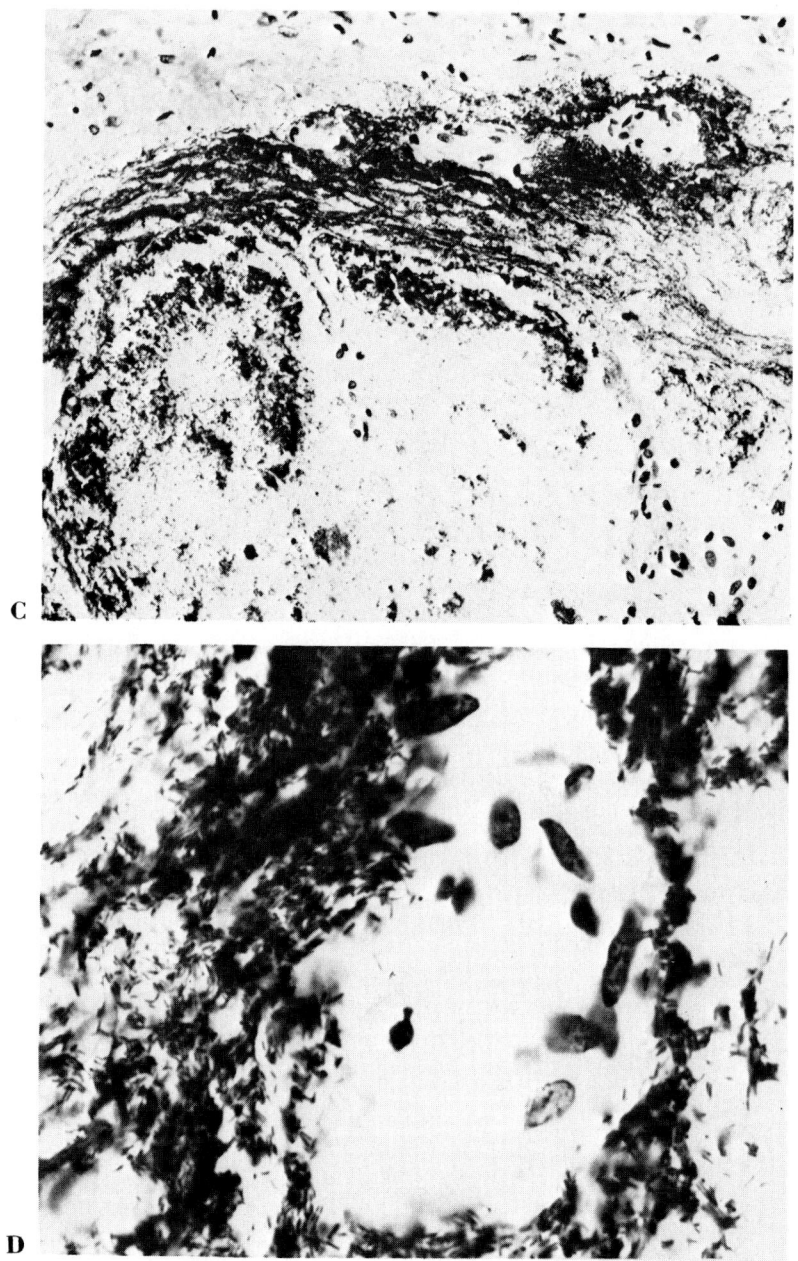

C

D

Fig. 14–15. Prostate specimen from patient with necrotizing infectious prostatitis. Totally necrotic glands adjacent to viable prostatic tissue. [(A): × 72; (B): × 342] Perivascular distribution suggestive of *P. aeruginosa* [(C): × 360; (D): × 1350]. The latter was cultured from the blood.

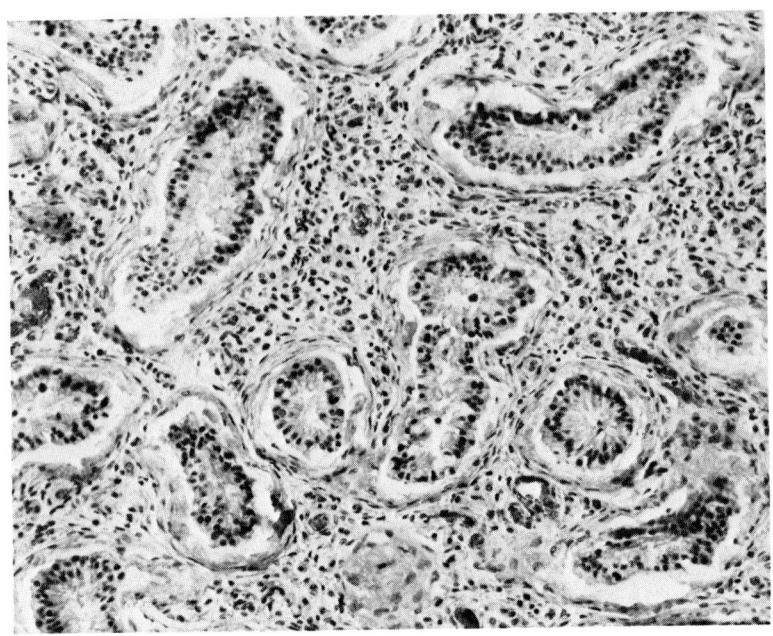

Fig. 14—16. Photomicrograph of testis showing markedly decreased spermatogenesis in a man who sustained scrotal burns. Note mononuclear inflammatory infiltrate within interstitial tissue, preservation of seminiferous tubules which contain almost exclusively Sertoli's cells. (× 180)

PERINEAL BURNS

Patients with extensive burns frequently have perineal burns. Significant bacteriuria (75 percent of patients) and septicemia (50 percent) often intervene. Complications of perineal burns include prostatic and periurethral abscesses.[22]

References

1. Shakespeare, P. G., Coombes, E. J., Hambleton, J., and Furness, D.: Proteinuria after burn injury. Ann. Clin. Biochem. 18:353–360, 1981.
2. Yu, H., Cooper, E. H., Settle, J. A. D., and Meadows, T.: Urinary protein profiles after burn injury. Burns 9:339–349, 1983.
3. Finley, J., Van Beek, A., and Glover, J. L.: Myonecrosis complicating carbon monoxide poisoning. J. Trauma 17:536–539, 1977.
4. Eklund, J., Granberg, P. O., and Liljedahl, S. O.: Studies on renal function in burns. I. Renal osmolal regulation, glomerular filtration rate and plasma solute

composition related to age, burned surface area and mortality probability. Acta Chir. Scand. 136:627–640, 1970.

5. Eklund, J.: Studies on renal function in burns. II. Early signs of impaired renal function in lethal burns. Acta Chir. Scand. 136:735–740, 1970.

6. Eklund, J.: Studies on renal function in burns. III. Hyperosmolal states in burned patients related to renal osmolal regulation. Acta Chir. Scand. 136:741–751, 1970.

7. Davis, D. M., Pusey, C. D., Rainford, D. J., Brown, J. M., and Bennett, J. P.: Acute renal failure in burns. Scand. J. Plast. Reconstr. Surg. 13:189–192, 1979.

8. Sevitt, S.: A review of the complications of burns, their origin and importance for illness and death. J. Trauma 19:358–369, 1979.

9. Heptinstall, R. H.: Acute renal failure. *In* Pathology of the Kidney, 2nd ed. Boston, Little, Brown & Co., 1974, pp 781–820.

10. Waugh, D., Schlieter, W., and James, A. W.: Infraglomerular epithelial reflux. An early lesion of acute renal failure. Arch. Pathol. 77:93–96, 1964.

11. Sevitt, S.: Reactions to Injury and Burns and Their Clinical Importance, 1st. ed. Philadelphia, J. B. Lippincott Co., 1974, pp 185–187.

12. Bohle, A., Jaenicke, J., Meyer, D., and Schubert, G. E.: Morphology of acute renal failure: Comparative data from biopsy and autopsy. Kidney Internatl. 10:S9–S16, 1976.

13. Mason, A. D., Jr., Teschan, P. E., and Muirhead, E. E.: Studies in acute renal failure. III. Renal histologic alterations in acute renal failure in the rat. J. Surg. Res. 3:450–456, 1963.

14. Jaenike, J. R., and Schneeberger, E. E.: The renal lesion associated with hemoglobinemia. II. Its structural characteristics in the rat. J. Exp. Med. 123:537–546, 1966.

15. Teplitz, C.: The pathology of burns and the fundamentals of burn wound sepsis. *In* Burns. A Team Approach, 1st. ed. C. P. Artz, J. A. Moncrief, and B. A. Pruitt, Jr., eds. Philadelphia, W. B. Saunders Co., 1979, pp 45–94.

16. Baldwin, D. S., Levine, B. B., McCluskey, R. T., and Galto, G. R.: Renal failure and interstitial nephritis due to penicillin and methicillin. N. Eng. J. Med. 279:1245–1252, 1968.

17. Solez, K., Kramer, E. C., Fox, J. A., and Heptinstall, R. H.: Medullary plasma flow and intravascular leukocyte accumulation in acute renal failure. Kidney Internatl. 6:24:37, 1974.

18. Baker, S. B. de C.: Intravascular haemopoiesis in the renal medulla in shock. J. Pathol. Bacteriol. 75:421–428, 1958.

19. Teplitz, C.: Pathogenesis of *Pseudomonas* vasculitis and septic lesions. Arch. Pathol. 80:297–307, 1965.

20. Sevitt, S.: Coronary thrombosis following injury and burns. Med. Sci. Law 13:185–191, 1973.

21. Rosse, W. F.: Differentiation of hemoglobin and myoglobin in urine. *In* Hematology, 2nd ed. W. J. Williams, E. Beutler, A. J. Erslev, and R. W. Runnels. NY, McGraw-Hill Book Co., A Blakiston Publication, 1977, pp 1613–1614.

22. McDougal, W. S., Peterson, H. D., Pruitt, B. A., Jr., and Persky, L., The thermally injured perineum. J. Urol. 121:320–323, 1979.

Chapter 15

ENDOCRINE SYSTEM

The extreme stress, fluid imbalance, changes in metabolism, and hormonal fluctuations[1-3] that occur in burn patients have led us and other observers to scrutinize the endocrine system for morphologic evidence of degenerative or hyperplastic changes. Significant morphologic findings have been seen only in the adrenal and pituitary glands.

ADRENAL GLAND

Thermal injury is associated with a variety of morphologic changes in the adrenal gland. These are most often related to the profound degree of stress that occurs with severe thermal injury and its subsequent complications. The pivotal role played by the adrenal gland is best exemplified by a recent report of an adrenalectomized patient who sustained a 36 percent total body surface thermal injury. During his postburn course, he manifested severe fluid and electrolyte problems, profound nutritional abnormalities, severely impaired wound healing, and a markedly defective immunologic function.[4] Significant morphologic changes in the adrenal glands of burn patients are discussed below.

Adrenal Hemorrhage

Severe vascular congestion is often seen at the corticomedullary junction within the first 24 hr post burn. (Fig. 15–1). Later, microscopic areas of hemorrhage may develop in the adrenal cortex. Massive adrenal hemorrhage is occasionally (2.0 percent of burn autopsies) identified (Fig. 15–2). Some cases of severe adrenal hemorrhage are associated with disseminated intravascular coagulopathy resulting in thrombi occluding the adrenal medullary veins (Fig. 15–3) (See Chapter 11).[5]

Severe thermal injury is one of many (Table 15–1) conditions that may

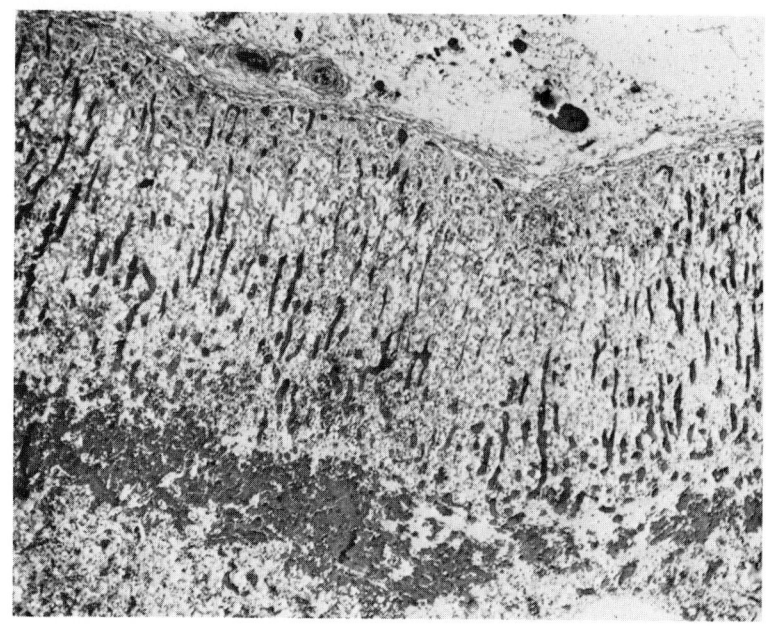

Fig. 15–1. Adrenal gland with severe corticomedullary congestion in a burn patient who succumbed shortly after injury. (H & E, × 50.4)

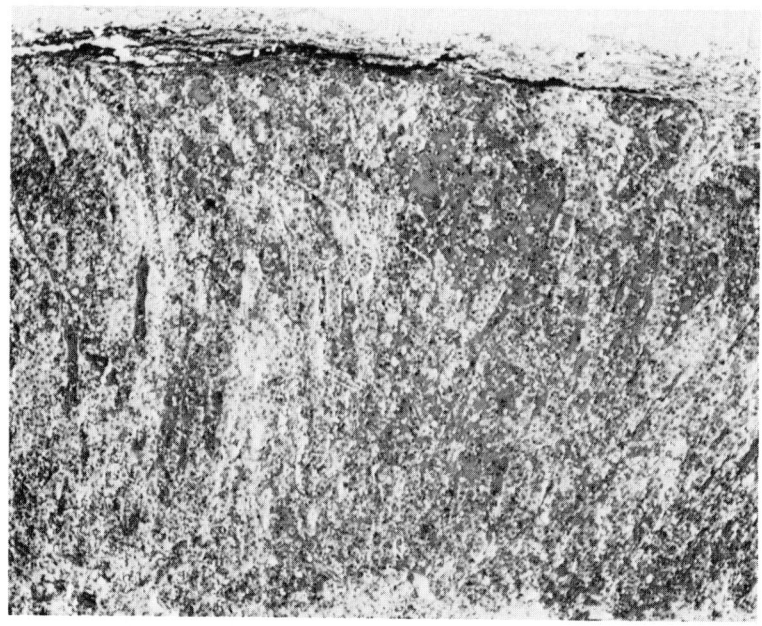

Fig. 15–2. Massive adrenal hemorrhage is occasionally seen in patients with severe burns. (H & E, × 50.4)

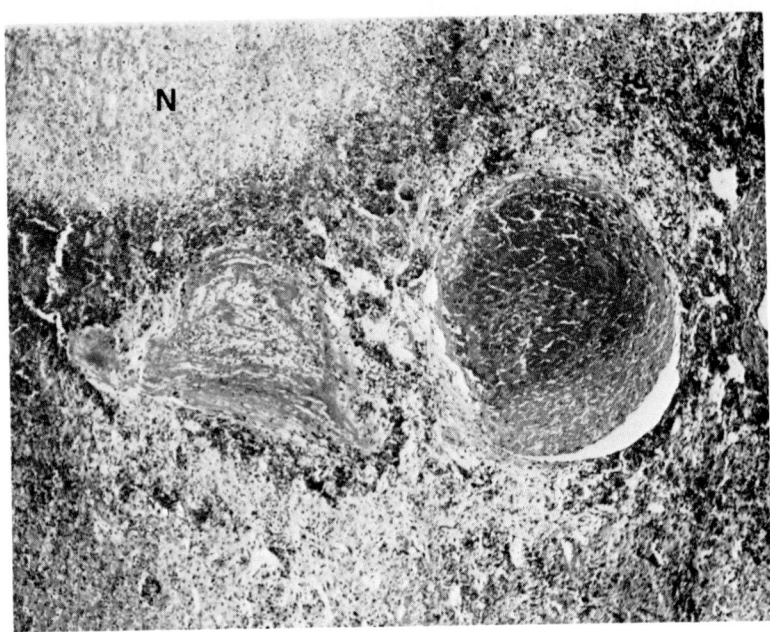

Fig. 15–3. Severe adrenal hemorrhage may also be associated with disseminated intravascular coagulation, as in this case in which thrombi were found in the adrenal medullary veins. A zone of infarction necrosis (N) is also present. (H & E, × 50.4)

contribute to, or be associated with, massive adrenal hemorrhage and necrosis. Severe burns often obscure clinical signs and symptoms of adrenal apoplexy. These clinical findings include abdominal pain, pallor, prostration, cyanosis, hypotension, and ileus.[5] Laboratory features which may predate adrenal hemorrhage are nonspecific; these include leukopenia with a subsequent rise in eosinophils. Sodium and chloride loss and potassium retention, hypoglycemia, and hemoconcentration develop following adrenal injury.[6] In some cases of this type, a complex interaction of shock, excess ACTH,[7] and disseminated intravascular coagulopathy occurs to a variable degree and may cause severe adrenal hemorrhage.

Another significant cause of severe adrenal hemorrhage ("adrenal apoplexy") is direct trauma. Sevitt[8] found severe adrenal hemorrhage in one-quarter to one-third of cases with severe closed injuries to the chest or abdomen. The injury was usually unilateral, affected the right adrenal more often than the left, and was related to "crushing" of the gland against the vertebral column. Some cases were noted in thermally injured patients with head trauma.

Table 1 *Conditions Associated with Severe Adrenal Hemorrhage*

Infections

    Pneumonia
    Meningitis, especially meningococcal
    Perforated viscus with peritonitis
    Septicemia, NOS
    Viral infections

Cardiovascular disease

    Atherosclerosis
    Myocardial infarct
    Hypertension
    Arteriosclerosis
    Thrombophlebitis
    Inflammation of adrenal veins

Renal failure

    Uremia

Severe trauma

    Skeletal fractures
    Extensive burns

Major surgery

    Gastrointestinal resection

Therapy

    ACTH
    Anticoagulants
    Insulin intoxication

Hematopoietic system

    Disseminated intravascular coagulopathy

Miscellaneous

    Pancreatitis

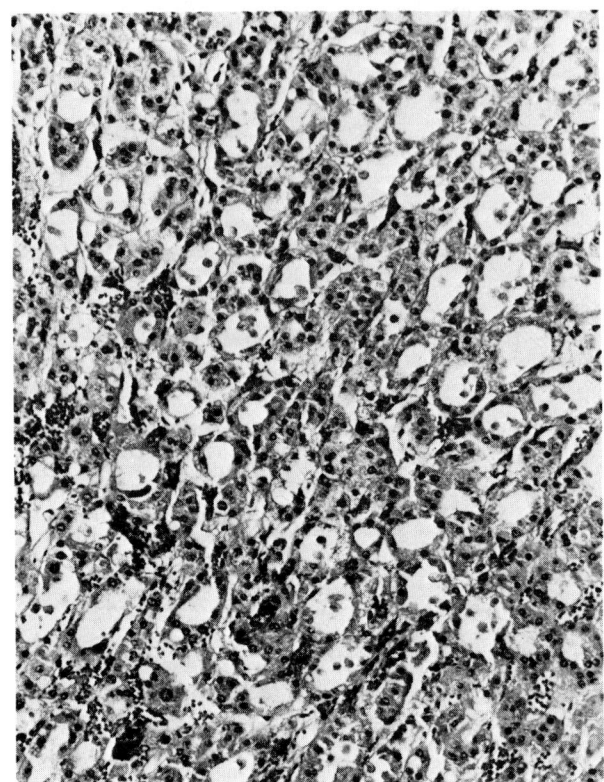

Fig. 15–4. The zona fasciculata cord cells of the adrenal cortex may show a cylindrical arrangement. This is thought to represent cortical regeneration. (H & E, × 126)

"Degenerative Changes" of the Adrenal Cortex

Severe stress may cause hyaline droplet formation in the adrenal cortical cells and necrobiosis. A hollow cylindrical appearance of the zona fasciculata cell cords (Fig. 15–4) is thought to represent regeneration of cortical cells around foci of cytolysis. This phenomenon has been termed Rich's tubular degeneration[9] and has been seen in a variety of severe infections. Despite the term "degeneration," such changes are usually associated with adrenocortical hyperactivity.[8]

Lipid Depletion of the Adrenal Cortex

Stress-related lipid depletion of the adrenal cortex is a well-recognized phenomenon. Most severely burned patients have yellow-brown to brown, thin adrenal cortices. The initial microscopic change is multifocal lipid depletion starting in the inner part of the zona fasciculata and extending into the outer adrenal cortex. Oil red O or Sudan Black stains of frozen sections of adrenal tissue document the marked lipid depletion in severely burned patients. In contrast, special stains demonstrate abundant RNA and alkaline phosphatase in these lipid-depleted adrenal cortical cells and indicate that these cells not only are active but may be hyperactive. In some patients dying a week or later after thermal injury, lipid depletion is more prominent in the outer adrenal cortex. This finding is thought to be related to the tendency for the inner cortex to reestablish more quickly normal lipid levels.[8]

Following severe burns, maximal adrenal cortical secretory (corticosteroids) activity has been measured. In contrast to postoperative patients, severely burned patients have no increment in endogenous corticosteroid secretion following parenteral injections of adrenocorticotrophic hormone.[10] The principal corticosteroids released by the adrenal cortex are the 17-hydroxycorticosteroids.[11] It has been shown in animal experiments that denervation of the area of burn trauma blocks the adrenal cortical response.[12]

Infectious Complications Involving the Adrenal Gland

Microscopic foci of suppurative necrosis secondary to bacterial septicemia have been described in the adrenal.[8] As described previously in Chapter 8, disseminated cytomegalovirus infections may also cause focal, mild cortical necrosis. Cortical cells at the periphery of these lesions may have intranuclear and intracytoplasmic viral inclusions typical of cytomegalovirus.[13]

Lymphoid Infiltrates in the Adrenal Gland

Lymphoid infiltrates in the zona reticularis and adjacent adrenal medulla are frequently seen in burn patients who have survived more than 10 days post burn. When these lymphoid infiltrates are prominent in the adrenal gland, they are often seen in the hepatic portal triads and corticomedullary junction of the kidney as well. This distribution suggests a nonspecific immune response to a systemic abnormality, probably infection.

Adrenal Medulla

Abundant data suggest that catecholamines are the principal mediators of the hypermetabolic response to thermal injury.[14,15] No gross or light microscopic changes are noted in the adrenal medulla following thermal injury. However, biochemical and ultrastructural data in experimental animals suggest that adrenal medullary hypersecretion occurs within 15 min following trauma, and that this sudden mobilization of secretory products continues for at least 12 hr. At 1–3 days post burn, increased compensatory synthesis of catecholamines is noted. Later in these animals, there may be functional exhaustion of the adrenal medulla. Other animal studies demonstrate a late morphologic and functional "normalization" of the adrenal medulla at about 28 days post burn.[16] Although the exact ultrastructural alterations in the burn patient are not known, they are very likely similar to the above findings in experimental animals.

In a study involving 14 fatal human burn cases in which the urine and adrenal tissue were assayed for adrenalin and noradrenalin, two-thirds of the patients were found at the time of death to have both a low adrenalin output and a low adrenalin content.[17]

PITUITARY GLAND

In burn patients who die within the first few days post burn, the only changes in the pituitary gland may be marked hyperemia of the tissue. Later, Crooke's hyaline degeneration (degranulation of PAS-positive mucoid cells) may occur in the anterior pituitary "basophilic" cells. Such alterations have been associated with increased secretion of adrenocorticotrophic hormones and lipid depletion of the adrenal cortex.[8] Sequential ultrastructural studies of the pituitary gland demonstrate edema and degenerative changes in the immediate postburn course. Later, features of increased metabolic activity predominate.[18]

Fat Embolism

Sevitt[19] has reported fat emboli in the pituitary. These fat emboli, which appear more commonly in the posterior lobe, are *not* associated with infarction. Occasional acute microhemorrhages were the only sequel to fat emboli in the pituitary noted by Sevitt.

Infarction

Pituitary infarcts were observed in 6.3 percent of autopsies (15 of 238 cases) at the Institute of Surgical Research.[20] In some of these cases, the infarcts almost totally obliterated the gland. Other cases had distinct subcapsular infarcts (Fig. 15–5). Hemorrhage was inconspicuous and infectious lesions were observed in only one case. The latter appeared to be an infarct which became secondarily infected by *Staphylococcus* sp. Specifi-

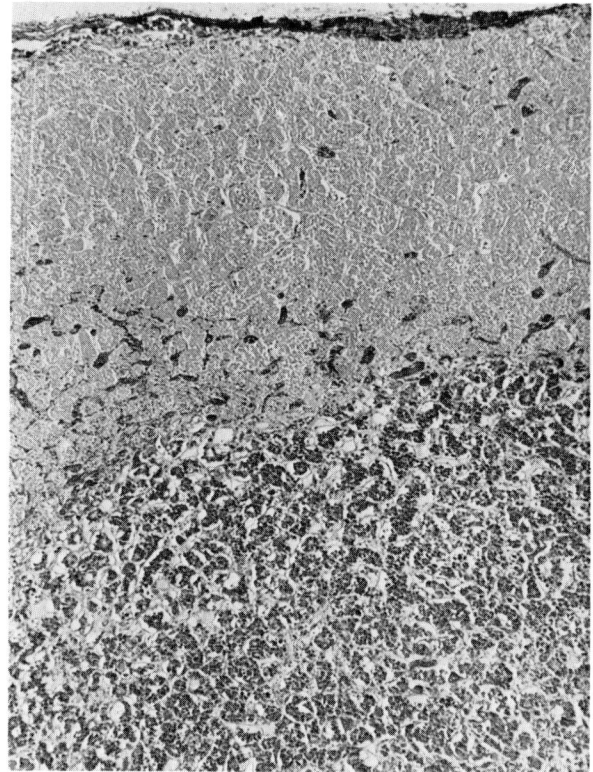

Fig. 15–5. Severe subcapsular infarction of the pituitary gland in a burn patient. (H & E, × 31.5)

cally, no abscess was found despite the presence of abundant bacteria within the infarcted tissue. Severe hypotensive episodes very likely contribute to the development of these infarcts. No thrombi were identified in these 15 cases of pituitary infarction. Morphologically, the duration of pituitary infarcts appeared to vary from one to several days. Patients with pituitary infarcts usually (87 percent) had extensive burns (45 percent or more of their total body surface) and relatively prolonged survivals.[20]

Hormonal Function

Numerous surveys have indirectly evaluated pituitary function in the burn patient. A sustained elevation of adrenocorticortrophic and antidiuretic hormones, marked depression of prolactin levels and (in young adults) follicular stimulating hormones, and minimal changes in thyrotrophic, luteotrophic, and growth hormones have been reported.[21–23] Furthermore, in the presence of hypoglycemia, the response of the pituitary gland with growth hormone secretion is depressed or completely unresponsive in the burn patient.[22,23] To our knowledge, clinical evidence of panhypopituitarism has not been identified in long-term survivors. As the above survey indicates, some patients with severe burns are susceptible to developing severe pituitary infarcts. Burn patients with severe burns complicated by hypotension should be carefully observed for subtle signs of impaired pituitary function.

THYROID GLAND

Evidence of hypermetabolic state in thermally injured patients[14,24] naturally focuses attention on the thyroid gland. Data on physiologic function, however, are controversial. Some investigators report increased thyroid activity while others find no change post burn.[25] Many investigators currently believe that the thyroid plays a minor role, if any, in the hypermetabolic response in the burn patient[14,26]; however, recent studies by Becker et al.[27] have shown that thermally injured patients have a significant decrease in serum triiodothyronine (T_3, the metabolically active hormone) and an increase in reverse T_3 levels. Similar alterations in peripheral thyroid hormone concentrations are found in other forms of critical illness.[28] Additionally, metabolism of thyroid hormones has been found to be accelerated during acute infections in man and experimental animals.[29] These findings suggest that low serum concentrations of active thyroid hormone result from increased hormonal clearance.

Morphologic changes consisting of peripheral colloid vacuolation, which suggests hyperactivity of the thyroid, have been found in the first 2 days post burn and in patients with infection.[8] Such evidence of increased activity cannot be confirmed by physiologic and endocrinologic studies.[26] We have observed no consistent pattern of morphologic change in the thyroid gland of patients who die either early or late post burn.

The presence of an "inhalation injury" or infectious pulmonic complications in severely burned patients may lead to the performance of tracheostomy. Tracheostomy stomal infections are relatively common; not only may these lead to staphylococcal or gram-negative bacterial tracheobronchitis and bronchopneumonia, but infection may also extend into the thyroid gland. One such case of a burn patient with staphylococcal abscesses involving the thyroid (Fig. 15–6) had recurrent staphylococcal septicemia. Perivascular and intravascular cocci within the thyroid represented the most likely source for septicemia in this patient.

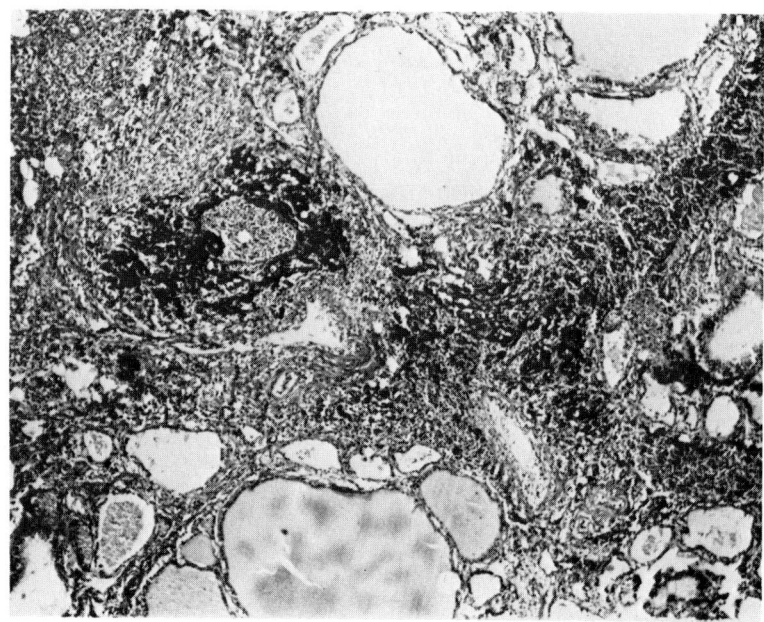

Fig. 15–6. Rarely, suppurative thyroiditis may result from extension of infections from the soft tissues around a tracheostomy wound. This patient had suppurative inflammation and numerous staphylococcal bacteria within the thyroid. (H & E, × 50.4)

Reference

1. Becker, R. A., Wilmere, D. W., and Goodwin C. W., Jr.: Free T4, free T3 and reverse T3 in critically ill, thermally injured patients. J. Trauma 20:713–721, 1980.
2. Brizio-Molteni, L., Molteni, A., Warpeha, R. L., Angelats, J., Lewis, N., and Fors, E. M.: Prolactin, corticotropin, and gonadotropin concentrations following thermal injury in adults. J. Trauma 24:1–7, 1984.
3. Pruitt, B. A., Jr.: Forces and factors influencing trauma care: 1983 A.A.S.T. presidential address. J. Trauma 24:463–470, 1984.
4. Stratta, R. J., Saffle, J. R., Holliman, C. J., et al.: Thermal injury in an adrenalectomized patient. J. Trauma 23:934–948, 1983.
5. Foley, F. D., Pruitt, B. A., Jr., and Moncrief, J. A.: Adrenal hemorrhage and necrosis in seriously burned patients. J. Trauma, 7:863–870, 1967.
6. Thrash, A. M., and Iri, H.: Adrenal infarction. Six case reports. Arch. Pathol., 75:538–542, 1963.
7. Greendyke, R. M.: Adrenal hemorrhage. Am. J. Clin. Pathol. 43:210–215, 1965.

8. Sevitt, S.: Reactions to Injury and Burns and Their Clinical Importance, 1st. ed. Philadelphia, J. B. Lippincott Co., 1974, pp 148, 154–157, 1966.

9. Sommers, S. C.: Adrenal glands. *In* Pathology, 6th. ed., Vol. 2, Chapter 36. W. A. D. Anderson, ed. St. Louis, C. V. Mosby Co., 1971, p 1473.

10. Markley, K., Bocanegra, M., Ego-Aguirre, E., Chiappori, M., and Morales, G.: Adrenocortical function after major surgical operations and thermal trauma in man. Surgery. 47:389–398, 1960.

11. Wise, L., Margraf, H. W., and Ballinger, W. F.: Adrenal cortical function in severe burns. Arch. Surg. 105:213–220, 1972.

12. Hume, D. M., and Egdahl, R. H.: The importance of the brain in the endocrine response to injury. Ann. Surg. 150:697–712, 1959.

13. Nash, G., Asch, M. J., Foley, F. D., and Pruitt, B. A., Jr.: Disseminated cytomegalic inclusion disease in a burned adult. J. A. M. A. 214:587–588, 1970.

14. Wilmore, D. W., Long, J. M., Mason, A. D., Jr., Skeen, R. W., and Pruitt, B. A., Jr.: Mediator of the hypermetabolic response to thermal injury. Ann. Surg. 180:653–669, 1974.

15. Vaughan, G. M., Becker, R. A., Allen, J. P., Goodwin, C. W., Pruitt, B. A., Jr., and Mason, A. D., Jr.: Cortisol and corticotrophin in burned patients. J. Trauma 22:263–272, 1982.

16. Zyss, R., and Gajkowska, B.: Changes in the function and fine structure of the chromaffin cells of rat adrenal medulla after burn. Pol. Med. J. 11:405–414, 1972 (Pathologia Polska 22, 1971).

17. Goodall, M. C., and Haynes, B. W., Jr.: Adrenal medullary insufficiency in severe thermal burn. J. Clin. Invest. 39:1927–1932, 1960.

18. Wegiel, J., Zyss, R., Kaszczynka, M., and Nka, M.: Ultrastructure of the rat adenohypophysis in burn disease. Acta Med. Pol. 12:289–292, 1971.

19. Sevitt, S.: Reactions to Injury and Burns and Their Clinical Importance, 1st. ed. Philadelphia, J. B. Lippincott Co., 1974, pp 209–210.

20. Panke, T. W., McLeod, C. G., Jr., and Langlinais, P. C.: Pituitary infarcts in thermally injured patients. Unpublished data.

21. Popp, M. B., Srivastava, L. S., Knowles, H. C., Jr., and MacMillan, B. G.: Anterior pituitary function in thermally-injured male children and young adults. Surg. Gynecol. Obstet. 145:517–524, 1977.

22. Dolecek, R., Zavada, M., Adamkova, M., and Leikep, K.: Endocrine rsponses in burned subjects: Insulin, somatotrophin, renin, angiotensin II, ACTH and LH. Burns 1:43–46, 1974

23. Wilmore, D. W., Orcutt, T. W., Mason, A. D., Jr., and Pruitt, B. A., Jr.: Alterations in hypothalamic function following thermal injury. J. Trauma 15:697–703, 1975.

24. Wilmore, D. W., and Aulick, L. H.: Metabolic changes in burned patients. Surg. Clin. N. Am. 58:1173–1187, 1978.

25. Cope, O., Nardi, G. L., Quijano, M., Rovit, R. L., Stanbury, J. B., and Wight, A.: Metabolic rate and thyroid function following acute thermal trauma in man. Ann. Surg. 137:165–174, 1953.

26. Trofimov, G. A.: Changes of thyroid function in burn disease. Fed. Proc. 22:1181–1183, 1963.

27. Becker, R. A., Wilmore, D. W., and Johnson, D. W.: Depressed serum triiodothyroinine (T$_3$) levels following thermal injury. Fed. Proc. 35:216, 1976.

28. Cavalieri, R. R., and Rapoport, B.: Impaired peripheral conversion of thyroxine to triiodothyronine. Ann. Rev. Med. 28:57–65, 1977.
29. DeRubertis, F.R. and Kasch, P.C.: Accelerated host metabolism of L-thyroxine during acute infection: Role of the leukocyte and peripheral leukocytosis. J. Clin. Endo. Metab. 40:589–600, 1975.
30. Tedeschi, L. G., and Peabody, C. N.: Cortical necrosis of the adrenal gland. Arch. Pathol. 73:18–24, 1962.

Chapter 16

NERVOUS SYSTEM

Cutaneous thermal injury has been associated with a variety of central and peripheral nervous system disorders[1] and alterations of EEGs.[2] Often, clinical manifestations cannot be correlated with pathologic findings. The following represents a compilation of known pathologic changes in the central nervous system that are related to thermal injury and a summary of unexplained clinical and laboratory findings.

POSTBURN DELIRIUM AND RESIDUAL NEUROLOGICAL DEFICITS

Immediate postburn agitation or obtundation may relate to hypovolemia or hypoxemia, and may be easily corrected while exercising care not to promote cardiac failure or cerebral edema by overhydration. Electrolyte disturbances (hyponatremia) and a variety of other conditions (Table 16–1) may cause seizures, especially in children. Unfortunately, cognitive and behavioral disorders may follow severe burns[1] and complicate the evaluation of other more significant neurological signs. As in other patients, septicemia is often associated in burn patients with mental disturbances characterized by delirium or altered state of consciousness.

In addition to the above, a vague clinical entity, "burn encephalopathy," has been mentioned in the early literature. This syndrome has no consistent clinical or morphologic characteristics, as indicated in a historical review of the subject.[3] In one review study, Andreasen et al.[2] noted that 30 percent of 32 burned patients less than 65 years old developed symptoms of delirium; a list of probable causative factors was proposed (Table 16–1). Whereas most studies have demonstrated no residual neurologic deficits in survivors, these authors described residual neurological abnormalities in two of their patients. These abnormalities included apathy, impaired memory, sexual impotence, and labile mood. From the above patients, a single case came to autopsy, and only cerebral edema and mild acute cerebral

Table 16–1 *Conditions Associated with Burns, Encephalopathy and Seizure[1,2,21]*

Hyponatremia or hypernatremia

Edema

Hypoxia

Fluid overload

Hypertension

Sepsis

Previous history of seizure disorder

Preexisting cerebrovascular disease

Hypovolemia

Blood loss

Dehydration

Pseudodiabetes

Renal failure

Meningitis

---

hemorrhage were noted.[2] Microscopically in other burn patients, degeneration of Purkinje's cells of the cerebellar cortex has been observed.[4] The nonspecific nature of these lesions suggests that they may be related to hypoxia, hemoconcentration, sepsis, and acidosis rather than to a metabolic disturbance.[5]

PERIPHERAL NERVOUS SYSTEM

A high incidence (29 percent) of peripheral neuropathy has been reported in acute burn patients.[6] Some of these neuropathies are secondary to deep (fourth degree) burns, especially the ulnar nerve at the epicondylar process and the distal radial nerve at the wrist.[7] Other nerves may be injured during escharotomy or debridement of vital tissue or intramuscular injections. Compression of blood vessels and nerves by circumferential burns in the extremities is a common postburn complication. However, compression of nerves by heterotopic bone formation around joint spaces (e.g., elbow) is uncommon and may not be recognized. Electrical injuries may destroy nerves primarily or secondarily through unappreciated fluid accumulation in fascial planes which compresses nerves or vascular structures or both. Finally, peripheral neuropathies may occur through improper positioning of extremities either during the application of splints or because of improper suspension causing injurious traction on the brachial or lumbosacral plexus.[6,7] Abnormal medial, ulnar, and peroneal nerve conduction has been demonstrated in patients of all age groups with significant burns (72 percent had burns covering more than 20 percent of their total body surface).[6] These neural deficits could not be related to uremia, septicemia, or use of magnesium sulfate in this study,[6] despite implication of such agents in peripheral nerve damage by other authors.

ELECTRICAL INJURY TO THE SPINAL CORD AND PERIPHERAL NERVES

The peripheral nervous system and spinal cord are particularly susceptible to electrical (thermal) injury. Injuries may be functionally divided into two types, immediate or delayed, according to the time of onset. In the *immediate* type, electric shock may cause a rare but potentially serious syndrome which is expressed by the immediate loss of consciousness followed by transient paralysis. This type of injury is exemplified by the patient struck by lightning. Clinically, the damage occurring immediately after electric injury is expressed by neurologic deficits which more commonly occur in motor nerves. Paraplegia, quadraplegia, or focal paralysis may develop, but complete recovery often occurs within hours or days after injury.[9]

The second or *delayed* form of spinal cord injury may result in a more serious outcome. Symptoms including quadraplegia, hemiplegia, or localized neurological loss may have a delayed (2 days to 2 years) onset. Signs of ascending paralysis, myelitis, or an amyotrophic lateral sclerosis-like syndrome may be seen. In 40 cases of this type reviewed by Levine et al.,[9] only two had even partial recovery. Wallerian degeneration, partial loss of neurons, and mild astrocytic gliosis were the prominent histologic findings in the spinal cord of the fatal case studied by Levine et. al.[9]

Neurologic deficits following peripheral nerve injury usually present as part of a polyneuritic syndrome involving nerves far removed from the apparent point of contact. It has been speculated that the late onset of neurologic deficit is related to microvascular thrombosis, but whatever the mechanism, late-appearing symptoms following spinal cord injury are more apt to be permanent than neurologic deficits occurring at the time of injury.[11]

CARBON MONOXIDE POISONING

Carbon monoxide poisoning is common when the thermal injury has occurred within a confined space, and burn victims who die within the first few days following injury may have morphologic lesions suggesting that severe exposure has occurred. Hemoglobin has a 300-fold greater affinity for carbon monoxide than oxygen.[12,13] The carboxyhemoglobin that forms also inhibits the release of oxygen from other hemoglobin molecules. This phenomenon greatly increases tissue hypoxia. Lethal levels of carboxyhemoglobin may occur rapidly in a fire (2–15 min).[12] No symptoms are associated with carboxyhemoglobin (COHb) levels less than 10 percent. Mild dilatation of cutaneous blood vessels and minimal headache occur with 10–20 percent COHb. This headache becomes moderately severe and throbbing at 20–30 percent and very severe at 30–40 percent with weakness, dizziness, dimmed vision, nausea, and vomiting. At 40–50 percent COHb, collapse is quite possible and increased respiration and pulse rates are noted. Coma with intermittent convulsions and Cheyne-Stokes respiration occur at 50–60 percent, suppressed cardiorespiratory function at 60–70 percent, and death when the COHb exceeds 70 percent.[2,12] Other clinical findings with carbon monoxide toxicity include speech, auditory, vestibular, and visual disturbances (amblyopia, hemianopsia, amaurosis, nystagmus, and pupillary dilation or constriction) and many other manifestations.[12]

The "cherry red" skin is characteristic of carbon monoxide poisoning and persists in the skin and organs for days after death. Occasionally, grayish discoloration is seen instead. Tissue often retains its cherry red color for a few days when immersed in formalin. Normal tissue loses its color within a few hours.[12,13] Extensive myonecrosis is highly suggestive of carbon monoxide poisoning.[14] Focal myocardial hemorrhage and necrosis have been induced in animal experiments in which dogs were exposed to high concentrations of carbon monoxide.[15] These lesions occur in humans by 10 days post exposure, and the left ventricle has subendocardial necrosis of papillary muscles. Lipid stains demonstrate fatty change in myocardial fibers and lipophages.[12,13]

Since the brain is the organ most sensitive to hypoxia, it is the site where most significant lesions are observed. Ischemic lesions vary from random microscopic foci to areas of macroscopic necrosis and cavitation of the pallidum and putamen. Initial lesions of the brain are petechiae in the cerebral cortex within a few hours. Ischemic changes of neurons consist of cytoplasmic eosinophilia, disappearance of Nissl substance, and nuclear pyknosis. Purkinje's cells of the cerebellum may be affected; 3 weeks later, they may be replaced by Bergmann's glia. Sommer's sector of Ammon's horn of the hippocampus is exquisitely sensitive to this hypoxia. Other areas sensitive to hypoxia include the cerebral cortex, striatum, pallidum, and thalamus. The third layer of the cerebral cortex, especially at the depths of the gyri, is very sensitive, and selective hypoxia may result in laminar necrosis of this layer.[16]

Probably the most characteristic, although not pathognomonic, lesion is hemorrhagic necrosis, which is most conspicuous in the white matter of the cerebrum, cerebellum, and basal nuclei of the brain (Fig. 16–1A). Highly suggestive is the finding of hemorrhagic lesions within the corpus callosum (Fig. 16–1B).[16] Rings of hemorrhage around small blood vessels occurring within a few days are also quite characteristic[13] of carbon monoxide poisoning. Further evolution of these lesions is generally similar to ischemic lesions of the brain resulting from vascular occlusion or stenosis. Finally, perivascular lymphocytic infiltrates may occur as nonspecific ischemic changes, and should not be misinterpreted as an encephalitis.[12,13]

CYANIDE POISONING

Hydrogen cyanide is released when upholstery and other materials containing polyurethane are combusted. Cyanide binds with cytochrome oxidase in mitochondria but produces clinical symptoms similar to those produced by carbon monoxide poisoning. Only chemical analysis for cyanide can make the diagnosis.[17] Winek[18] reports that embalming the body with formaldehyde may destroy the cyanide. He further indicates that cyanide content in fire victims is "considerably" less than the amount found in suicide victims. It may be, however, that a patient with severe thermal injury may tolerate far less cyanide. Recognizing cyanide poisoning and separating it from carbon monoxide poisoning is important, as treatment of the former is with sodium nitrite while treatment of the latter is oxygen with or without assisted ventilation.[17]

HEMATOGENOUS INFECTIOUS LESIONS

Episodes of severe septicemia may result in multiple "metastatic" bacterial lesions within the central nervous system. These foci have been identified within the meninges, choroid plexus, cerebellum, and cerebrum, and, if *Pseudomonas* species are the causative organisms, have the characteristic necrotizing features of metastatic *Pseudomonas* lesions seen elsewhere in the body (Fig. 16–2). Similar cerebral lesions have been observed in the *Pseudomonas*-infected burned rat. In patients with staphylococcal septicemia, microscopic cerebral abscesses have also been identified.[19]

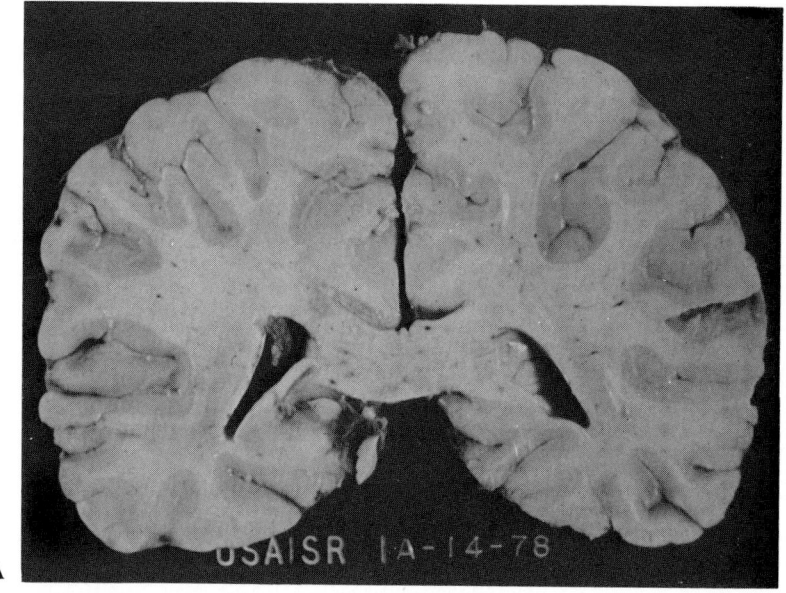

A

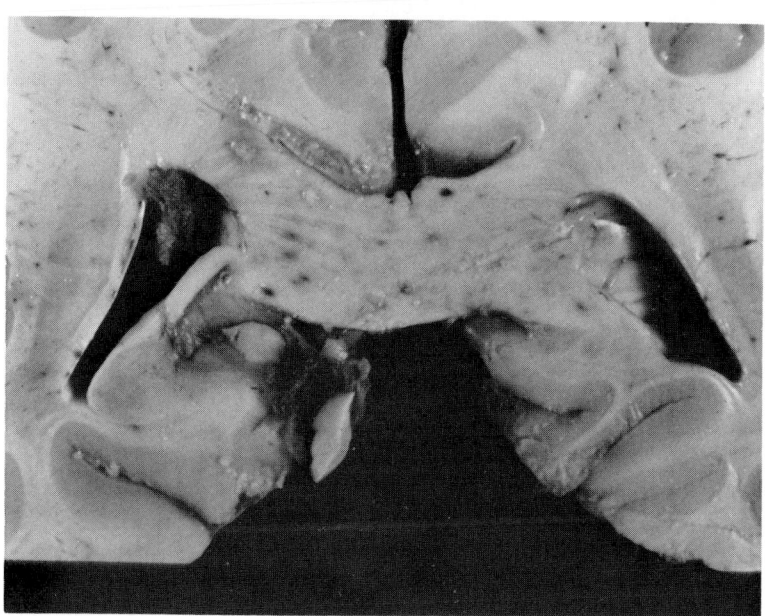

B

Fig. 16–1. (A): Multiple small areas of hemorrhage and necrosis within the white matter of the cerebrum and basal nuclei of the brain may be found in patients who succumb to carbon monoxide poisoning. (B): Hemorrhagic necrosis within the corpus callosum is highly suggestive of carbon monoxide poisoning, although similar lesions may be seen in patients who suffer terminal anoxia.

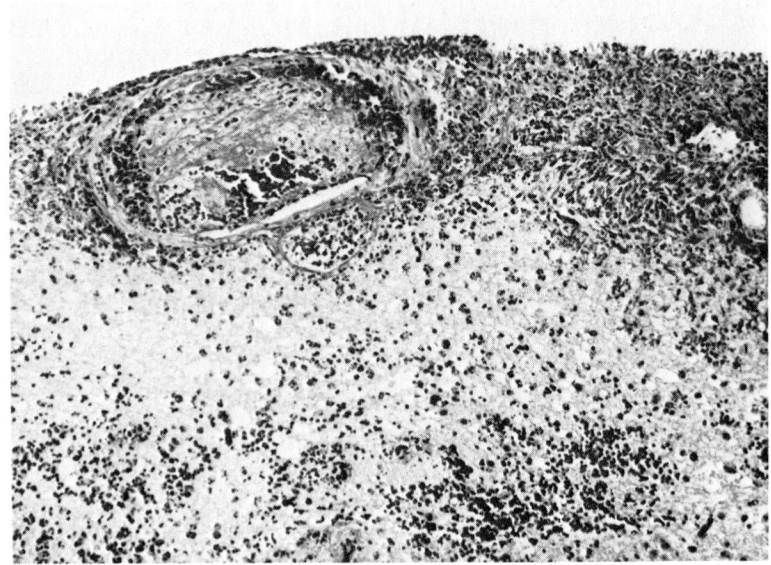

Fig. 16–2. Rarely, septicemic patients may develop embolic bacterial lesions in small vessels of the brain. Bacterial tissue stains will reveal the causative organism, usually *Pseudomonas aeruginosa* or *Staphyloccus* sp. in and adjacent to the affected vessels. (H & E, × 150)

COMPLICATIONS OF INTRAVENOUS HYPERALIMENTATION WITH FAT EMULSIONS

Recent reports have documented an infrequent complication of hyperalimentation with fat emulsions. "Milky-white lumps" have been noted in the pulmonary artery and adherent to the endocardial surfaces of the heart. (See Chapter 9.) Additional accumulations were identified in the microvasculature of the lungs, cerebrum, and kidney. Specific neurologic symptoms were not apparent clinically.[20]

FAT EMBOLISM, CENTRAL NERVOUS SYSTEM

In patients with multiple skeletal fractures, one must be concerned about fat embolism. Although a marked functional reserve and a double arterial blood supply protect the lung from apparent physiologic derangement in most cases, some fat emboli may pass through the pulmonary microvasculature into the cerebral microvasculature. Micro infarcts, "ball-and-ring" hemorrhagic perivascular lesions, and subsequent impairment in central nervous system function may result. (See Chapter 17.)

References

1. Hughes, J. R., Cazaffa, J. J., Pruitt, B. A., Jr., Boswick, J. A., McManus, W. F., Bruck, H. M., and Borges, J.: Seizures following burns of skin. II. Conditions associated with these seizures. Dis. Nerv. Syst. 34:347–353, 1973.
2. Andreasen, N. J. C., Hartford, C. E., Knott, J. R., and Carter, D. A.: EEG changes associated with burn delirium. Dis. Nerv. Syst. 38:27–31, 1977.
3. Mettler, F. A.: Burn encephalopathy as a "diagnosis." J. Med. Soc. N. J. 71:817–823, 1974.
4. Teplitz, C.: The pathology of burns and the fundamentals of burn wound sepsis. *In* Artz, C. P., Moncrief, J. A., and Pruitt, B. A., Jr. Burns, A Team Approach, 1st. ed. Philadelphia, W. B. Saunders Co., 1979, p 91.
5. Mohnot, D., Snead, O. C., and Benton, J. W. Jr.: Burn encephalopathy in children. Ann. Neurol. 12:42–47, 1982.
6. Helm, P. A., Johnson, E. R., and Carlton, A. M.: Peripheral neurological problems in the acute burn patient. Burns 3:123–125, 1976.
7. Salisbury R. E., and Dingeldein, G. P.: Peripheral nerve complications following burn injury. Clin. Orthop. Rel. Res. 163:92–97, 1982.
8. Dayan, A. D., Gardner-Thorpe, C., Down, P. F., et al.: Peripheral neuropathy in uremia. Neurology 20:649–658, 1970.
9. Fleming, L. W., Lenman, J. A. R., and Stewart, W. K.: Effect of magnesium on nerve conduction velocity during regular dialysis treatment. J. Neurol. Neurosurg. Psychiatry 35:342–355, 1970.
10. Levine, N. S., Atkins, A., McKeel, D. W., Jr., et al.: Spinal cord injury following electrical accidents: Case reports. J. Trauma 15:459–463, 1975.
11. Pruitt, B. A., Jr.: Other Complications of Burn Injury. *In* Moncrief, J. A., and Pruitt, B. A., Jr., Burns. A Team Approach, 1st. ed. Philadelphia, W. B. Saunders Co., 1979, pp 523–552.
12. Finck, P. A.: Exposure to carbon monoxide. *In* Tedeschi, C. G., Eckert, W. G., and Tedeschi, L. G., Forensic Medicine. A Study in Trauma and Environmental Hazards, Chapter 29. Philadelphia, W. B. Saunders Co., 1977, pp 840–849.
13. Finck, P. A.: Exposure to carbon monoxide. Review of the literature and 567 autopsies. Milit. Med. 131:1513–1539, 1966.
14. Finley, J., Van Beek, A., and Glova, J. L.: Myonecrosis complicating carbon monoxide poisoning. J. Trauma 17:536–539, 1977.
15. Ehrich, W. E., Bellet, S., and Lewey, F. H.: Cardiac changes from CO poisoning. Am. J. Med. Sci. 208:511–523, 1944.
16. Blackwood, W., and Corsellis, J. A. N.: Greenfield's Neuropathology, 3rd. ed. Chicago, An Edward Arnold Publication, Year Book Medical Publishers, Inc. (distributors), 1976, pp 68–71.

17. Cahalane, M., and Demling, R. H.: Early respiratory abnormalities from smoke inhalation. J. A. M. A. 251:771–773, 1984.

18. Winek, C. L.: Injury by chemical agents. In Tedeschi, C. G., Eckert, W. G., and Tedeschi, L. K., Forensic Medicine. A study in Trauma and Environmental Hazards, 1st. ed. Vol. 3. Philadelphia, W. B. Saunders Co., 1977, p 1581.

19. Teplitz, C.: The pathology of burns and the fundamentals of burn wound sepsis. *In* Burns. A Team Approach, 1st. ed. Artz, C. P., Moncrief, J. A., and Pruitt, B. A., Jr., eds. Philadelphia, W. B. Saunders Co., 1979, p 91.

20. Hessov, I., Melsen, F., and Haug, A.: Postmortem findings in three patients treated with intravenous fat emulsions. Arch. Surg. 114:66–68, 1979.

21. Sevitt, S.: A review of the complications of burns, their origin and importance for illness and death. J. Trauma 19:358–369, 1979.

Chapter 17

OTHER COMPLICATIONS AND CONSIDERATIONS OF THERMAL INJURY

Several important topics in burn pathology have been condensed into this single chapter. Although these subjects do not interrelate smoothly, they are vitally needed for a complete discussion of the spectrum of burn pathology. This chapter provides the reader with an overview of such topics as periocular and auricular burns, thermal complications involving bone and joints, late complications of burns (keloids and malignant neoplasms), a survey of lesions and important clinical data that might lead the physician to a diagnosis of child abuse, and other important yet less common complications of trauma or thermal injury or both.

SKELETAL SYSTEM

A less common, but significant, site of injury following burns is the skeletal system. Such injuries may occur during a fall (fracture), may represent a complication of infection of the burn wound (costochondritis, osteomyelitis), or may be a late complication of the healing process (periarticular calcification). A summary of these and other lesions follows.

Costochondritis

Costochondritis and associated perichondritis (Fig. 17–1) are found in three distinct clinical settings in the burn patient: (1) Deep electrical burns of the chest; (2) thermal injury with associated soft tissue injury of the chest;

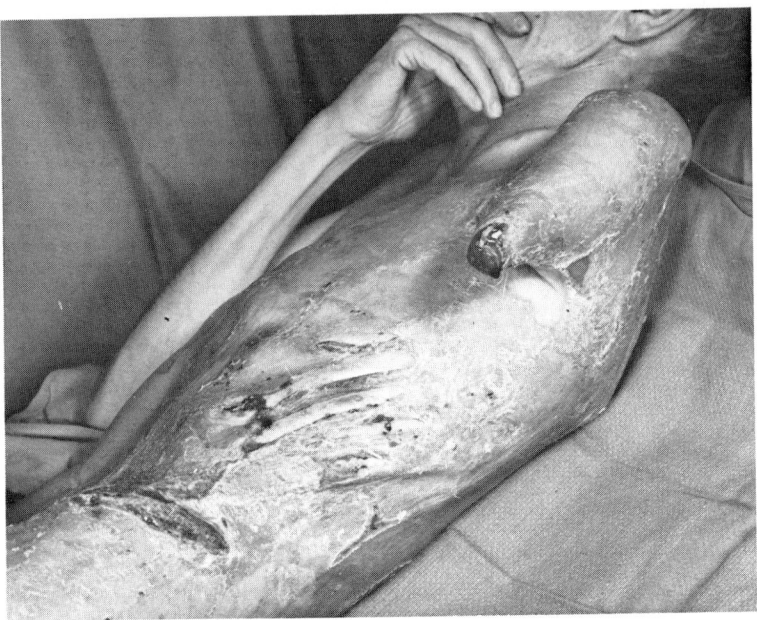

Fig. 17—1.   Thermal injury of the chest complicated by costochondritis. Exposed ribs are surrounded by granulation tissue with focal suppuration.

and (3) as a complication of abdominal surgery.[1] Burn patients are similar to other patients who develop costochondritis;[2] both groups classically experience persistent draining sinuses of the chest wall accompanied by tenderness, pain, edema, erythema, and induration.[1] Involvement of a single costal cartilage often requires resection of contiguous cartilages to prevent recurrence.

Microbiologically, *P. aeruginosa* is the most common isolate in burn patients with costochondritis, and may be seen in combination with *Staphylococcus aureus*, *Proteus mirabilis*, and *Providencia* sp.[1] In nonburn patients, the spectrum of microorganisms responsible for costochondritis also includes *Salmonella typhi*, *Actinomyces* sp., *Streptococcus* sp., *Blastomyces* sp., *Mycobacterium tuberculosis*, and *Escherichia coli*.[3]

Microscopically, the acute and/or chronic inflammatory process is located peripherally along apparent sinus tracts and involves perichondrium which is usually lysed. Where the inflammatory process affects cartilage, the chondrocytic nuclei are pyknotic or lysed, and pale-stained collagen is common.[3] A cystic alteration and "frayed" appearance of the edge of involved cartilage are also noted.

Periarticular Injuries

A variety of skeletal and periarticular changes may occur following thermal injury, as summarized in Table 17–1. Such lesions occur in 3–14 percent of burn patients; the incidence of each depends on the severity of burn and the thoroughness of the subsequent evaluation.[4-7] Few of these lesions heal well without surgical intervention.[4-6,8,9]

Periarticular calcification and heterotopic ossification usually involve the hip, elbow, and shoulder, and more commonly occur in children than adults. Although only a single joint area is involved, severe periarticular calcification may involve several joints,[10] It varies from minimal impairment (decreased extension by 10–15 degrees) to a "frozen" joint from an extraarticular bony bridge which may also be associated with ulnar nerve compression.[11] Trauma with damage to tendons and periarticular tissue may play some role in heterotopic ossification because third degree burns at the site are so common. Other conditions, such as prolonged immobilization of the joint and vigorous physical therapy, have been casually associated with heterotopic ossification by some,[10,11] but not others.[12] It is well known that immobilization causes a shift of calcium from bone, and this calcium shift is augmented by high protein diets.[13] Mobilized calcium may also play a role in periarticular heterotopic calcification. This periarticular heterotopic calcifi-

Table 17–1 *Summary of Skeletal, Articular and Periarticular Alterations Following Thermal Injury*

Bone

    Osteoporosis–generalized
              –local
    Periosteal new bone formation
    Osteophytes
    Fractures

Joints

    Chondrolysis
    Ankylosis
    Septic Arthritis

Periarticular

    Calcification
    Osteophytes
    Heterotopic, periarticular ossification

---

cation and ossification may extend into adjacent muscle and produce a histologic picture suggesting myositis ossificans. These changes are quite nonspecific and may not actually be caused by trauma, as they have been noted in poliomyelitis with paraplegia.[4] Plaque-like bony outgrowths may extend from the bone cortex (osteophytes) at the articular margin of the olecranon or coronoid processes. These osteophytes are similar to those occurring with degenerative joint disease.[4]

Articular Changes

Several alterations may occur in the *joint spaces,* and these are related to erosive articular lesions which heal by fibrosis and eventual bony *ankylosis.* Initial clinical changes are painful joints with fusiform swelling. If articular destruction is severe, there may be radiographic evidence of effacement of articular structures; and septic arthritis may be suggested (Fig. 17–2). This preceding septic arthritis may be asymptomatic.[11] The elbow, hip, and ankle are most often involved, and children appear to be more severely affected than adults. In addition to the articular calcification that may subsequently occur, periosteal new bone formation may also be seen.[4]

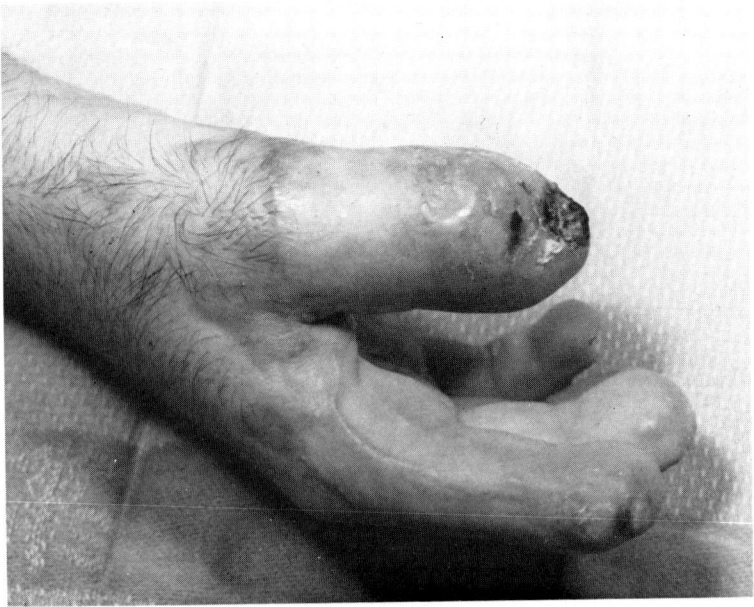

Fig. 17–2. Greatly enlarged thumb secondary to septic arthritis with marked soft tissue reaction.

Chondrolysis in the absence of infection is unusual and has been reported in a young man who developed chondrolysis of both hips. The lesions were first noted 5 weeks after hospitalization and were thought to be related to trauma (3.7 meter jump from plane following a crash) and the subsequent 5-week immobilization of his lower body.[14]

Skeletal Changes

Explosions, high voltage electrical injuries, motor vehicle and aircraft accidents, and other forms of trauma may result not only in extensive burns but also in skeletal *fractures,* usually of the long bones. Such fractures present major problems for therapy if they are in an area of thermal injury. Because of the necrotic burned tissue and progressive bacterial colonization, a cast cannot be applied and the patient must be treated with traction to allow for topical antibiotic application and wound inspection.[15] Later in the postburn course, hematoma at the site of bone fracture may become secondarily infected if septicemia occurs.

Periosteal new bone formation may suggest previous bone destruction, either directly caused by deep burns or secondarily related to the associated hyperemia. Serum alkaline phosphatase may be used to determine the age of the bone.[9]

Generalized osteoporosis is a consistent finding in all severely burned patients who are generally immobilized and have prominently elevated adrenocorticosteroid secretion. Localized osteoporosis may occur in bone underlying deep burns. Some degree of bone atrophy and demineralization may persist up to 3 years post burn.[4]

Thermal Injury of Bone

Deep thermal injuries affecting the periosteum and outer bone cortex occurs in 2–3.3 percent of patients sustaining burns.[16] Such injuries are secondary to prolonged involuntary contact with flaming material, explosions, and electrical injuries.[5] Necrotic periosteum and bone may subsequently become infected and separate as a sequestrum. However, if sterile, the dead bone may act as a graft and eventually become revascularized. Osteomyelitis has its own infrequent but very grave secondary consequences, including such disseminated infections as endocarditis, pneumonia, and cerebral abscesses.

Thermal Injury to Skull

Partial thickness thermal injury to the skull is associated with injury to the outer bone table and intervening diploë. The inner bone table is generally intact. The first changes of bone necrosis occur in the bone marrow, where the nuclei of hematopoietic cells assume a ghost-like appearance. Bone necrosis is marked by the disappearance of osteocytes from bone lacunae, often in an irregular and patchy distribution which precludes accurate determination of the actual depth of bone necrosis. In *full thickness* thermal injury of the skull, both inner and outer tables of the skull are destroyed.[17]

Repair is slow in onset and is generally prolonged. Healing from full thickness lesions frequently occurs from the underlying dural membrane. Granulation tissue and fibrosis extend through Volkmann's canals into dead bone. New bone proliferation appears at 4 weeks and is most prominent at 7 weeks post burn. Complete healing may take months.[17] This slow rate of skull bone healing in adults contrasts with that seen in other sites. For example, rib fractures have new medullary bone within 10–14 days. This slow rate of skull bone healing may be related to the loss of vasculature on the site of osseous repair.

Pyogenic infections are a constant threat to necrotic areas of the skull (Fig. 17–3). Such infections are most common in patients with full thickness injury to skull bone. Microorganisms are generally gram-positive cocci and less frequently gram-negative bacilli. Infectious agents may invade through the outer bone table of the skull to involve the diploë. The inner bone table usually restricts the spread of infection. It has been suggested that early skin closure over the skull burn may decrease or prevent pyogenic bone infection.[17,18] The benefits of early closure may outweigh risks of subsequent infection from unsuspected contaminated skull wounds.[19]

Joint Dislocation

Although very unlikely to occur in the closely observed patient, scar contracture may cause several joints to become dislocated. These include: thumb, hip, elbow, and foot (vertical talus deformity).[11] Simple surgical intervention is usually sufficient to prevent evolution of these dislocations.

FAT EMBOLISM

Significant fat embolism (FE) in burn patients is rare, except in those who have also sustained multiple skeletal fractures. It is *not* due to hyperlipemia[20] but is actually related to fragments of intramedullary bone, sometimes including bone spicules, and large fat globules which embolize to the pulmonary microvasculature (Fig. 17–4A) or subsequently to other systemic organs (Fig. 17–4B).[20–22] Trauma that precedes FE usually involves multiple fractures of long bones (tibia, femur). The incidence is variable, but it can be found in 90–100 percent of patients who die shortly after sustaining serious skeletal fractures.[20]

Pulmonary FE has no clinical importance in the great majority of patients because it is usually mild. Several factors may contribute to this

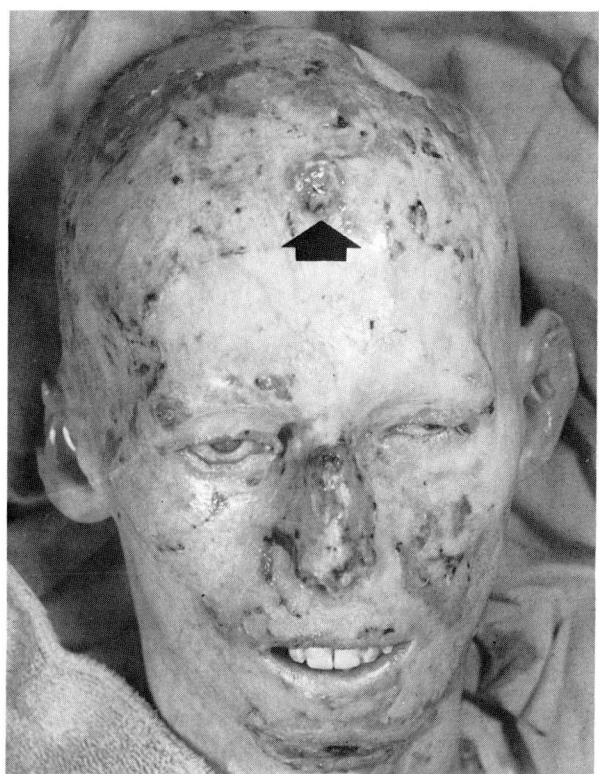

Fig. 17–3. Severe thermal injury of scalp showing one of several foci of suppurative infection secondary to *Staphylococcus aureus* (arrow). Patient also has ectropion involving both eyelids.

lack of symptoms: (1) an enormous pulmonary capillary bed, (2) a large functional reserve of the lung, and (3) the fluid and deformable nature of the emboli, which may cause only partial occlusion. In contrast, even mild fat embolization to the brain may have great clinical consequences and may be responsible for the patient's death.[21]

Although less common, significant direct trauma to subcutaneous adipose tissue may also cause FE. The injury may be caused by accidental trauma or major surgical procedures.[22]

Pathogenesis

Intramedullary adipose tissue enters the venous system of bone at the site of fracture during the first 24–48 hr following traumatic injury. Although less common, significant direct surgical or accidental trauma to subcutane-

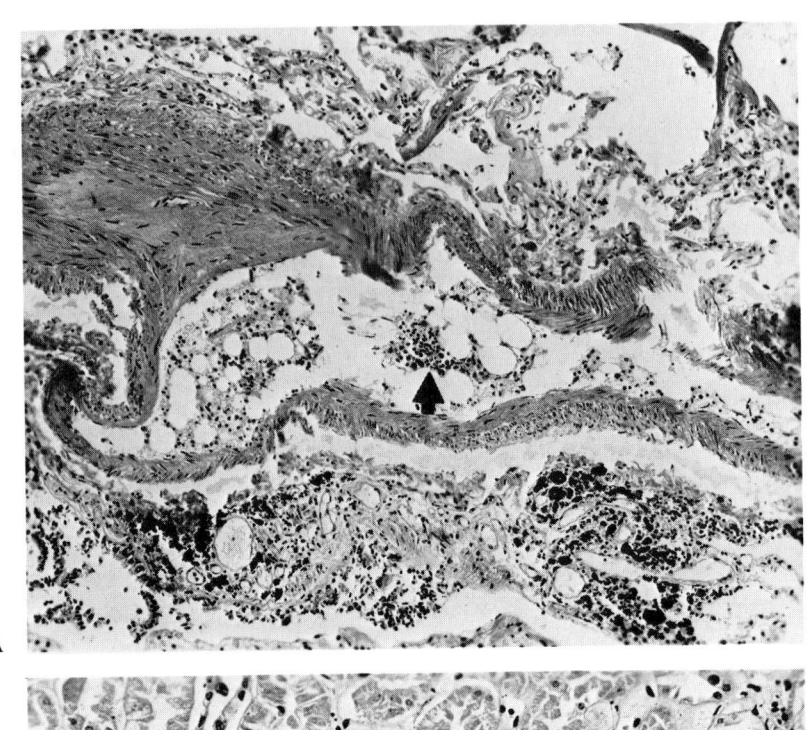

A

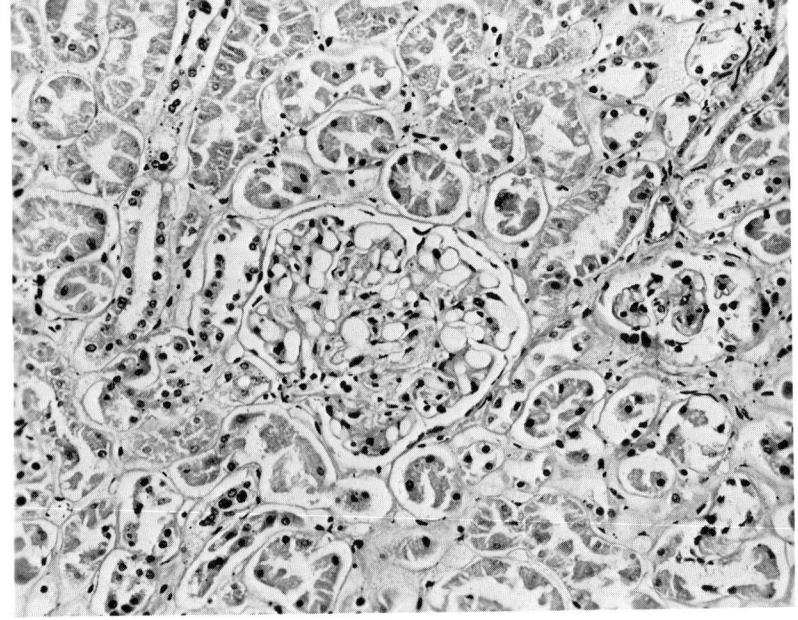

B

Fig. 17–4 A, B.

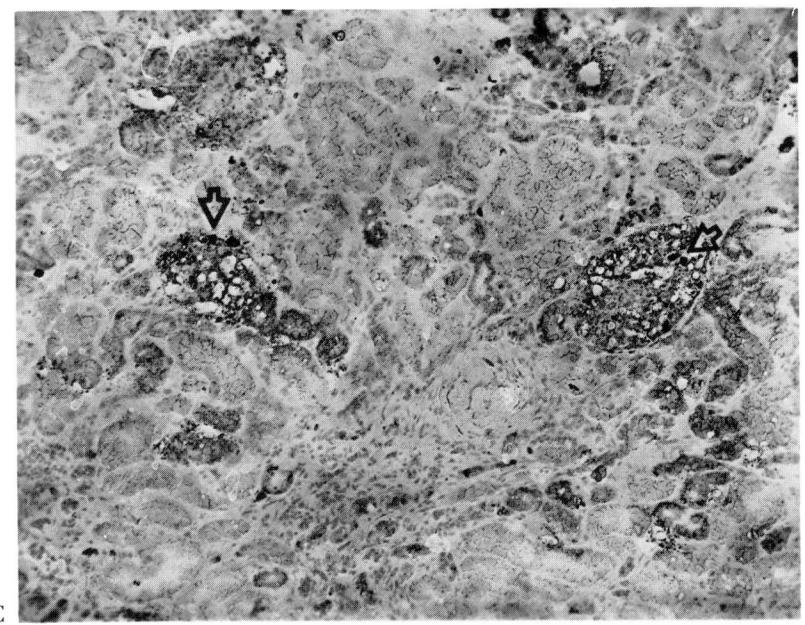

Fig. 17–4. Photomicroscopic demonstration of fat emboli in **(A)** lungs and **(B)** kidney and **(C)** kidney with Oil Red O stain.

ous adipose tissue may also cause FE. Patients with significant FE have several common features: (1) numerous fat emboli typically involving small arteries and arterioles of the lung and brain (see Chapter 16 on Central Nervous System) rather than capillaries and venules, (2) multifocal pulmonary intraalveolar hemorrhage, and (3) possible death within hours to 2 days.[21–23]

Hydrolysis of fat emboli causes macroglobules of fat to form without histologic evidence of identifiable fat cells from which they arose. Such fat globules, which appear to coalesce with erythrocytes and platelets, may be associated with fibrin thrombi and may explain the laboratory findings associated with FE (see below). Probably both this physiochemical and the above mechanical block of the microvasculature play a role in "fat embolism."[24,25] Traumatic lipemia, commonly encountered after severe hemorrhage, burns and, cold injury, is related to elevated plasma chylomicrons and serum lipoproteins. The composition of lipids contrasts with fat embolism, where the triglyceride composition is similar to bone marrow fat.[26]

Clinical Presentation

Symptoms of significant FE usually appear 12–72 hr following trauma. Manifestations of pulmonary involvement include progressive pulmonary insufficiency (arterial pO_2 less than 60 mmHg), fever and, tachycardia. Serial chest roentgenograms demonstrate fine stippling and hazy infiltration in both lung fields that rapidly progress to fluffy, dense mottling and microatelectasis if severe FE has occurred.[22,23,27]

Systemic fat embolism is much less frequent and may present as a characteristic petechial rash involving the anterior shoulders, upper chest, and neck on the second or third day after injury.[22,28,29] Fat globules may be detected in retinal blood vessels.[23,30] Generalized central nervous system involvement may present as somnolence and disorientation and progress to stupor, coma, and decerebrate rigidity. Localized neurologic deficits may include focal epileptic seizures, expressive aphasia, or hemiparesis.[22] Occasional renal involvement may occur secondary to FE.[31]

Patients with significant (symptomatic) FE were previously associated with up to a 35 percent mortality rate. This latter group of patients often progressed into coma terminally.[23] Current therapy has markedly improved the prognosis for patients with FE.[32]

Laboratory Findings

Patients with FE have a sudden, unexplained fall in the hematocrit which may be associated with thrombocytopenia. Laboratory evidence may suggest a disseminated intravascular coagulopathy (See DIC in Chapter 11, the Hematopoietic System). Hypocalcemia, frequently associated with severe FE, may be related to the release of free fatty acids following hydrolysis of fat which then binds to calcium to form "soap."[33] No change is seen in serum triglyceride levels, but serum lipase is elevated. Neutral fat may be detected in urine,[23] cerebrospinal fluid,[30,34] and frozen sections of peripheral blood and tracheal aspirates.[29,35,36] However, the high incidence of neutral fat in sputa of patients without FE negates the diagnostic value of this observation. Skin biopsies of petechiae and possibly a renal biopsy may demonstrate systemic FE.[28,37]

It is important to realize, however, that fat globules are not a pathognomonic finding indicating FE, and may be seen in a variety of conditions other than trauma (Table 17–2).[20,22,26,38] Even serum lipases may be markedly elevated by the administration of morphine, which is frequently used in traumatized patients.[23]

Fat embolism must be distinguished from recently described complications associated with the use of high concentrations of fat emulsions in hyperalimentation. The latter is usually recognized grossly. The lack of bony

Table 17–2 *Nontraumatic Causes of Fat Embolism*

Diabetic acidosis and coma

Acute pancreatitis

Miscellaneous infections

  Clostridium welchii
  Bacterial osteomyelitis
  Pneumonia, occasionally

Excessive alcoholism

Fatty nutritional cirrhosis

Chronic venous congestion

Terminal cardiac failure

Carbon tetrachloride

Phosphorus poisoning

Salvarsan administration

Acute, severe osteomyelitis

---

spicules and fat cells by microscopic examination further distinguishes fat emulsions from fat emboli.[39] The topic is discussed in Chapter 9, the Cardiovascular System.

OCULAR THERMAL INJURY

Asch et al.[40] reported ocular complications in 7.5 percent of 1400 burn patients admitted to a major burn unit. These injuries consisted of globe (corneal ulcers, lacerations, burns, and penetrating injuries) and oculoadnexal (conjunctivitis and ectropion) lesions. Although none of these complications is fatal, significant morbidity is common.

Ectropion is one of the most common complications of *oculoadnexal* injury (Fig. 17–4) and frequently (47/123 eyelids) requires surgical release to prevent subsequent persistent conjunctivitis, keratitis, and corneal ulcers. Ectropion is an eversion of the edge of the eyelid related to contraction of excessive scar tissue in the late "healing" phase of burn injury. Flame injuries are the most common cause of ectropion.[40]

Conjunctivitis is another form of *oculoadnexal* injury which may occur at any time during the postburn course. Chemical burns usually result in conjunctival lesions in the early postburn course which are successfully treated medically. Late in the postburn course ectropion may cause conjunctivitis secondary to exposure of oculoadnexal tissues.

Corneal ulcers and lacerations are the most common injuries to the ocular globe.[40] Trauma to the cornea or progression of a burn involving the cornea usually causes the corneal ulcer. Since *P. aeuruginosa* often is the predominant significant pathogenic microorganism in the burn wound, one must vigorously treat such corneal ulcers, which may become rapidly colonized. *Pseudomonas* infections of the eye are well known to cause rapid and fulminant necrosis of the entire globe within hours. There may be only minimal evidence of this devastating progression.[41]

AURICULAR CHONDRITIS

Approximately one of four patients with facial burns involving the ear develops auricular chondritis unless meticulous care is initiated when the burn patient first enters the hospital.[42] Burns cause damage to the ear in two ways. *Severe, direct thermal injury* may cause full thickness burn with subsequent autoamputation (autochondrectomy). The helical rim of the ear is most commonly involved; however, the entire ear may be so injured.

The above process must be differentiated from a *suppurative chondritis* in which the burned ear becomes secondarily infected (Fig. 17–5). Suppurative chondritis is usually seen between the third and fifth postburn weeks and may complicate a partial or full thickness burn. Clinically, the ear is painful, tender, erythematous, moist, and swollen. The inflammation is initially confined to the anthelix or helix. An abscess may form on the anterior surface of the ear and may spontaneously drain, but usually must be excised. Most cultures (74 percent) yield a mixed bacterial flora, with *P. aeruginosa* noted in 95 percent of cultures. *Staphylococcus aureus* and *Proteus* sp. are the next most common microorganisms identified. Topical sulfamylon (Winthrop) has reduced (29–19 percent) the incidence of aural suppurative chondritis.[42]

Pathology: Mild early lesions of auricular chondritis consist of a mild acute perichondritis. When the infection becomes suppurative, the involved cartilage diffusely loses its basophilic staining. Erosions of cartilage are common, and chondrocytic lacunae frequently lack nuclei. Colonies of bacteria are commonly identified. The overlying epithelium shows varied stages or repair, and ulcerated areas may be identified.

Thermal injury to the *inner ear* is uncommon. Frenkiel and Alberti reported 11 cases, mostly involving steelworkers in whom hot "slag" perforated the tympanic membrane. The resultant injury was associated with a recurrent suppurative inflammatory process that responded poorly to medical and surgical therapy. Prevention measures (earplugs) were strongly recommended.[43]

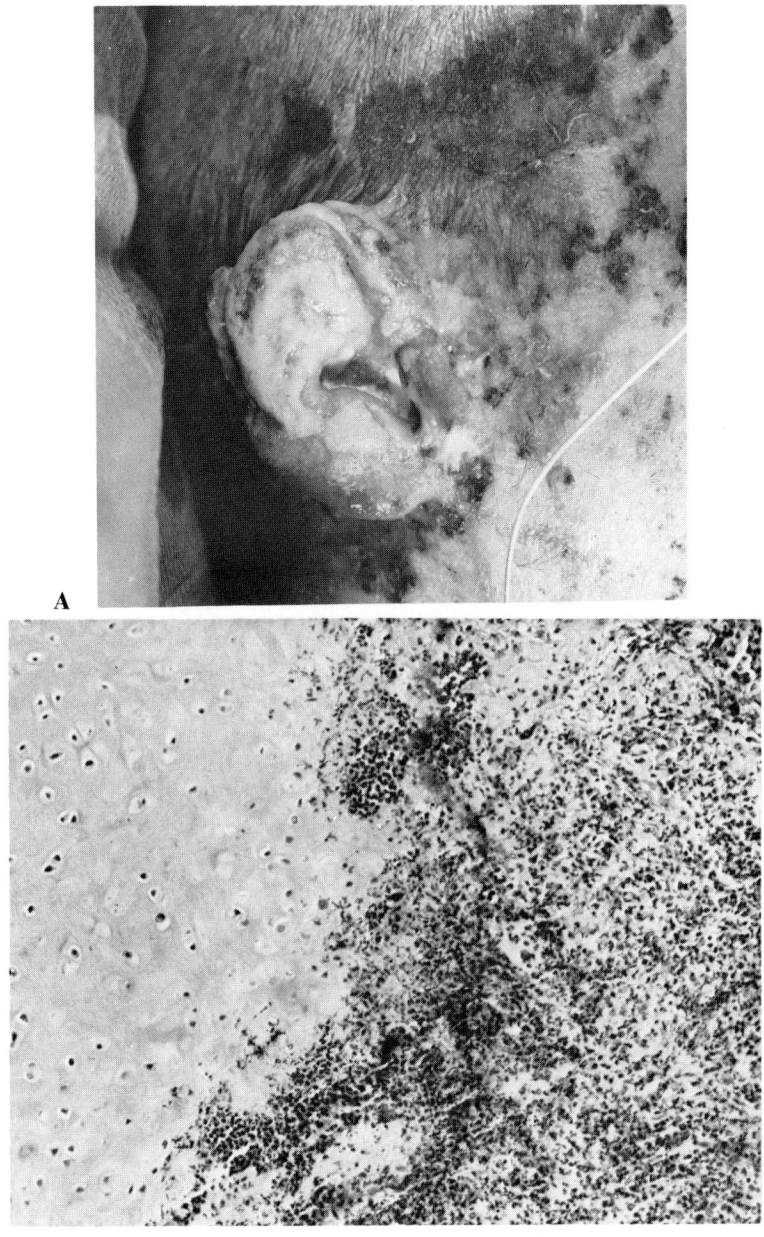

Fig. 17–5. (A): Severe thermal injury to the ear complicated by infectious auricular chondritis. (B): Photomicrograph of auricular chondritis. Acute and chronic inflammatory infiltrate extending into and eroding edge of cartilage. (× 135).

291

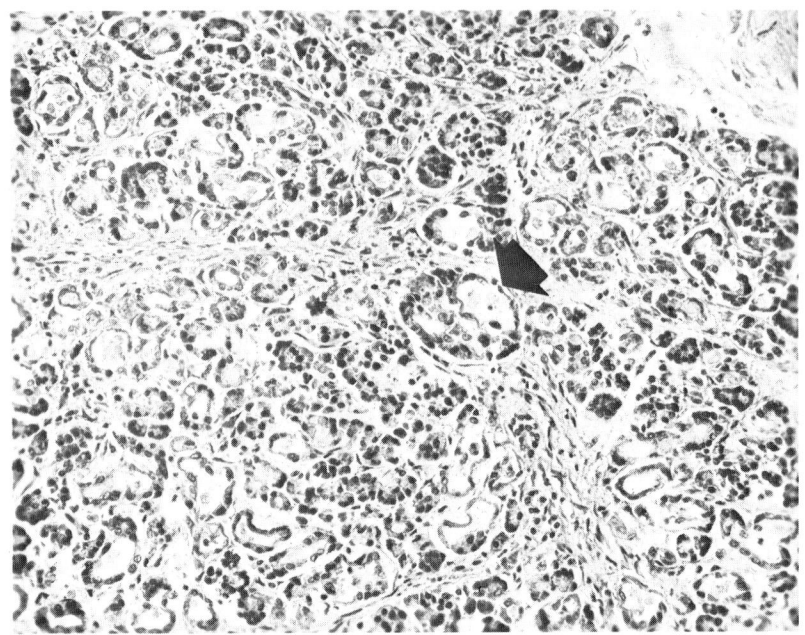

Fig. 17–6. Photomicrograph of pancreas in a burn patient with severe renal disease demonstrating distended pancreatic acini filled with an eosinophilic inspissated material.

PANCREATITIS

Few significant findings are seen in the pancreas of the burn patient. The occasional small foci of fat necrosis and mild inflammation in the pancreas are usually insignificant. Hemorrhagic necrosis of the pancreas occurs in less than 1 percent of burn autopsies and may be fatal.

In most burn patients with serious renal failure, changes described by Baggenstoss[44] are noted in the pancreas. The pancreatic acinar glands are variably dilated and are filled with an inspissation of pale, amorphous secretion (Fig. 17–6). There may also be a scant lymphocytic infiltrate. The pathogenesis of this lesion is unknown.

PSEUDODIABETES MELLITUS OF BURNS

Hyperglycemia following severe thermal injury, a common sequel in the early postburn period, may be related to high levels of circulating corticosteroids and catecholamines.[45,46] High levels of blood glucose in the "normal" patient with severe burns during the second postburn week are not associated with ketosis and may thus be distinguished from hyperglycemia of true diabetic patients. Glucagon levels are prominently elevated and return to "normal" when the burn wound is healed. In mild burns, levels of both insulin and glucose are elevated, suggesting insulin resistance. In severe burns, insulin is decreased while glucose is elevated.[47] When the burn wound is closed, insulin returns to "normal." The islets of Langerhans appear to participate in the metabolic response to burns. It has been speculated that catecholamines may direct islet-cell response.[46]

Pseudodiabetes mellitus of burns (PDMB) may be associated with marked hyperglycemia (800–1000 mg/dl). Cerebral symptoms in PDMB are most commonly secondary to dehydration due to glycosuria and obligatory polyuria.[48] PDMB has been associated with pyknosis, hyperchromatism, and other postmortem findings in pancreatic islet cells suggesting injury. Mitoses within the islet cells and normoglycemia during the later postburn course suggest that the damage is reversible.[48]

Patients with true diabetes mellitus have a similar predisposition to developing insulin-resistent hyperglycemia. However, these patients have additional problems, including delayed and poor wound healing because of a defective inflammatory response[49] and a markedly increased risk for infections, especially mycotic (see Chapter 11, the Hematopoietic System).

HYPERTROPHIC SCAR FORMATION AND KELOIDS

Exuberant desmoplasia within the burn wound is the most frequent late complication in the burn patient. Severe hypertrophic scars most commonly occur in children and young adults (Fig. 17–7A). Deep partial thickness burns with prolonged spontaneous healing appear to be particularly susceptible to excessive scarring.[50,51] Keloids are defined as cutaneous scars that grow beyond the confines of the original wound, in contrast to hypertrophic scars, which are raised scars which remain within the boundary of the original wound.[52] The formation of hypertrophic scars and keloids is related to the unexplained development of an exuberant intradermal proliferation of fibrous connective tissue which becomes increasingly hypovascular. Whorls of dense collagen fibers become hyalinized (Fig. 17–7B). The overlying epidermis plays a passive role in keloid and hypertrophic scar formation, merely conforming to the distorted outline of the underlying nodules of connective tissue. Hypertrophic scars and keloids require approximately one year (a few months to several years) to develop.[53] In contrast to keloids, hypertrophic scars may undergo partial to complete spontaneous resolution.

Deitch et al.[51] have evaluated several variables associated with the formation of hypertrophic scars. In their analysis of 100 children and adults

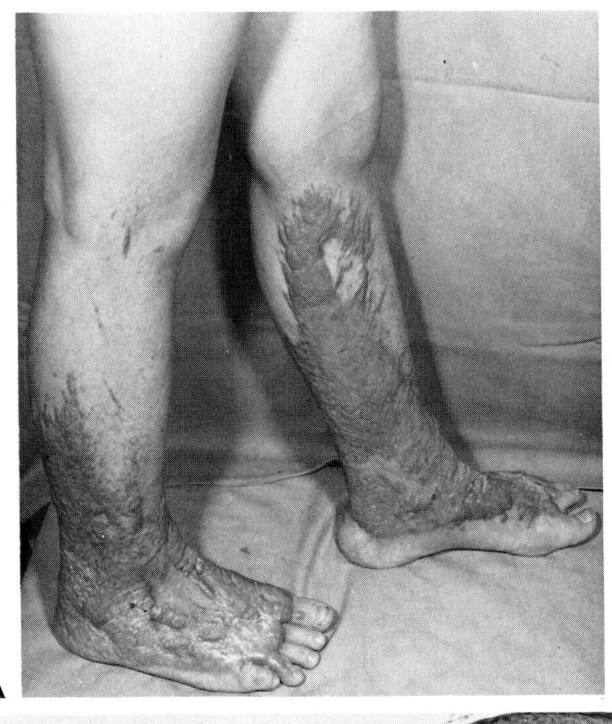

A

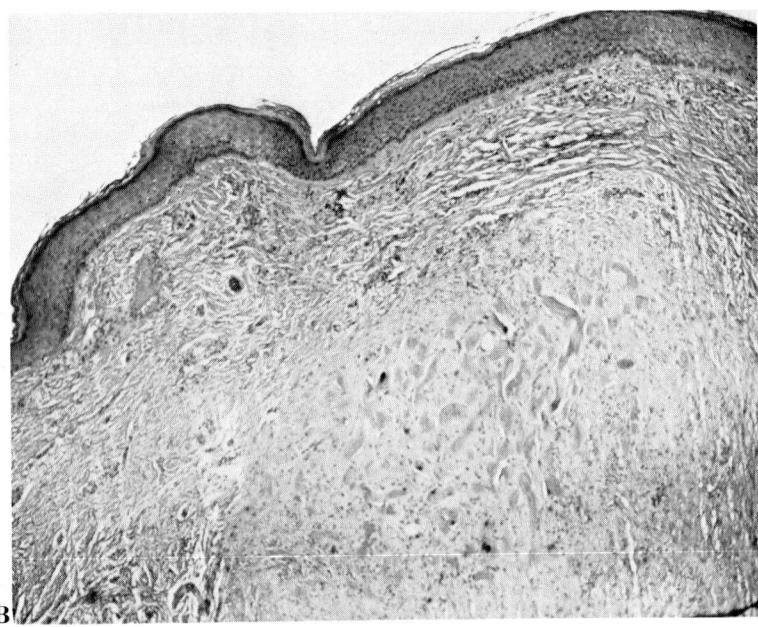

B

Fig. 17–7. (A): Hypertrophic scars following severe thermal injury. (B): Photomicroscopic slide of hypertrophic scar showing thick hyalinized bundles of connective tissue that replace adnexal structures and distort overlying epidermis ($\times$ 54).

with thermal injury, they demonstrated that the rate of burn wound healing was paramount for the development of hypertrophic scars. Healing in less than 14 days was optimal. Healing between 14 and 21 days was associated with a 33 percent incidence of hypertrophic scars; this incidence rose to 78 percent if healing was prolonged beyond 21 days. Black patients only had an increased incidence of hypertrophic scars if the time for healing exceeded 14 days. Certain sites of burn injury (chest, back, shoulders, buttocks) were associated with a slightly increased incidence of hypertrophic scars because of the increased skin tension. No correlation was found between patient age and hypertrophic scars. These findings would undoubtedly be useful in any protocol utilizing selective and individualized prophylactic pressure therapy in burn patients.[51]

The pathogenesis of hypertrophic scars and keloids is an enigma. Increased levels of a serum alpha globulin component have been associated with an increased collagen synthesis rate in hypertrophic scars[54] and with suppression of collagenase activity and decreased collagen turnover.[55] Evidence of active collagen turnover has been demonstrated by the identification of "fibroclasts" and "myofibroclasts" several months after scar tissue formation.[55] Also, increased lysl oxidase activity has been identified in postburn scars.[56] It is currently uncertain how these findings might be manipulated to affect appropriate wound healing.

The most effective therapy for patients who form hypertrophic scars in their healing burn wounds is the application of circumferential pressure via the utilization of elastic garments.[52] Such pressure has been associated with some resolution of the hypertrophic scar; ultrastructural evaluation has demonstrated reorganizational changes in the fibrous connective tissue.[57] Corticosteroids and surgery have also been beneficial therapeutic adjuncts.[52,58,59]

NEOPLASTIC TRANSFORMATION WITHIN THE BURN WOUND

With the passage of several (10–40 or more) years, neoplasms develop in some burn scar wounds.[60–66] Neoplasms developing in burn scars are usually squamous cell carcinomas, also known as Marjolin's ulcers (Fig. 17–8); less frequently, basal cell carcinomas and malignant melanomas occur.[61–63,66–69] Also, a variety of sarcomas and reactive epithelial proliferations have been noted in burn scar, including keratoacanthoma, osteogenic sarcoma, fibrosarcoma, and poorly differentiated sarcoma.[60,64,70–72] Just recently a contiguous squamous cell carcinoma and malignant melanoma were reported to develop synchronously in a burn scar.[73] These neoplasms rarely occur in sites of successful primary skin grafting, and they are seldom multiple. Although squamous cell carcinomas arising in radiated skin may be

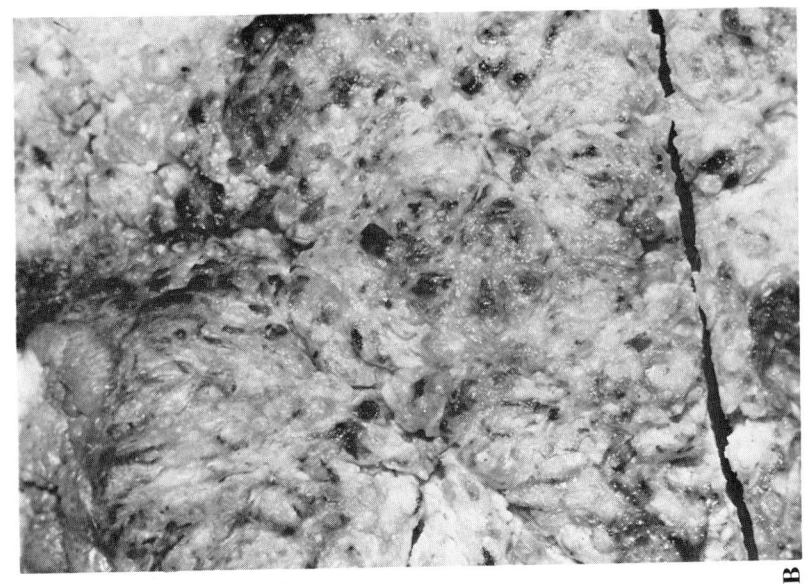

Fig. 17-8 A, B.

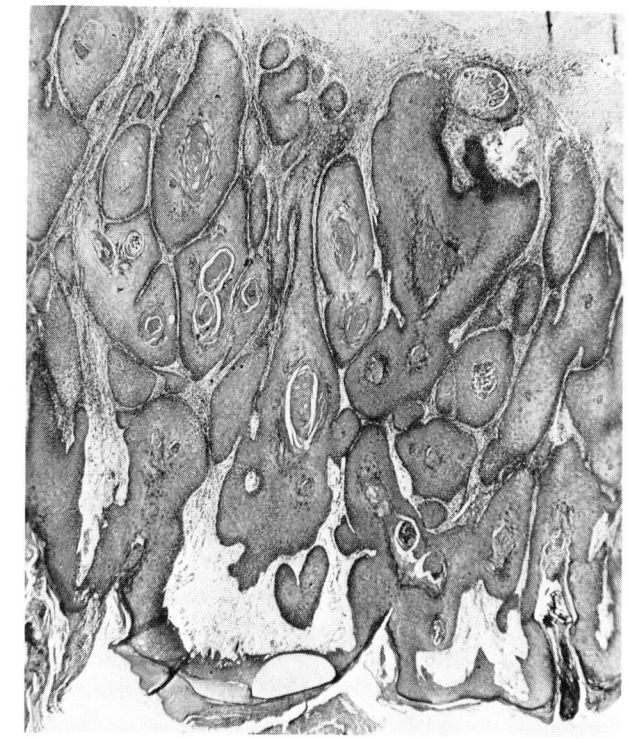

D

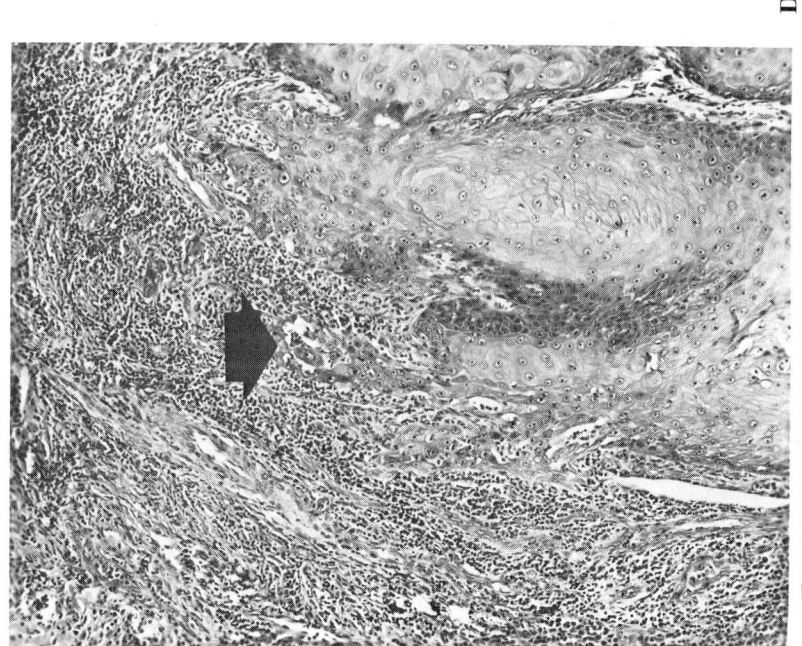

C

Fig. 17–8. (**A,B**): Cross-section and "close-up" photograph of a squamous cell carcinoma of the leg in a 61-year-old man who sustained thermal injuries more than 15 years previously. (**C,D**): Photomicroscopic slides of well-differentiated squamous cell carcinoma with areas of possible early invasion [(**D**): arrow].

composed of spindle cells, only typical round-to-polygonal squamous cell carcinomas have been reported in the burn wound scar.

Burn scar malignancies have been predominantly found in sites of poor healing, poor vascularity, repetitive trauma, or mistreatment.[63] A variety of noxious agents may elicit a neoplastic alteration in the skin.[62] There is experimental evidence that the temporary state of immunological depression in the burn patient may permit accelerated neoplastic transformation.[74,75] However, the effect is transient and completely reversed by 10 days post burn.[75] Increased resistance to tumor growth appears during burn wound healing.[76] Others believe that the hypertrophic burn scar wound represents an immunologically privileged site which may permit tumor growth.[77]

Although the prognosis of burn wound malignancy is generally good, squamous cell carcinoma does metastasize to regional lymph nodes. Novick et al.[66] reported a series of 46 cases of patients with neoplastic transformation of burn wound scar. One-third of these patients developed metastases, and most of these latter patients died from their disease. These malignant neoplasms occurred on all body surfaces; however, the lower extremities were more prone to develop metastatic disease (54 percent of patients developed carcinoma of skin of the legs) than carcinoma at other sites of the body.[66] The above indicates the need for careful observation of patients with the development of ulcers in burn scars. Such ulcers should be immediately excised, along with a wide margin of normal tissue surrounding the ulcer. Bostwick et al.[77] presented a group of patients with well-differentiated squamous cell carcinoma originating in burn wound scar that manifested an extraordinary aggressive course once the neoplasms were "released from their scar milieu."

Three cases of keratoacanthoma have recently been reported in burn patients (ages 30–43 years, all Caucasian). These tumors appear within a few weeks to months after thermal injury; this rapid development of keratoacanthoma sharply contrasts with the prolonged latent period for malignant neoplasms.[65] Previous reports of skin cancer which occurred within a few months of thermal injury and healed spontaneously may represent keratoacanthoma.

Nonneoplastic pseudoepitheliomatous hyperplasia in the burn scar may also be confused with squamous cell carcinoma and may be found at the edge of chronic ulcers in the burn wound. This reactive epithelial proliferation lacks the atypical cytologic appearance, abnormal mitoses, and invasive characteristics of squamous cell carcinoma.[60]

CHILD ABUSE

Any young child with burns must be considered a possible victim of child abuse. The persuasive influence of the parents may mislead the examining physician into believing that the burn is accidental, and the pathologist may be the last and only unbiased observer to examine the patient for evidence of child abuse.

Recognition of child abuse by the pathologist is vital for the protection of other siblings in the family. In addition to this moral responsibility, it recently has become a legal responsibility in many states to report cases of *suspected* child abuse.

Most abused children are infants and toddlers (67–79 percent), and 24–25 percent are less than 1 year old.[78–80] Battered children have a 60 percent incidence of additional battering and a 10–15 percent subsequent mortality. Permanent neurological sequellae were seen in up to 15 percent of battered children.[80] Males are more commonly abused, and the children are most often burned by hot liquids. The youngest child of the family tends to be most abused. If the first-born child is abused, the subsequent siblings have a 13-fold increased chance of abuse.[79]

All social classes may seriously abuse children, but semiskilled workers of the lower classes are the most common offenders. The parents have average intelligence, are usually young (early 20s), may be itinerant factory workers, are often unemployed, and frequently are isolated from grandparents and relatives. There is usually family discord, and the abused child is usually unwanted. The male parent is very often a stepfather or male consort of the mother, is more likely to demonstrate abnormal (64 percent) or even aggressive psychopathic disorders (33 percent), and frequently has a criminal record (29 percent).[81] Battering parents, who are often themselves abused children, manifest a character disorder rather than a psychotic abnormality. The parental action is almost always precipitated by misdeeds of the child; however, the parental response is unduly violent.[79]

Many children are accidentally burned each year. However, certain burns are highly suggestive, if not diagnostic, of child abuse. Acute punctate burns with several scars of similar dimensions and appearance may suggest infliction by a lit cigarette. Burns confined to the buttocks are likewise suspicious.[79] In addition, immersion "doughnut" burns, "splash" burns, and particularly flexion ("zebra") burns may be inflicted on the patient.[82]

Burns from abuse involve less of the body surface area (usually less than 1 percent) than accidental burns.[83,84] Burns involving the perineum are more common with abuse, while accidental burns are more likely to affect the extremities.[83,84] Accidental scald burns are most often second degree burns if promptly treated. Third degree burns resulting from a scald burn suggest forced immersion. The soles and palms are covered by thick epidermis and are particularly resistent to third degree scald burns.[85]

The pathologist and physician must look beyond the burn for other evidence of child abuse (See Tables 17–3, 17–4). Most injuries of other kinds are inflicted with the hand. The head, face, or neck of the child are the most common sites for abrasions, bruises, and lacerations. Careful search of the scalp may uncover several bruises. More serious injuries include skull

Table 17–3 *Clinical and Pathoeologic Features Suggesting Child Abuse*[78–87]

Clinical History

Prior hospitalization for "accidental" trauma

Unexplained delay (>24 hr) between time of injury and first attempt to obtain medical attention, e.g., burns older than alleged day of accident

Parent departs abruptly after child admitted to the hospital and telephones for progress reports (? fears reprisal for child abuse)

Account of incident not compatible with age or ability of patient—6-month-old child crawling into bathtub filled with hot water

Account of incident not compatible with distribution of burn. Hot water, third degree burns from upper chest downward without burns of arms—not compatible with accidental fall into tub

Alleged no witness to the "accident"

Relatives other than parents bring burned child to hospital

Burn attributed to action of a sibling or other child (although this does occur)

Changes or inconsistencies in parent and parental story

Child taken to a different doctor or accident center after each injury

Clinical Findings

Child excessively withdrawn, submissive, "flattened affect," or overly polite

Evidence of neglect, i.e., malnutrition, anemia, growth failure, poor hygiene

Child does not cry during painful procedures

Remains isolated in bed; avoids contact with other children

Unresponsive initially to friendly overtures by the staff

Scalds of hands or feet with symmetrical third degree burns suggesting forced immersion for prolonged period

Isolated burn of buttocks—unusual area for accidental burn

Ocular injuries: subconjunctival subhyaloid and retinal hemorrhages, retinal separation, lens displacement

Oral cavity lesions: bruises of the lip, especially upper lip, with lacerations of inner side; ruptured frenulum, detachment of inner surface of lip from gum; fractured anterior teeth, tongue lacerations

Finger tip pressure marks, necks, sides of chest

Bruises of abdomen associated with ruptured abdominal organs

"Bites," almost always inflicted by mother (rule out other children by size of bite)

Punctate ulcers or scars suggesting cigarette burns

Multiple skeletal fractures of variable ages

Autopsy Findings

Subdural hematomas, acute, recent, old

Cerebral contusions

Lacerations of liver, intestine, mesentery

Multiple hematomas, scars in varying stages of healing

As in Clinical Findings*

---

* Mentioned under "Autopsy findings" are *internal* abnormalities not seen clinically. The pathologist should also carefully describe the *external* "clinical findings."

Table 17-4 *Comparison of "Inflicted" Abuse and "Accidental-Neglect" Abuse in Children*[a]

	Inflicted Abuse	Accidental-Neglect Abuse
Age	Infant, toddler	Toddler to young child
Maternal background	Educationally or culturally deprived	Educationally or culturally deprived
Maternal characteristics	Withdrawn or hostile	Overwhelmed, manipulative, defensive
Marital status	Single and boyfriend in home; separated; occasionally married	Divorced, separated, single, occasionally married
Employment	Often not employed	All long-term unemployment
Burn type	Usually all scald	Contact, scald, or flame
Other injuries	Scars, whip marks, rib or skull fractures, bruises, bite marks, rope burns	Malnourished, growth failures, poor hygiene, anemia
Acute precipitating factors	Often acute marital conflict	Neglect, abuse

Features Common to Both Groups: Wife battery, young mothers (17-25), and previous reports of child abuse are common

[a] Adapted from Ayoub, C., and Pfeiffer, D.: Burns as a manifestation of child abuse and neglect. Amer. J. Dis. Child. 133: 910–914, 1979.

fractures (especially parietal), subdural hematoma, and cerebral contusions. Subdural hematoma may occur without skull fractures or scalp bruises. A child who is violently shaken may develop a "whiplash" injury, and the only evidence may be faint finger- and thumb-mark bruises on the trunk and arms. Intraocular hemorrhage is strong presumptive evidence of a battered child.[80] Abdominal injuries, suggested by the presence of abdominal bruises, consist mostly of ruptures of the liver, intestine, and mesentery.[79]

Ayoub and Pfeifer[65] have further delineated child abuse by creating two categories: inflicted abuse and accidental-neglect abuse. The above classification clearly describes two entirely different modes of abuse, active and passive, respectively. Distinguishing features are summarized in Table 17–4.

When a multiplicity of factors is present (See Table 17–2), one should be highly suspicious of child abuse, regardless of the circumstances. The interested reader is referred to Stone et al.,[78] who have outlined an approach for investigating such incidents.

References

1. Reckler, J. M., Flemma, R. J., and Pruitt, B. A., Jr.: Costal chondritis: An unusual complication in the burned patient. J. Trauma 13:76–80, 1973.
2. Moschcowitz, A. V.: The treatment of diseases of the costal cartilages. Tr. Amer. Surg. Assoc. 36:327–349, 1918.
3. Pontius, J. G., Clagett, O. T., and McDonald, J. R.: Costal chondritis and perichondritis. Surgery 45:852–857, 1959.
4. Evans, E. B., and Smith, J. R.: Bone and joint changes following burns. J. Bone Jt. Surg. 41A:785–799, 1959.
5. Asch, M. J., Curreri, P. W., and Pruitt, B. A., Jr.: Thermal injury involving bone: Report of 32 cases. J. Trauma 12:135–139, 1972.
6. Munster, A. M., Bruck, H, M., Johns, L. A., von Prince, K., Kirkman, E. M., and Remig, R. L.: Heterotopic calcification following burns: A prospective study. J. Trauma 12:1071–1074, 1972.
7. Kubacek, V., Fait, M., and Poul, J.: A case of heterotopic ossification in the hip joint area following skin burn. ACTA Chir. Plast. (Praha) 19:209–214, 1977.
8. Evans, E. B.: Orthopedic measures in the treatment of severe burns. J. Bone Jt. Surg. 48A:643–669, 1966.
9. Hoffer, M., Brody, G., and Ferbic, F.: Excision of heterotopic ossification about elbows in patients with thermal injury. J. Trauma 18:667–670, 1978.
10. Heslop, J. H.: Heterotopic periarticular ossification in burns. Burns 8:436–438, 1982.
11. Jackson, D., and Mac, G.: Destructive burns: Some orthopedic complications. Burns 7:105–122, 1980.
12. Pruitt, B. A., Jr.: Complications of thermal injury. Clin. Plast. Surg. 1:667–691, 1974.
13. November-Dusansky, A., Moylan, J. A., Linkswiler, H., and Elson, C.: Calciuretic response to protein loading in burn patients. Burns 6:198–201, 1980.
14. Pellicci, P. M., and Wilson, P. D., Jr.: Chondrolysis of the hips associated with severe burns. A case report. J. Bone Jt. Surg. 61A:592–596, 1979.
15. Pruitt, B. A., Jr.: Other complications of burn injury. In Burns: A Team Approach, 1st. ed., Chapter 36. C. P. Artz, J. A. Moncrief, and B. A. Pruitt, Jr., eds. Philadelphia, W. B. Saunders Co., 1979, pp 523–552.
16. Trapnell, D. H., and Jackson, D.: Bone and joint changes following burns. Clin. Radiol. 16:180–186, 1965.
17. Sevitt, S.: Healing of burns of the skull: A histological study. Burns 3:133–149, 1977.

18. Worthen, E. F.: Regeneration of the skull following a deep electrical burn. Plast. Reconstr. Surg. 48:1–4, 1971.
19. Jackson, D.: Burns of bones: Can these bones live. II. Results of 98 cases and discussion of treatment. Burns 1:356–372, 1975.
20. Sevitt, S. Reactions to Injury and Burns and Their Clinical Importance, 1st. ed. Philadelphia, J. B. Lippincott Co., 1974, pp 188–217.
21. Sevitt, S.: The boundaries between physiology, pathology and irreversibility after injury. Lancet 2:1203–1210, 1966.
22. Tedeschi, L. G., and Hechtman, H. B.: Posttraumatic embolism. *In* Forensic Medicine. A Study in Trauma and Environmental Hazard, 1st. ed. C. G. Tedeschi, W. G. Eckert, and L. G. Tedeschi, eds. Philadelphia, W. B. Saunders Co., 1977, pp 409–413.
23. Thomas, J. E., and Ayyar, R.: Systemic fat embolism. A diagnostic profile in 24 patients. Arch. Neurol. 26:517–523, 1972.
24. Dines, D. E., Burgher, L. W., and Okazaki, H.: The clinical and pathologic correlation of fat embolism syndrome. Mayo Clin. Proc. 50:407–411, 1975.
25. Oh, W. H., and Mital, M. A.: Fat embolism: Current concepts of pathogenesis, diagnosis and treatment. Orthop. Clin. N. Amer. 9:769–779, 1978.
26. Spencer, H. Pathology of the Lung, 3rd. ed. Philadelphia, W. B. Saunders Co., 1977, pp 554–555.
27. Wiener, L., and Forsyth, D.: Pulmonary pathophysiology of fat embolism. Amer. Rev. Resp. Dis. 92:113–118, 1965.
28. Stephens, J. H., and Fred, H. L.: Petechiae associated with systemic fat embolism. Arch. Dermatol. 86:515–517, 1962.
29. Lahiri, B., and Zu Wallack, R.: The early diagnosis and treatment of fat embolism syndrome: A preliminary report. J. Trauma 17:956–959, 1977.
30. Cross, H. E.: Examination of cerebrospinal fluid in fat embolism: Report of a case. Arch. Intern. Med. 115:470–474, 1965.
31. Wagman, E., Barbara, A., Marquis, J., Chirls, M., and Falla, A.: Renal fat embolization and urostealith formation complicating femoral fracture. J. A. M. A. 198:721–723, 1966.
32. Guenter, C. A., and Braun, T. E.: Fat embolism syndrome. Changing prognosis. Chest 79:143–145, 1981.
33. Peltier, L. F., Adler, F., and Lai, S.: Fat Embolism: The significance of an elevated serum lipase after trauma to bone. Amer. J. Surg. 99:821–825, 1960.
34. Robb-Smith, A. H. T., Hung, A. H., Russell, D., and Greenfield, J. G.: Discussion on fat embolism and brain. Proc. R. Soc. Med. 34:639–642, 1941.
35. Pollak, R., and Meyers, R. A. M.: Early diagnosis of the fat embolism syndrome. J. Trauma 18:121–123, 1978.
36. Chan, K. M., Tham, K. T., Chiu, H. S., Chow, Y. N., and Leung, P. C.: Post-traumatic fat embolism—Its clinical and subclinical presentations. J. Trauma 24:45–49, 1984.
37. Sevitt, S.: The boundaries between physiology, pathology and irreversibility after injury. Lancet 2:1203–1210, 1966.
38. Tedeschi, C. G., Castelli, W., Kropp, G., and Tedeschi, L. G.: Fat macroglobulinemia and fat embolism. Surg. Gynecol. Obstet. 126:83–90, 1968.
39. Hessov, I., Melsen, F., and Haug, A.: Postmortem findings in three patients treated with intravenous fat emulsions. Arch. Surg. 114:66–68, 1979.

40. Asch, M. J., Moylan, J. A., Jr., Bruck, H. M., and Pruitt, B. A., Jr.: Ocular complications associated with burns: Review of a five year experience including 104 patients. J. Trauma 11:857–861, 1971.
41. Hazlett, L. D., Rosen, D., and Berk, R. S.: *Pseudomonas* eye infections in cyclophosphamide-treated mice. Invest. Ophthal. Visual Sci. 16:649–652, 1977.
42. Dowling, J. A., Foley, F. D., and Moncrief, J. A.: Chondritis in the burned ear. Plast. Reconstr. Surg. 42:115–122, 1968.
43. Frenkiel, S., and Alberti, P. W.: Traumatic thermal injuries of the middle ear. J. Otolaryngol. 6:15–22, 1977.
44. Baggenstoss, A. H.: The pancreas in uremia, histopathologic study. Amer. J. Pathol. 24:1003–1017, 1948.
45. Munster, A. M.: Management of diabetic patients with thermal injury. Surg. Gynecol. Obstet. 134:483–484, 1972.
46. Wilmore, D. W., Lindsey, C. A., Moylan, J. A., Faloona, G. R., Pruitt, B. A., Jr., and Unger, R. H.: Hyperglucagonaemia after burns. Lancet 1:73–75, 1974.
47. Volenec, F. J., Clark, G. M., Mani, M. M., Kyner, J., and Humphrey, L. J.: Metabolic profiles of thermal trauma. Ann. Surg. 190:694–698, 1979.
48. Sevitt, S. Reactions to Injury and Burns and Their Importance, 1st. ed. Philadelphia, J. B. Lippincott Co., 1974, pp 129–130.
49. Goodson, W. H., III, and Hunt, T. K.: Wound healing and the diabetic patient. Surg. Gynecol. Obstet. 149:600–608, 1979.
50. Pruitt, B. A., Jr.: The burn patient: II. Later care and complications of thermal injury. Curr. Prob. Surg. 16:1–95, 1979.
51. Deitch, E. A., Wheelahan, T. M., Rose, M. P., Clothier, J., and Cotter, J.: Hypertrophic burn scars: Analysis of variables. J. Trauma 23:895–898, 1983.
52. Ketchum, L. D., Cohen, I. K., and Masters, F. W.: Hypertrophic scars and keloids. Plast. Reconstr. Surg. 53:140–154, 1974.
53. Parks, D. H., Evans, E. B., and Larson, D. L.: Prevention and correction of deformity after severe burns. Surg. Clin. N. Amer. 58:1279–1289, 1978.
54. Diegelmann, R. F., Bryant, C. P., and Cohen, I. K.: Tissue alpha-globulins in keloid formation. Plast. Reconstr. Surg. 59:418–423, 1977.
55. Baur, P. S., Jr., Barratt, G. F., Brown, G. M., and Parks, D. H.: Ultrastructural evidence for the presence of "fibroclasts" and "myofibroclasts" in wound healing tissues. J. Trauma 19:744–756, 1979.
56. Hayakawa, T., Hino, M., Fuyamada, H., Nagatsu, T., Aoyama, H., and Izawa, Y.: Lysyl oxidase activity in human normal skins and postburn scars. Clin. Chim. ACTA 71:245–250, 1976.
57. Bauer, P. S., Barratt, G., Linares, H. A., Dobrkovsky, M., de la Houssaye, A. J., and Larson, D. L.: Wound contractions, scar contractures and myofibroblasts: A classical case study. J. Trauma 18:8–22, 1978.
58. Ohmori, K.: Application of microvascular free flaps to burn deformities. World J. Surg. 2:193–202, 1978.
59. Sharzer, L. A., O'Brien, B. M., Horton, C. E., Adamson, J. E., Meadick, R. A., Carraway, H. H., Hayhurst, J. W., and McLeod, A.: Clinical applications of free flap transfer in the burn patient. J. Trauma 15:766–771, 1975.
60. Arons, M. S., Lynch, J. B., Lewis, S. R., and Blocker, T. G., Jr.: Scar tissue carcinoma. Part I. A clinical study with special reference to burn scar carcinoma. Ann. Surg. 161:170–188, 1965.

61. Stone, N. H., and Montiel, M. M.: Multiple basal cell carcinomas arising in radiated burn scars. Case report. Plast. Reconstr. Surg. 46:506–509, 1970.
62. Treves, N., and Pack, G. T.: Development of cancer in burn scars. Surg. Gynecol. Obstet. 51:749–782, 1930.
63. Montani, D., Hartl-Prpic, V., Vladovic-Relja, T., and Prpic I.: Cancer in old burn scars. Chir. Plast. 2:203–210, 1974.
64. Drut, R., and Barletta, L.: Osteogenic sarcoma arising in an old burn scar. J. Cutan. Pathol. 2:302–306, 1975.
65. Monafo, W. W., and Bohling, C.: Keratoacanthoma arising in newly healed burn scars. Burns 1:172–174, 1974.
66. Novick, M., Gard, D. A., Hardy, S. B., and Spira, M.: Burn scar carcinoma: A review and analysis of 46 cases. J. Trauma 17:809–817, 1978.
67. Nancarro, J. D.: Malignant melanoma arising in an unstable burn scar. Br. J. Plast. Surg. 32:135, 1979.
68. Gellin, G. A., and Epstein, W. L.: Malignant melanoma from thermal burn scar. Arch. Dermatol. 111:1214–1215, 1975.
69. White, S. W.: Basal cell carcinoma arising in a burn scar: Case report. J. Dermatol. Surg. Oncol. 9:159–160, 1983.
70. Moule, R.: Cancers in burn scars ("Marjolin's ulcers"). In Transactions to the International Society of Plastic Surgeons, Second Congress. A. B. Wallace, ed. London, E & S Livingstone, 1960, pp 130–142.
71. Barletta, L., Laguens, R., and Wamba, N.: Hemangioendotelioma maligno del dorso desarrollado en cicatriz de antigua guemadura. Prensa Med. Argent. 63:1163–1166, 1966.
72. Fleming, R. M., and Rezek, P. R.: Sarcoma developing in an old burn scar. Amer. J. Surg. 54:457–465, 1941.
73. Muhlemann, M. F., Griffiths, R. W., and Briggs, J. C.: Malignant melanoma and squamous cell carcinoma in a burn scar. Br. J. Plast. Surg. 35:474–477, 1982.
74. Castillo, J. L., and Goldsmith, H. S.: Burn scar malignancy in possible depressed immunologic setting. Surg. Forum 19:511–513, 1968.
75. Munster, A. M., Gale, G. R., and Hunt, H. H.: Accelerated tumor growth following experimental burns. J. Trauma 17:373–375, 1977.
76. Minster, J. J.: Decreased resistance to tumor cells after stress followed by increased resistance. Proc. Soc. Exp. Biol. Med. 113:377–379, 1963.
77. Bostwick, J., III, Pendergrast, W. J., Jr., and Vasconez, L. O.: Marjolin's ulcer: An immunologically privileged tumor? Plast. Reconstr. Surg. 57:66–69, 1976.
78. Stone, N. H., Rinaldo, L., Humphrey, C. R., and Brown, R. H. Child abuse by burning. Surg. Clin. N. Amer. 50:1419–1424, 1970.
79. Knight, B.: The battered child. In Forensic Medicine. A Study in Trauma and Environmental Hazards, 1st. ed., Chapter 10. C. G. Tedeschi, W. C. Eckhert, and L. G. Tedeschi, eds. Philadelphia, W. B. Saunders Co., pp 500–509, 1977.
80. Smith, S. M., and Hanson, R.: 134 battered children: A medical and psychological study. Br. Med. J. 3:666–670, 1974.
81. Smith S. M., Hanson, R., and Noble, S.: Parents of battered babies: A controlled study. Br. Med. J. 4:388–391, 1973.

82. Lenoski, E. F., and Hunter, K. A.: Specific patterns of inflicted burn injuries. J. Trauma 17:842–846, 1977.

83. Kumar, P.: Child abuse by thermal injury—A retrospective survey. Burns 10:344–348, 1984.

84. Keen, J. H., Lendrum, J., and Wolman, B.: Inflicted burns and scalds in children. Br. Med. J. 4:268–269, 1975.

85. Hight, D. W., Bakalar, H. R., and Lloyd, J. R.: Inflicted burns in children. Recognition and treatment. J. A. M. A. 242:517–520, 1979.

86. Ayoub, C., and Pfeifer, D.: Burns as a manifestation of child abuse and neglect. Amer. J. Dis. Child. 133:910–914, 1979.

87. Woolridge, E. D., Jr.: Forensic odontology. *In* Forensic Medicine. A Study in Trauma and Environmental Hazards, 1st. ed. C. G. Tedeschi, W. C. Eckert, and L. G. Tedeschi, eds. Philadelphia, W. B. Saunders Co., pp 1137–1139, 1977.

Chapter 18

PREDICTING SURVIVAL AND ASSIGNING THE CAUSE OF DEATH IN BURN PATIENTS

In previous chapters, we have explored and analyzed innumerable, and sometimes seemingly unrelated, pathophysiological events that were directly or indirectly associated with thermal injury. The physician, the surgeon, and certainly the pathologist may find these many direct and indirect effects of thermal injury difficult to evaluate or arrange in any order of importance. In this final chapter, we would like to present a time-related framework of events and complications as they might be anticipated in the postburn course of a seriously injured patient. It is suggested that this framework would delineate both the typical and the unusual events and indicate ancillary findings that one might expect and search for at particular junctures in the postburn course.

In addition, tables will be provided to aid the physician in determining probability for survival based on the extent and severity of burns and age of the patient. Factors that might affect interpretation of these tables are discussed.

SURVIVAL POTENTIAL OF THE BURN PATIENT

For the purpose of conveying a meaningful prognosis to the patient or his relatives, it is important to have criteria for evaluation of the patient's clinical condition and his probability of survival. Such information is equally

307

important to the pathologist, who must analyze whether the burn size and severity sufficiently explain the hospital course and terminal complication(s). Several methods are available for predicting the prognosis of a thermally injured patient.

Simple Approach—Burn Size and Age

This most common method utilizes a small number of criteria, such as the extent of second and third degree burn and the patient's age, to assess the severity of injury and prognosis for survival. Utilizing a system of probit analysis, Bull and Fisher[1] have prepared a grid of approximate mortalities based on patient's age and burned surface area. In terms of large general burn populations, this method provides highly accurate data[2-9] and has greatest utility because it may be easily and immediately employed shortly after the patient's admission. It depends only on the accuracy with which the extent of the patient's thermal injury is determined. This method is currently in use at most burn units. Survival data for burn patients are included (see Fig. 18–1, 18–2).

Universal Data Gathering

A more complex and complete set of criteria utilizes the above variables as well as the assessment of preexisting disease.[10] In addition, daily evaluation of clinical status will reveal new complications or unexpected improvements. All such elements may be utilized to reassess prognosis in a highly individualized manner which is most useful for predicting survival of the individual burn victim with unusual complicating health factors.

Initial attempts to evaluate the complicating factors in thermally injured patients are discussed briefly by Bull and Fisher.[1] These investigators attempted to compensate for complicating disease in older patients (where other, preexisting diseases are obviously more common) and determined that their compensatory technique could be applied to their "Grid of Approximate Mortality Probabilities." However, other factors of obvious medical importance (previous myocardial infarcts, diabetes mellitus) may occur in younger patients, and no method for evaluating these factors is currently available.

Fisher et al.[10] have recently demonstrated how knowledge of preexisting disease, coupled with survival data based on age and extent of burn, may provide more meaningful information about successful management of burn patients. Zawacki et al.[7] have included evaluations of previous bronchopulmonary disease, blood gas studies, and presence of airway edema, in addition to extent of burns and patient age, and noted significant improvement in estimating mortality. A careful history may not only increase

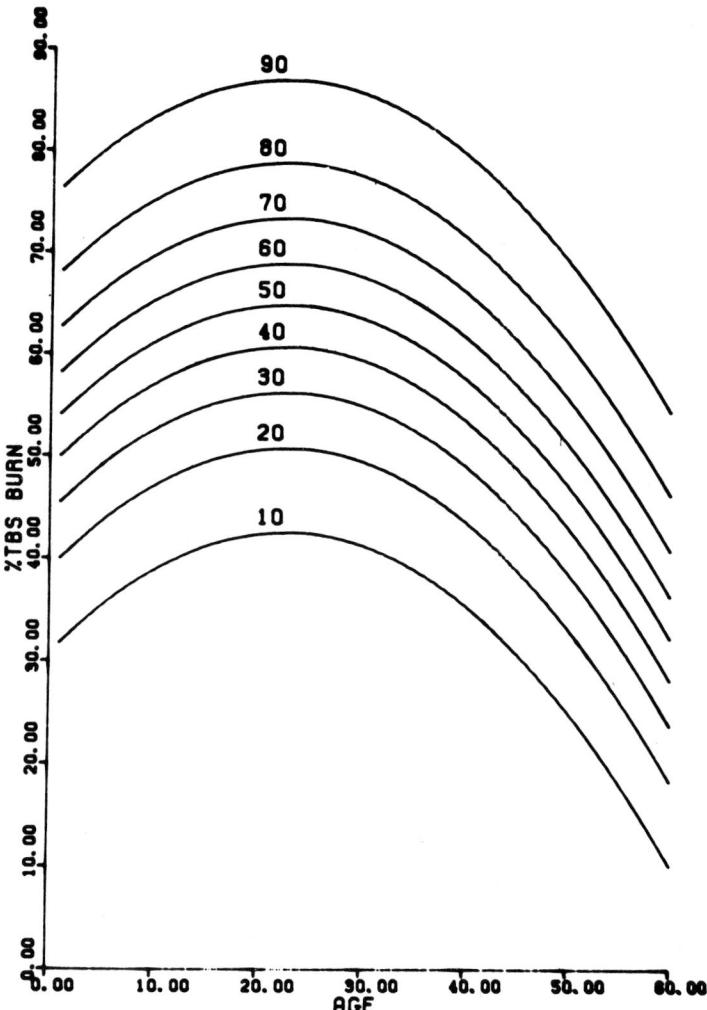

Fig. 18–1. Mortality contours obtained from probit analysis of 937 consecutive patients treated at the New York Hospital–Cornell Medical Burn Center. Reprinted with permission from Feller, I., Flora, J. D., and Bawol, R.: Baseline results of therapy for burned patients. J. A. M. A. 236, 1,946, 1976.

accuracy of prognostication, but also help the physician to anticipate possible complications, e.g., diabetic ketoacidosis and profound hyperglycemia, myocardial infarction, renal failure, respiratory failure, etc. Unfortunately, some important clinical factors may be unknown at the time of the patient's admission (inhalation injury, asymptomatic severe atherosclerotic cardiovascular disease, renal disease, diabetes mellitus, etc.). Stress from

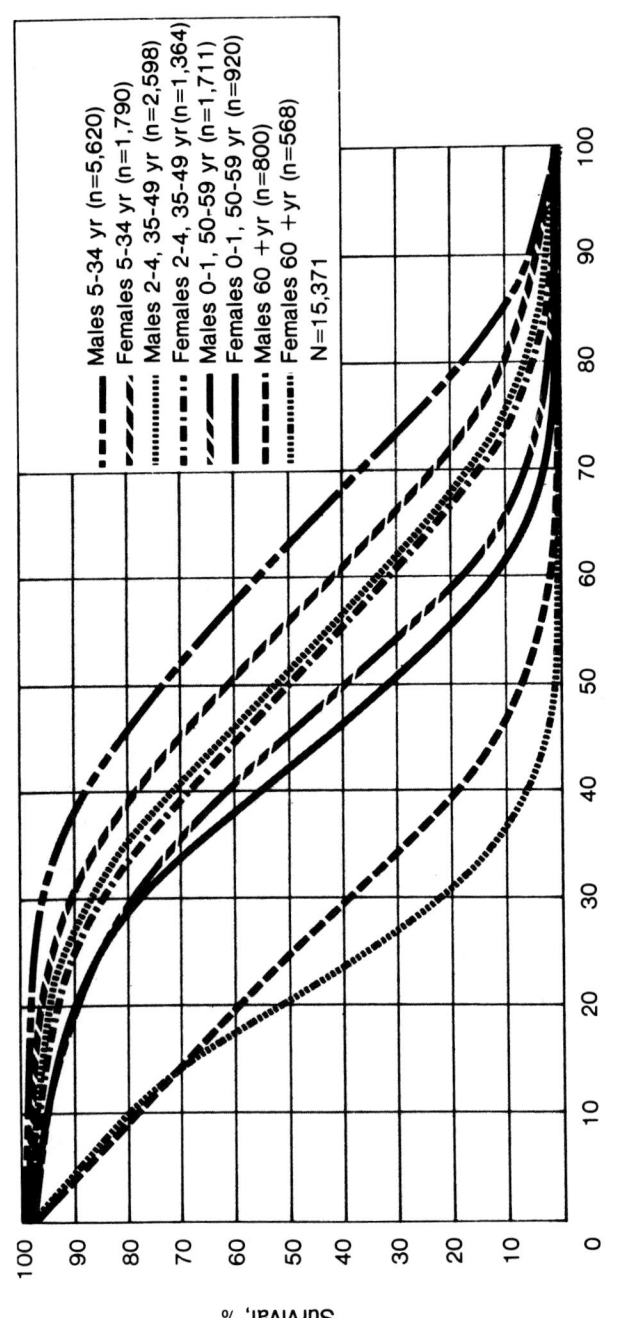

Fig. 18–2. Survival curves estimated by probit analysis for specified age and sex groups. Reprinted with permission from Curreri, P. W., Lutherman, A., Brown, D. W., and Shires, G. T.: Burn Injury. Analysis of survival and hospitalization time for 937 patients. Ann. Surg. 192:473, 1980.

Table 18–1 *Definition of Subsystem Failure in the Burn Patient*

Subsystem	Definition of Failure
Cardiac	Requires catecholamines for support of blood pressure
Pulmonary	Requires mechanical ventilatory support
Renal	Needs hemodialysis
Immunologic	Clinical presence of septicemia and/or bacteremia

thermal injury severely taxes functional reserves in all body organs and often discloses these underlying disorders.

During the hospital course, additional complications often occur in the severely burned patient which obviously affect the prognosis of the patient. Accurate daily recording of such complications, together with the above clinical information, would provide the most complete analysis from which one could daily evaluate the patient's progress. Recently, a plethora of articles have evaluated complications of severe thermal injury and monitored the significance of cardiac, respiratory, renal, and other systemic failure.[11–14] Marshall and Dimick attempted to quantitate failure of these major subsystems (Table 18–1) in 168 severely burned patients (> 40 percent total body surface second degree or third degree burns or both).[13] Universal mortality was associated with severe cardiac or renal failure. A lower mortality (21 percent) was noted with no subsystem or a single subsystem failure in contrast to multiple subsystems (98 percent). Their definition of subsystem failure (Table 18–1) lists highly identifiable events[13] which are rather extreme, and are universally recognized as grave departure points for the severely burned patient. However, this article and others point the direction for further research which might elicit more sensitive and predictable parameters.

In summary, the "simple" approach for estimating survival offers very few parameters to be evaluated and may be immediately utilized at the time of the patient's admission. It depends only upon the patient's age and a careful and accurate determination of the extent of second and third degree burn. In contrast, the "universal data gathering" approach represents, at this time, only an ideal to be worked toward. No universally accepted system is available; the estimate requires daily updating; and the system is greatly dependent upon a complete medical history for prognostic accuracy.

DETERMINING THE CAUSE OF DEATH

Although the exact pathophysiologic events leading to death or to significant morbidity may be difficult to detail, an appreciation of typical time-related changes in burn patients and the classical patterns of complications may be helpful. To achieve these ends, we have summarized and categorized the essential features of complications associated with thermal injury.

Severity of Burns

Prognosis is directly correlated with the extent of second and third degree burns.[2] Burns covering 40 percent or more of the total body surface are associated with a mortality rate of approximately 30 percent.[15] Patients with burns involving more than 85–90 percent of the total body surface rarely survive, and almost all die within the first postburn week. Tables with data estimating survival in burn patients are readily available.[2]

Age of Burn Patient

It is well known that age is a major prognostic indicator in the burn patient. The mortality rate is exceedingly high in very young and elderly burn patients. Presumably, the increased mortality in very young children is at least partially related to the increased body surface area to weight ratio.[16] In addition, the infant has a combination of immature organs and a high rate of growth that may be associated with an attenuated physiologic and biochemical reserve.

In the elderly, several factors[16] apparently impair the ability to respond to thermal injury. Some of these factors include

1. Gradual thinning and decreased elasticity of skin provide a surface more susceptible to full thickness destruction. In addition, wound healing in the elderly is poor because of delayed eschar separation and impaired ability to form granulation tissue, which is necessary for reepithelialization or successful skin grafting. Delayed wound healing also leads to an increased incidence of infection.
2. Decreased dexterity and response time lead to more severe burns in the elderly.
3. A decreased reserve capacity in vital organs, especially the respiratory

and cardiovascular systems, diminishes the physiologic capacity of these elderly patients to respond to the hypermetabolic state which follows severe thermal injury.

4. A diminished renal capacity in the elderly more readily results in renal failure when excess or insufficient fluid is presented to the kidney. Increased nitrogenous wastes may not be appropriately excreted.

Presence of these and other severe diseases predating the thermal injury is more common in older patients, and survival data (mortality figures) are heavily weighted by the factors described above.

Infection

Most commonly, the burn wound is the source of bacterial and occasional mycotic infection. Subsequent hematogenous dissemination from the burn wound to viscera and unburned skin often occurs. When burn exposure is sustained in a closed space, partially combusted substances and gases may cause inhalation injury to the tracheobronchial tree which predisposes the lung to "airborne" bronchopneumonia. Cannula-associated suppurative thrombophlebitis and endocarditis are additional sources of serious infection. Severe ascending pyelonephritis is an uncommon source of septicemia.

Insulin-Dependent Diabetes Mellitus

In burn patients who are continuously threatened by invasive microorganisms, severe diabetes mellitus appears to increase significantly both the number and the severity of infections which may be fatal in patients who have relatively small burns. (See Chapter 17.)

Myocardial Infarcts

Burn patients must be closely observed for subtle signs of ischemic heart disease. (See Chapter 9, Cardiovascular System.) Despite their lack of cardiac symptoms, elderly burn patients often have advanced coronary vascular disease. These patients are unusually susceptible in the early postburn period (first 3 weeks) to the marked cardiac stress caused by the hypermetabolic state.

Young burn patients without significant coronary vascular disease may likewise have infarcts complicating severe multisystemic disease. Therefore, myocardial infarction in the young patient is only one of several causes leading to death. In contrast, myocardial infarction in the elderly burn patient is often an isolated event and may be the only significant complication of the thermal injury.

Inhalation Injury

Inhalation injury is a major complicating factor in burn patients. A patient with a serious inhalation injury often has a severe restrictive ventilatory defect probably related to increased pulmonary water and (circumferential) burns of the thorax.[17] The patient with severe pulmonary injury appears to be far more susceptible to gram-negative pulmonary infections. It is also known that uncomplicated inhalation injuries in patients without burns usually respond favorably to symptomatic treatment. This complex relationship between inhalation injury and burn size has been investigated in an animal model. Zawacki et al.[7] proposed that mortality was primarily a function of dosage of smoke and burn size. The inhalation injury in burn patients usually causes a chemical necrotizing tracheobronchitis which destroys the protective respiratory mucosa. This tissue destruction drastically affects the lungs' capacity to clear necrotic debris, bacteria, and mucus plugs and often leads to atelectasis and pneumonia.

Renal Failure

Morphologic findings in the kidney at autopsy most commonly support the clinical diagnosis of renal failure and allow the pathologist to quantify, in a general sense, the injury to the kidney. However, morphologic changes are not always present, and clinical data must be utilized to determine the effect of the renal failure on the patient's course. Some evidence of possible toxic injury (nephrotoxic type of acute tubular necrosis) suggests that unknown substances, possibly from the burn wound, may affect the renal function of severely burned patients.

CAUSES OF DEATH RELATED TO TIME ELAPSED POST BURN

The following text represents a general guide to factors causing significant morbidity and mortality at specific times in the postburn period (Table 18–2).

Table 18–2 *Cause of Death as Related to Time Post Burn*

Death in the first 3 days post burn

Patients with extensive burns
"Burn wound shock"
Myoglobinuric nephrosis in severe electrical injury
Constrictive problems with soft tissue fluid accumulation
    Entremities
    Thorax
    Abdomen
Carbon monoxide poisoning

Death during the fourth through seventh days post burn

Burn wound infections, bacterial
Airborne bronchopneumonia secondary to "inhalation injury"
Pulmonary hemorrhage
±Renal failure
±Aspiration pneumonia

Death after 7 days post burn—common causes[a]

Burn wound infection, bacterial (occasionally mycotic, especially in patients
    with diabetes mellitus)
Hematogenous pneumonia from infected burn wound
Airborne bronchopneumonia
Pulmonary hemorrhage
Pulmonary hyaline membrane disease
Renal failure, expecially acute renal tubular necrosis
Acute myocardial infarct[b]
Suppurative thrombophlebitis and endocarditis
Aspiration pneumonia
Severe ischemic enterocolitis[c]
Severe disseminated intravascular coagulopathy[c]
Severe liver failure with cholestatic "hepatitis"[c]
Necrotizing prostatitis[c]
Hematogenous pyelonephritis[c]

Death after 7 days post burn—uncommon causes[d]

Extensive pulmonary thromboembolism
Ascending pyelonephritis
Extensive ischemic necrosis, liver
Apoplectic adrenal hemorrhage
Extensive pituitary infarct
Hemorrhagic necrotizing pancreatitis
Gastrointestinal stress ulcer

---

[a] Usually develops after 2 weeks postburn.
[b] When infarcts are found in young patients, they usually occur after the second postburn week in patients with severe multisystemic disease.
[c] Less common, but seen annually in at least one postmortem examination (50–55/year).
[d] Seen in 1% or less of burn patients at postmortem examination.

Immediate Postburn Period
(Time of Burn to 3 Days Post Burn)

In the early postburn course, causes of death are often difficult to ascertain by morphologic studies. Patients who die within 3 days following thermal injury usually have sustained extensive, severe burns and often died in "burn shock."[18] The latter is a term used to describe cardiovascular failure from a variety of causes in the early postburn period. Adequate fluid therapy has markedly reduced these fatalities. Patients dying in the immediate postburn period usually have severe metabolic, biochemical, and physiologic derrangements not readily quantified by morphological evaluation of tissues. Deaths during this period are often termed "resuscitation failures."[17]

Patients with severe electrical injuries may succumb to myoglobinuric nephrosis during this period. Constrictive effects of fluid accumulation beneath extensive circumferential burns of the neck, thorax, and abdomen may impair air conduction of the trachea, respiratory excursion, and cardiovascular return, respectively. Carbon monoxide intoxication must be suspected in any patient who has been burned in a confined space and dies within a few hours post burn.

Acute Postburn Period
(Fourth Through Seventh Postburn Days)

During this peirod, progressive bacterial colonization of the burn wound leads to burn wound infection. In fact, burn wound infections, although uncommon during the first postburn week, may be the only recognizable event during this period. Similarly, in patients with inhalation injury, severe injury to the tracheobronchial tree provides an appropriate milieu for bacterial growth (airborne bronchopneumonia) or pulmonary hemorrhage. Renal failure, with early evidence of acute renal tubular necrosis, also may be encountered during this period.

Subacute and Late Postburn Period
(After 7 Days Post Burn)

After the seventh postburn day, morphologic evidence of multisystemic failure is common. Most complications relate directly or indirectly to infections or to cardiovascular insufficiency. (See Table 18–3.) *Infections* which initially start in the burn wound or lung in the acute postburn period may, in the late postburn period, arise in several different organs such as kidney, urinary bladder, prostate, thyroid (as a consequence of an infected tracheostomy site), major veins and cardiac valves (as a consequence of

Table 18–3 *Categorization by Disease Entity in the Late (Beyond the First Week) Postburn Period*

Infectious

Burn wound infection—bacterial, mycotic, viral
Hematogenous/airborne pneumonia
Bacterial thrombophlebitis/endocarditis with septic pulmonary thromboembolus
Hematogenous pyelonephritis
Ascending pyelonephritis
Extensive viral infections
Disseminated intravascular coagulopathy, related to gram-negative bacterial infections
Necrotizing prostatitis
Hematogenous intramyocardial abscesses
Acute noncalculous cholecystitis due to hematogenous infection

Cardiovascular insufficiency[a]

Myocardial infarct due to increased cardiovascular demands with or without coronary vascular disease
Renal failure due to acute tabular necrosis or renal cortical necrosis
Ischemic enterocolitis
Extensive pituitary infarcts
Extensive hepatocellular necrosis

Miscellaneous

Pulmonary hyaline membrane disease
Liver failure with severe cholestatic ''hepatitis''
Aspiration pneumonia
Pulmonary hemorrhage
Drug reactions, liver, and bone marrow
Adrenal hemorrhage
Disseminated intravascular coagulopathy
Pulmonary embolism

[a] Usually develops after 2 weeks postburn.
[b] When infarcts are found in young patients, they usually occur after the second postburn week in patients with severe multisystemic disease.

intravenous cannulae), etc. Since the latter sites for infection are less common and more difficult to identify, they often remain undetected during life.

Cardiovascular insufficiency is commonly related to marked fluid loss in the early postburn period. Intravascular hemolysis and gastrointestinal and burn wound hemorrhage may also play an important role in the evolving hypovolemia. However, later in the postburn course, gram-negative bacterial septicemia often plays a central role in hypotensive episodes, and may give rise to ischemic disease of several organ systems (ischemic enterocolitis, acute renal tubular necrosis, extensive infarcts of the pituitary). Endotoxemic shock occasionally may play a role in the development of ischemic heart disease. (See Chapter 9, Cardiovascular System.)

The *miscellaneous* disease entities which may occur in the late postburn period often relate to complications of therapy, e.g., pulmonary hyaline membrane disease, or not unexpected sequelae in the seriously ill patient, e.g., aspiration pneumonia. Also included are some entities such as severe cholestatic "hepatitis" which are currently often difficult to explain by clear pathogenetic mechanisms.

Several significant systemic alterations are not readily quantifiable by morphologic measurements, and, therefore, totally elude the pathologist in his interpretation of the case. Currently, the pathologist must rely on premortem historical and laboratory information to determine the significance of hypovolemia or an excessively stimulated cardiovascular, endocrine, or metabolic system. Abnormalities in function, (hematopoiesis, coagulation, immunologic system) may not be apparent by evaluation of morphologic changes alone.

While the number of morphologic alterations in the *early* postburn period is limited to a few recognizable histopathologic processes in a small number of organ systems, virtually every organ system may be involved in the *late* postburn course. Less than a thorough and extensive postmortem examination may lead to an incomplete and potentially misleading analysis of the cause of death. For example, the patient who has documented invasive infection of the burn wound by *Pseudomonas aeruginosa* on the seventh postburn day but survives until the sixteenth postburn day may have none of the expected complications of his initial *Peudomonas* infection. Hypothetically, his burn wound may now have an invasive mycotic infection, and he may also have a hematogenous pneumonia secondary to *Staphylococcus aureus*. If careful examination of cannulated veins has not been undertaken, the origin for staphylococcal septicemia (suppurative thrombophlebitis in a cannulated vein) may never be detected. Assiduous investigation of the myocardium may reveal small (less than 3 mm) abscesses which would further suggest staphylococcal thrombophlebitis. By the twelfth postburn day, the patient may also have developed severe pulmonary hyaline membrane disease and acute renal tubular necrosis. Terminally, he may have suffered an acute myocardial infarct despite the lack of significant coronary vascular disease. Such complex patterns of disease development in the late postburn period are by no means uncommon or rare. Careful evaluation of time-related changes and correlation with the clinical course can usually be carried out, and meaningful data may emerge from such studies which will direct the course of future therapy of other patients.

In summary, this chapter presents methods for determining potential for survival, outlines the spectrum of postburn complications, and relates the complications to both pathogenetic mechanisms and time of occurrence post burn. Since complications of thermal injury are constantly changing and often differ among the major treatment centers, burn physicians are encouraged to study carefully several of the comprehensive autopsy reviews which emphasize the many variables associated with this "universal trauma model" and which address pathogenesis and categorization of the causes of death.[19-22]

References

1. Bull, J. P., and Fisher, A. J.: A study of mortality in a burn unit: A revised estimate. Ann. Surg. 139:269–274, 1954.
2. Feller, I., and Crane, K. H.: National burn information exchange. Surg. Clin. N. Am. 50:1425–1446, 1970.
3. Feller, I., Flora, J. D., and Bawol, R.: Baseline results of therapy for burned patients. J. A. M. A. 236:1943–1947, 1976.
4. McCoy, J. A., Micks, D. W., and Lynch, J. B.: Discriminant function probability model for predicting survival in burned patients. J. A. M. A. 203:644–646, 1968.
5. Stern, and Waisbren, B. A.: A method by which burn units may compare their results with a baseline curve. Surg. Gynecol. Obstet. 142:230–234, 1976.
6. Stern, M., and Waisbren, B. A.: Comparison of methods of predicting burn mortality. Scand. J. Plast. Reconstr. Surg. 13:201–204, 1979.
7. Zawacki, B. E., Azen, S. P., Imbus, S. H., and Chang, Y. C.: Multifactorial probit analysis of mortality in burned patients. Ann. Surg. 189:1–5, 1979.
8. Curreri, P. W., Lutherman, A., Brown, D. W., Jr., and Shires, G. T.: Burn injury. Analysis of survival and hospitalization time for 937 patients. Ann. Surg. 192:472–478, 1980.
9. Roi, L. D., Flora, J. D., Jr., Davis, T. M., and Wolfe, R. A.: Two new burn severity indices. J. Trauma 23:1023–1029, 1983.
10. Fisher, J. C., Wells, J. A., Fulwider, B. T., and Edgerton, M. T.: Editorial: Do we need a burn severity grading system? J. Trauma 17:252–255, 1977.
11. Tweed, A., and Ross, J. F.: A review of the mortality in the burns units at the Victoria General Hospital and the Isaac Walton Killam Hospital, January 1967 to April 1977. Ann. Plast. Surg. 2:491–498, 1979.
12. Bowser, B. H., Caldwell, F. T., Baker, J. A., and Wells, R. C.: Statistical methods to predict morbidity and mortality: Self assessment techniques for burn units. Burns Incl. Therm. Inj. 9:318–326, 1983.
13. Marshall, W. G., Jr., and Dimick, A. R.: The natural history of major burns with multiple subsystem failure. J. Trauma 23:102–105, 1983.
14. Hamit, H. F.: Factors associated with deaths of burned patients in a community hospital. J. Trauma 18:405–418, 1978.

15. Pruitt, B. A., Jr., Mason, A. D., Jr., and Hunt, J. L.: Burn injury in the aged or high-risk patient. *In* The Aged and High Risk Surgical Patient, 1st. ed. J. H. Siegel and P. D. Chodoff, eds. New York, Grune & Stratton, Inc., 1976, pp 523–546.

16. Pruitt, B. A., Jr., Tumbusch, W. T., Mason, A. D., Jr., and Pearson, E.: Mortality in 1,100 consecutive burns treated at a burns unit. Ann. Surg. 159:396–401, 1964.

17. Baxter, C. R.: Problems and complications of burn shock resuscitation. Surg. Clin. N. Am. 58:1313–1322, 1978.

18. Davies, M. R. Q., Cywes, S., van der Riet, R. I. S., Davies, D., and Rode, H.: A review of deaths in a paediatric burns unit. S. Afr. Med. J. 50:1479–1483, 1976.

19. Sevitt, S.: A review of the complications of burns, their origin and importance for illness and death. J. Trauma 19:358–369, 1979.

20. Teplitz, C.: The pathology of burns and the fundamentals of burn wound sepsis. *In* Burns. A Team Approach, 1st. ed. C. P. Artz, J. A. Moncrief, and B. A. Pruitt, eds. Philadelphia, W. B. Saunders Co., 1979, pp 45–94.

21. Linares, H. A.: A report of 115 consecutive autopsies in burned children: 1966–1980. Burns. 8:263–270, 1982.

22. Wartman, W. B.: Mechanism of death in severe burn injury: The need for planned autopsies. *In* Research in Burns. C. P. Artz, ed. Philadelphia, Amer. Inst. Biol. Sci. & F. A. Davis Co., 1960, pp 6–14.

INDEX